T5-ARY-141

Invitation to the Mathematics of Fermat–Wiles

Dedicated to André Warusfel

Invitation to the Mathematics of Fermat–Wiles

Yves Hellegouarch

University of Caen, France

This work has been published with the help of
the French Ministére de la Culture – Centre national du livre

San Diego San Francisco New York
Boston London Sydney Tokyo

This book is printed on acid-free paper.

English translation by Leila Schneps of the original French second edition of the author's INVITATION AUX MATHÉMATIQUES DE FERMAT-WILES (© Masson, 1st edition, 1997, © Dunod, 2nd edition, 2001)

Academic Press
A division of Harcourt, Inc.
Harcourt Place, 32 Jamestown Road, London NW1 7BY, UK
http://www.academicpress.com

Academic Press
A division of Harcourt, Inc.
525 B Street, Suite 1900, San Diego, California 92101-4495, USA
http://www.academicpress.com

Translated by Leila Schneps

ISBN 0-12-339251-9

Library of Congress Catalog Number: 2001089851

A catalogue record of this book is available from the British Library

Typeset by Newgen Imaging Systems (P) Ltd., Chennai, India
Printed and bound in Great Britain by MPG Books Ltd, Bodmin, Cornwall

02 03 04 05 06 07 MP 9 8 7 6 5 4 3 2 1

CONTENTS

FOREWORD

Some time in the middle of the 17th century, Fermat wrote, somewhere, that for every $n \geq 3$, the equation

$$a^n + b^n = c^n \tag{1}$$

had no solution in the set of strictly positive integers, and he asserted several times that he had proved this result for $n = 3$ and $n = 4$ (see Chapter 1). Assuming this it is easy to see that to prove Fermat's assertion in general, it suffices to prove it for every odd prime number $n \geq 5$. If we set

$$x = -\frac{a}{c}, \qquad y = -\frac{b}{c},$$

it is equivalent to showing that the only rational points on the affine curve

$$x^n + y^n + 1 = 0 \tag{2}$$

are the points $(-1, 0)$ and $(0, -1)$.

As we have known since the 19th century that the natural framework for solving problems in planar algebraic geometry is the projective plane, we need to show that the projective curve F_n given by

$$X^n + Y^n + Z^n = 0 \tag{3}$$

has exactly three rational points, namely $(0, 1, -1)$, $(-1, 0, 1)$ and $(1, -1, 0)$.

We will say that these points are the "trivial" points of F_n, since they satisfy the equation

$$XYZ = 0.$$

Although the main theorem of this book is Fermat's assertion, we may also introduce a secondary theorem, which is very close to the main one. Let us state this secondary result.

Denes' Conjecture: For every prime number $n \geq 3$, the only rational solutions of the projective curve G_n given by

$$X^n + Y^n + 2Z^n = 0 \tag{4}$$

are the points

$$(1, -1, 0), \qquad (1, 1, -1).$$

This book was written at the very time when this conjecture had just been **proved**, by H. Darmon and L. Merel.

In order to understand the meaning of Fermat's assertion, let us first consider a fundamental result due to Faltings.

In 1983, Faltings succeeded in proving the famous Mordell conjecture, which stated that every smooth projective algebraic curve defined over a given number field and of genus ≥ 2 has only a **finite** number of rational points.

By this result, we see that for $n \geq 5$, the projective curve F_n has only a finite number N_n of rational points.

However, it does not say whether the number N_n is independent of n, and even less that $N_n = 3$. Furthermore, it can be applied just as well to the curves G_n, and thus cannot serve to show that the number M_n of their rational points is independent of n and equal to 2! Moreover, Faltings' result disappears for $n = 3$, since the curves F_n and G_n are then of genus one!

Although Faltings' theorem is truly marvellous, it is not precise enough to attack Fermat's assertion decisively[1].

This is not the case for the approach developed by A. Wiles, which is so precise that it cannot even be applied without modification to the curves G_n. This approach uses three major notions:

(1) the absolute Galois group $G_{\mathbb{Q}}$ and its representations of degree 2, of which we give a fundamental example in Chapter 4,

(2) certain rather special elliptic curves, known as the curves $E_{A,B,C}$, which are attached to an ABC relation, i.e. a relation of the form

$$A + B + C = 0$$

where $(A, B, C) \in \mathbb{Z}$ are relatively prime integers such that $ABC \neq 0$.

The simplest of these curves is $E_{1,1,-2}$, whose equation is given by

$$y^2 = x(x + 1)(x - 1)$$

(3) certain functions of one complex variable, which are so extraordinarily rich in symmetries that their very existence would never have been suspected if they had not actually been discovered. These functions are studied in Chapter 5; one of the simplest

[1] Note, however ([Hel 3]), that if n is a large enough multiple of a prime number $p \geq 5$, then Faltings' theorem does imply that $N_n = 3$ (see Chapter 6, Exercise 17).

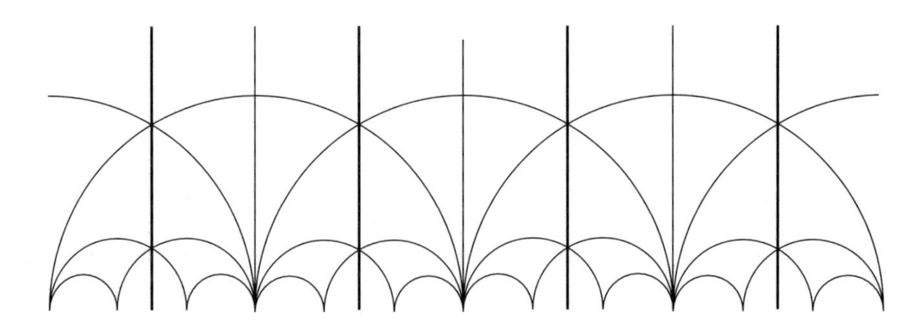

examples, called the modular invariant, which is defined for $\Im\tau > 0$ by a series

$$\tau \longmapsto q = e^{2i\pi\tau} \longmapsto j(q) = \frac{1}{q} + 744 + 196884q + \cdots ,$$

is invariant under the group of positive hyperbolic isometries of the tiling shown below:

Chapter 6 is devoted to describing the way in which these actors collaborate in a game which, given the fundamental push by Wiles, ends up by proving the Fermat–Wiles theorem.

In this book, we can only give a superficial description of this "glass bead game", because our intention is not to write a book for specialists. However, even at this level, the game displays a strange beauty which covers great depths and hard work.

Barring an unexpected surprise, this proof marks the conclusion of one of the great conceptual cathedrals which the 20th century will leave to its descendants. On this topic, it is difficult not to think of what Pascal wrote in the Preface to his Treatise on the Vacuum:

"*... From this, we see that by a particular prerogative, not only does each man advance day by day in the sciences, but all men together make continual progress as the universe ages, because the same thing happens in the aging of mankind as a whole as happens during the aging of a single man. Thus, the entire body of mankind, over many centuries, must be considered as a single man, who lives forever and continues to learn [...]. Those whom we call the ancients were truly new in all things, and form the childhood of mankind; as we have added to their knowledge the experience of the centuries which followed them, it is in ourselves that we should seek the antiquity which we dream of in others.*"

Three hundred years after Pascal, Fermat's theorem was proved, and we now perceive Fermat and Pascal themselves as men who were "*new in all things*". We are, indeed, old and forgetful at the end of the 20th century! This book was begun in 1994, as a university course, and I slowly became aware of the usefulness of transmitting all of these scattered, precious results which are no longer taught (I am thinking particularly of the "analytic geometry" of Fermat and Descartes), and I also saw that our students were happy to travel through centuries of mathematics, following an idea as single as Ariadne's thread, which served to provide them with a synthetic and transversal vision. Alas, in doing so, I forgot that our ministers desire to attach their names to current reforms; the latest of these marked the end of this experiment.

Throughout this text, the reader will perceive many traces of pedagogical choices consciously made during this interrupted teaching experience. As I could always count on my colleagues, or on books, to provide proofs of the general theorems, but as I was perfectly aware that they never had time to give applications, I made it my principal business to show the other side of the coin. This book introduces a host of mathematical objects each more fascinating than the other, and ... references for the proofs of the main structure theorems.

The book also contains non-classical proofs of certain results, using *ad hoc* ideas, which make it possible to appreciate the depths of the classical proofs by looking at certain suitably chosen special cases. This aspect is amplified in the problems contained in the volume, problems which were often given in examinations. As for the exercises, their goal is either to complete the text on certain minor points, or to attract the reader's attention to points of view not mentioned in the text which lead to new directions, or sometimes to add to the historical culture of the reader by including original texts, many of which were communicated to me by Didier Bessot and Jean-Pierre Le Goff.

This book is what survived of my interrupted teaching, and I must say that without the encouragements of André Warusfel, without a particularly opportune sabbatical semester, and without the patience and the talent of the secretaries of our mathematics department, it would never have seen the light of day.

I

PATHS

The goal of this book is to give a brief description of the steps leading to the solution of a problem which was first raised by Pierre de Fermat in the margin of his copy of Bachet de Meziriac's translation of Diophantus' *Arithmetica.*

The question Diophantus asked was the following:

"*Divide a given square into two other squares*" (and he gave an example of his method).

Let 16 *be the given square; I will call* N^2 *and* $16 - N^2$ *the squares to be determined. It remains to find N such that* $16 - N^2$ *is a square.*

I set $16 - N^2 = (2N - 4)^2$ *and I obtain* $N = 16/5$.

To quote from d'Alembert's article "Diophantus" in the Encyclopedia of 1750, "Diophantus' method consists in reducing the situation to an equation in one unknown via a sequence of substitutions". In other words, Diophantus did not seek to give an exhaustive solution to his problem. His methods were *ad hoc*, although they were very clever methods nonetheless (see Exercise 1.1).

When the following famous remark by Fermat burst upon the scene, it represented a real break with all that real been done before[1].

"*To decompose a cube into two other cubes, a fourth power, and generally an arbitrary power into two of the same powers above the second power, is an impossible thing and I have certainly found its admirable proof. This narrow margin would not contain it*".

This is the assertion which became known as "Fermat's last theorem"; it is the only one of his many theorems which remained unproven. In modern language, we state it as follows:

For every $n > 2$, *the equation*

$$\begin{cases} x^n + y^n = z^n \\ xyz \neq 0 \end{cases}$$

has no solution $(x, y, z) \in \mathbb{Z}^3$.

[1] Diophantus gave a recipe, whereas Fermat stated a general property of integers, in accordance with the Aristotelian concept stating that "there is no science except of that which is general".

1.1 DIOPHANTUS AND HIS *ARITHMETICA*

Diophantus is one of the great names of the "Silver Age" of Greek mathematics, and it is thought that he lived in Alexandria, together with other scholars such as Pappus and Proclus, between about 250 and 350 A.D.

He is known for his *Arithmetica* which contained 13 books: at Fermat's time only six were available, but Arabic translations of four more were discovered quite recently.

The mathematics of Diophantus' *Arithmetica* appear quite foreign to the Greek mathematics of the "Golden Age" of Euclid; in some aspects, they actually seem more closely related to the **Babylonian tradition**. However, unlike the Babylonians, Diophantus was interested by **exact** rational solutions of determinate or indeterminate equations. As far as we know, it was Diophantus who first introduced the indeterminate equations which led to the development of "Diophantine analysis", a domain of mathematics which lies within the realm of "number theory".

However, Diophantus also did the work of an algebraist, and he used notations close to those of Viète.

For example, Diophantus would have written the polynomial

$$2x^4 + 3x^3 - 4x^2 + 5x - 6$$

in the form

$$\Delta\Delta 2 \quad K3 \quad x5 \quad M \quad \Delta 4 \quad U6$$

where Δ denotes x^2, K denotes x^3, M denotes the minus sign and U is unity.

The following important identity, which was well-known in the Middle Ages, can be found in the work of Diophantus:

$$\begin{aligned}(a^2+b^2)(c^2+d^2) &= (ac+bd)^2 + (ad-bc)^2\\ &= (ac-bd)^2 + (ad+bc)^2.\end{aligned}$$

However, as we noted, Diophantus seems to have been more interested by the computation of a particular solution than by the exhaustive analysis of a given problem.

An evaluation of Diophantus as a mathematician

The first algebraist? lacks generality ...
The last Babylonian? too abstract ...
The first number theoretician? doesn't work in $\mathbb{N}$ ...

1.2 TRANSLATIONS OF DIOPHANTUS

In the 16th century, Diophantus' *Arithmetica* was an obscure text which, for all practical purposes, had been forgotten. R. Bombelli[2] rediscovered the book in 1570 and incorporated

[2] R. Bombelli 1526–1572, introduction to complex numbers.

it into his book *Algebra*, written in Italian in 1572. Then W. Holtzmann (alias Xylander[3]) gave a complete Latin translation of the *Arithmetica* in 1575.

F. Viète[4] transformed the book and incorporated it into his own works (*Isagoge*, 1591; *Zététique*, 1593), emphasizing its algebraic aspects.

C.G. Bachet de Meziriac[5], on the other hand, who was not a real algebraist, but the author of a work called *Pleasant and delectable problems to be solved by numbers* and the original discoverer of "Bézout's identity" in $\mathbb{Z}$ (1624), prepared a bilingual (Greek/Latin) edition of Diophantus' *Arithmetica* in 1621: it was this excellent version which was studied and annotated by Fermat.

1.3 FERMAT

Pierre de Fermat[6] was a jurist, councillor at the Parliament of Toulouse and known for his facility in writing verses in different languages (Latin, Greek, Italian and Spanish).

But in this book, we are interested in Fermat as an amateur mathematician of such extraordinary talent that professionals of the time bowed to his superior knowledge.

Although we are interested mainly in his number theoretic discoveries, we must mention here that Fermat did not only work in number theory.

In **geometry**, one of his very first results was to reconstruct the planar loci of Apollonius according to Pappus' analysis: his first book on the subject appeared before 1629, the second in 1636.

Then in 1636, independently of Descartes, Fermat invented analytic geometry and saw that if he had known it in 1629 he could have saved a great deal of time. However, he remarked that "there is, for science, a certain importance in not depriving posterity of the works of the mind even when they are yet not completely formulated; work which is at first simple and crude becomes stronger and grows by new inventions. It is even important, for study, to be able to fully contemplate the hidden progress of the mind and the spontaneous development of the art".

In **analysis**, Fermat was a precursor of differential and integral calculus.

In **combinatorics** and **probability**, he was the equal of Pascal.

In **optics** he introduced the calculus of variations to justify the laws of Snell–Descartes (nature always acts by the shortest paths).

Let us now turn to a detail of **number theory**, his favourite domain.

The work of Fermat concerned:

– Fermat's small theorem:
  For every prime number p and for every $a \in \mathbb{Z}$ not divisible by p, we have

$$a^{p-1} \equiv 1 \bmod p.$$

[3] Xylander 1532–1576.

[4] Viète 1540–1603.

[5] C.G. Bachet 1581–1638.

[6] Born in 1601 in Beaumont de Lomagne, died in 1665 in Castres, where he was a councillor in the Chamber of the Edict.

- *Fermat's equation* (usually mistakenly called Pell's equation):

$$x^2 - Ay^2 = 1.$$

- The *Fermat numbers*

$$F_n = 2^{2^n} + 1.$$

- The representation of prime numbers by quadratic forms, in particular by

$$X^2 + Y^2 \quad \text{and} \quad X_1^2 + X_2^2 + X_3^2 + X_4^2.$$

- Fermat's famous assertion or "last theorem", stating that for $n > 2$ and $(x, y, z) \in \mathbb{Z}^3$, we have

$$x^n + y^n = z^n \quad \Longrightarrow \quad xyz = 0.$$

- A great number of individual Diophantine equations.

1.4 INFINITE DESCENT

Infinite descent was one of Fermat's greatest discoveries. This technique allowed him to prove many results on the integers by a device which could not extend to rational or real numbers.

Superficially, the method of infinite descent resembles the induction method, or rather, its negation is a descending sort of induction. But the "mystery" of Fermat's method is difficult to define. Here is what Fermat himself said of it:

"Since the ordinary methods which are in books were insufficient to prove such difficult propositions, I finally discovered an altogether singular method to succeed.

I called this method of proof indefinite or infinite descent: at first I used it only to prove negative propositions, such as for example:

> That there is no number, less by a unit than a multiple of 3, which is composed of a square and the triple of another square;
> That there is no right triangle in numbers whose area is a square number.

The proof is done by reduction to the absurd in this manner:

If there were a right triangle with integral sides whose area was equal to a square, there would be another triangle smaller than this one which would have the same property. If there was a second, smaller than the first, with the same property, then by the same reasoning, there would be a third, smaller than the second, which would have the same property, and then a fourth, a fifth, and so on descending to infinity. However, given a number, there is no way to descend from it infinitely (here I only speak of integers). Hence we conclude that it is impossible to have a right triangle whose area is a square.

We infer from this that neither is it possible to have a triangle whose sides have fractional (rather than integral) lengths and whose area is a square. Because if we had such a triangle with fractional sides, then we could construct one with integral sides, which is in contradiction with what we proved above.

I do not add the reason for which I infer that if there were a right triangle of this nature, there would be another of the same nature smaller than the first, because the discourse would be too long and **therein lies the whole mystery of my method**. I would be greatly satisfied if Pascal and Roberval and many other scholars sought it upon my indications.

I remained for a long time unable to apply my method to positive assertions, because the proper angle to approach such questions is much more uncomfortable than the one which I use for negative assertions. Thus, when I needed to show that **every prime number which surpasses a multiple of 4 by a unit is composed of two squares**, I found myself in a quandary. But finally, much-renewed meditation provided the light which I lacked, and positive assertions became solvable by my method, with the help of some new principles which necessity forced me to add to it. This progress in my reasoning on such positive assertions runs as follows: if a prime number chosen at will, which surpasses a multiple of 4 by a unit, is not composed of two squares, there will be a prime number of the same nature, less than the given one, and then a third still less, etc., descending to infinity until one arrives at the number 5, which is the least of all primes of this nature, which shows that it cannot be composed of two squares, which yet it is. Hence we infer, by deducing the impossible, that all primes of this nature are, consequently, composed of two squares".

Let us try to illustrate this subtle difference between induction **on** n and Fermat's method of infinite descent, on a simple example of each.

(a) Induction

Definition 1.4.1 *A number is said to be triangular of root n, if it is the sum T_n of the first n integers:*

$$T_n = 1 + 2 + \cdots + n.$$

Formula 1.4.1 Inverting the order of the terms of the sum we have

$$T_n = n + (n-1) + \cdots + 1,$$

and adding term by term, we find

$$2T_n = (n+1) + (n+1) + \cdots + (n+1) = n(n+1),$$

hence

$$T_n = \frac{n(n+1)}{2}.$$

Definition 1.4.2 *A number is said to be pyramidal of root n, if it is the sum Π_n of the first n triangular numbers:*

$$\Pi_n = T_1 + T_2 + \cdots + T_n.$$

Theorem 1.4.1 *We have the identity*

$$\Pi_n = \frac{1}{6}\, n(n+1)(n+2).$$

Proof. Let us proceed by induction on n.

(1) The property holds for $n = 1$.
(2) Assuming it holds for n, we have

$$\Pi_{n+1} = \frac{1}{6}\, n(n+1)(n+2) + T_{n+1}.$$

As $T_{n+1} = (n+1)(n+2)/2$, we obtain

$$\begin{aligned}\Pi_{n+1} &= \frac{1}{6}\, n(n+1)(n+2) + \frac{1}{2}(n+1)(n+2)\\ &= \frac{1}{6}\,(n+1)(n+2)(n+3),\end{aligned}$$

so the property holds for $n+1$.
(3) As the property holds for $n = 1$, it also holds for $n = 2$; since it holds for $n = 2$ it also holds for $n = 3$, "and so on to infinity", as Pascal would say. □

(b) Infinite descent

Theorem 1.4.2 $\sqrt{3} \notin \mathbb{Q}$.

Proof. We give two different proofs.

(a) Let us first recall the classical solution due to Euclid.
If we had

$$\sqrt{3} = \frac{n}{d},$$

we would deduce that

$$3d^2 = n^2,$$

so the exponent of 3 in the decomposition into prime factors of this number should be both odd (left-hand expression) and even (right-hand expression). This would contradict the **uniqueness** of the decomposition into primes.

(b) We will now give a proof of the same result by the method of infinite descent.
Assume that $\sqrt{3} = a_1/b_1$, with $(a_1, b_1) \in \mathbb{N} \times \mathbb{N}_+$. Using the relation

$$\frac{1}{\sqrt{3}-1} = \frac{\sqrt{3}+1}{2}$$

we deduce that

$$\frac{\sqrt{3}+1}{2} = \frac{b_1}{a_1 - b_1},$$

hence $\sqrt{3} = (3b_1 - a_1)/(a_1 - b_1) = a_2/b_2$, with

$$\begin{cases} a_2 = 3b_1 - a_1 \\ b_2 = a_1 - b_1. \end{cases}$$

From $\sqrt{3} < 2$ we deduce that $a_1 < 2b_1$, so

$$a_2 = 3b_1 - a_1 > 0 \quad \text{and} \quad b_2 < b_1.$$

From $3/2 < \sqrt{3}$ we deduce that

$$b_2 = a_1 - b_1 > 0 \quad \text{and} \quad a_2 < a_1.$$

Thus, we see that

$$\sqrt{3} = \frac{a_1}{b_1} = \frac{a_2}{b_2} = \frac{a_3}{b_3} = \cdots$$

and the sequences of the numerators and denominators are strictly decreasing: this is absurd, so $\sqrt{3} \notin \mathbb{Q}$. □

A modern formulation of this descent argument

Let E denote the set of $b \in \mathbb{N}_+$ such that there exists $a \in \mathbb{N}$ such that $\sqrt{3} = a/b$. If E is not empty, then E contains a smallest element b_1, so $\sqrt{3} = a_1/b_1$ with $a_1 \in \mathbb{N}$.

Then the above argument shows that there exists $a_2/b_2 \in E$ with $b_2 < b_1$; this gives a contradiction, so E is empty. □

Remark 1.4.1 Unfortunately, the example of $\sqrt{3}$ is too simple to clearly reveal the "mystery" of infinite descent. Indeed, the property $\sqrt{3} \notin \mathbb{Q}$ is purely "local"; what the proof (a) really shows is that $\sqrt{3}$ does not belong to the field of 3-adic numbers (see Chapter 3). Fermat's method acquires its full importance when applied to theorems which are "everywhere locally true", but "globally false" (on this subject, see the article by J. Coates in [C-N]).

1.5 FERMAT'S "THEOREM" IN DEGREE 4

Let a **Pythagorean triangle** be a triple of natural numbers (a, b, c) such that

$$a^2 + b^2 = c^2.$$

Since Euclid's time[7], it has been known that if $(a, b) = 1$, then there exist relatively prime numbers p and q with $p > q$ such that if a is even, we have

$$a = 2pq, \qquad b = p^2 - q^2, \qquad c = p^2 + q^2.$$

[7] Euclid's Elements X29, lemma 1.

In a letter to Huygens, Fermat stated the following theorem

Theorem 1.5.1 *The area of a Pythagorean triangle cannot be a square.*

Proof. *(Fermat)* This area is given by $pq(p+q)(p-q)$, where the four factors are pairwise relatively prime. Thus, we have

$$p = x^2, \qquad q = y^2, \qquad p + q = u^2, \qquad p - q = v^2,$$

where u and v are odd and relatively prime. We have

$$u^2 = v^2 + 2y^2 \qquad \Longleftrightarrow \qquad 2y^2 = (u + v)(u - v);$$

as the greatest common divisor of $u + v$ and $u - v$ is 2, we see that

$$\begin{aligned} u + v &= 2r^2 \\ u - v &= 4s^2 \end{aligned}$$

or the contrary. Hence

$$u = r^2 + 2s^2, \qquad \pm v = r^2 - 2s^2$$

and $y = 2rs$. Consequently,

$$x^2 = p = \tfrac{1}{2}(u^2 + v^2) = r^4 + 4s^4,$$

and $(r^2, 2s^2, x)$ forms a Pythagorean triangle of area equal to $(rs)^2$ and hypotenuse strictly less than $\sqrt{x^4 + y^4}$: we can apply the descent. □

A modern formulation of this descent argument

Let E denote the set of $h \in \mathbb{N}_+^*$ such that there exists a Pythagorean triangle of hypotenuse h whose area is a square. If E is not empty, then E contains a smallest element $h = p^2 + q^2$ with $p > q$.

Then the above argument shows that $\sqrt{p} \in E$ and $\sqrt{p} < h$, which is a contradiction, so E is empty.

Corollary 1.5.1 *The equation*

$$\begin{cases} x^4 + y^4 = z^4 \\ xyz \neq 0; \quad x,\ y,\ z \text{ integral} \end{cases}$$

is impossible.

Proof. Assume that x, y, z form a "primitive" solution of the equation, i.e. a solution such that x, y, z are pairwise relatively prime. At least one of them must be even; let it be x. From the equation

$$x^4 = (z^2 - y^2)(z^2 + y^2),$$

we deduce that

$$z^2 - y^2 = 8a^4, \qquad z^2 + y^2 = 2b^4$$

with $(a, b) = 1$. We then have

$$4a^4 + b^4 = z^2,$$

which gives a Pythagorean triangle whose area is a square. □

Remark 1.5.1

(1) We can compare corollary 1.5.1 to the equation

$$x^4 + y^4 = 2z^4;$$

does it have any primitive solution other than $(\pm 1, \pm 1, \pm 1)$?

The answer is no, since otherwise we would have

$$z^8 - x^4y^4 = \left(\frac{x^4 - y^4}{2}\right)^2,$$

and the equation $X^4 - Y^4 = Z^2$ has only trivial solutions in $\mathbb{Z}^3$ (i.e. solutions with $XYZ = 0$) by Exercise 1.6.

(2) Clearly the same remark applies to the equation

$$x^4 + y^4 = 2z^2.$$

1.6 THE THEOREM OF TWO SQUARES

Like many others before him, Fermat became interested in the representation of an integer as the sum of two squares, and he began to construct the rigorous theory of such integers. More generally, he was interested in the representation of an integer by quadratic forms of the type $X^2 + AY^2$ with $A \in \{1, 2, 3\}$. As the method is the same in the three cases, we consider only the case $A = 1$. In this case, Fermat's result is as follows.

Theorem 1.6.1 *A positive integer n is the sum of two squares if and only if all the prime factors of n of the form $4m - 1$ have even exponents in the decomposition of n into prime factors.*

1.6.1 A Modern Proof

The modern method for proving this theorem is to work in the ring $\mathbb{Z}[i]$ of "Gaussian integers" [8].

Recall that the main property of this ring is that it is **Euclidean**. Indeed, if we define the norm of an element in this ring by $N(x+iy) = x^2 + y^2$, we obtain a homomorphism from the multiplicative monoid $\mathbb{Z}[i]\backslash\{0\}$ to the multiplicative monoid $\mathbb{N}^*$ of the integers ≥ 1; this follows from the fact that complex conjugation is an automorphism of $\mathbb{Z}[i]$, so we have

$$N(z_1z_2) = (z_1z_2)(\overline{z_1z_2}) = (z_1\bar{z}_1)(z_2\bar{z}_2) = N(z_1)N(z_2),$$

which is exactly Diophantus' identity.

The essential fact concerning $\mathbb{Z}[i]$ is that equipping it with the norm N, it becomes a **Euclidean ring**, i.e. we can perform Euclidean division in $\mathbb{Z}[i]$. Let $x \in \mathbb{Z}[i] \setminus \{0\}$; then for every $y \in \mathbb{Z}[i]$ there exist q and r in $\mathbb{Z}[i]$ such that

$$\begin{cases} y = xq + r \\ \text{with } N(r) < N(x). \end{cases}$$

It follows that $\mathbb{Z}[i]$ is principal, and every element of $\mathbb{Z}[i]$ can be written in an essentially unique way as a product of irreducible elements

$$x = u\pi_1^{n_1} \cdots \pi_r^{n_r},$$

where $\pi_1, \ldots, \pi_r$ are non-associated irreducible elements of $\mathbb{Z}[i]$, and u is a unit of $\mathbb{Z}[i]$, i.e. an element x of $\mathbb{Z}[i]$ such that $N(x) = 1$.

Apart from the prime number 2 in $\mathbb{Z}$ which can be written

$$2 = i^{-1}(1+i)^2$$

(thus it is a prime number in $\mathbb{Z}$ which is "ramified" in $\mathbb{Z}[i]$), the prime numbers in $\mathbb{Z}$ fall into two categories:

(i) prime numbers of the form $4n - 1$, which remain irreducible in $\mathbb{Z}[i]$ and which we call the "inert" primes.

(ii) prime numbers p of the form $4n + 1$, which split in $\mathbb{Z}[i]$, i.e. factor as

$$p = \pi\,\bar{\pi} = N(\pi),$$

where $\bar{\pi}$ denotes the number conjugate to π.

This is precisely the delicate point of the theory, the one which gave Fermat so much trouble. Let U and Σ denote the kernel and the image of the homomorphism of monoids:

$$N : \mathbb{Z}[i] \longrightarrow \mathbb{N}.$$

The set U is precisely the **group of units** of $\mathbb{Z}[i]$, i.e. the group of the invertible elements of this ring.

[8] C.F. Gauss 1777–1855.

Theorem 1.6.2 (Crucial Theorem) *Let $p \in \mathbb{N}^*$ be a prime. The following conditions are equivalent:*

(i) $p \in \Sigma$;
(ii) *p is reducible in* $\mathbb{Z}[i]$;
(iii) $p = 2$ *or* $p \equiv 1 \bmod 4$.

Proof. **(1)** Let us first show that (i) $\iff$ (ii).

Indeed, if $p \in \Sigma$, we have $p = (a + bi)(a - bi)$ in $\mathbb{Z}[i]$, and this is a non-trivial decomposition. Conversely, if $p = z_1 z_2$ with z_1 and $z_2 \notin U$, we see that $N(p) = N(z_1)N(z_2)$, i.e.

$$p^2 = N(z_1)N(z_2).$$

As $N(z_j) > 1$ for $j \in \{1, 2\}$, we obtain

$$p = N(z_1) = N(z_2),$$

so $p \in \Sigma$.

(2) Let us show that (ii) $\iff$ (iii): this is the subtle point!

We know that p is reducible if and only if $(p) = p\mathbb{Z}[i]$ is not prime.

Consider the isomorphisms

$$\begin{aligned}
\mathbb{Z}[i] &\cong \mathbb{Z}[X]/(X^2 + 1)\\
\mathbb{Z}[i]/(p) &\cong \mathbb{Z}[X]/(X^2 + 1, p)\\
&\cong (\mathbb{Z}[X]/(p))/(X^2 + 1)\\
&\cong \mathbb{F}_p[X]/(X^2 + 1).
\end{aligned}$$

If $p \neq 2$, we have

$$\begin{aligned}
(p) \text{ reducible} &\iff X^2 + 1 \ \text{ factors in } \mathbb{F}_p[X]\\
&\iff -1 \in (\mathbb{F}_p^*)^2, \ \text{ the group of squares in } \mathbb{F}_p^*\\
&\iff p \equiv 1 \bmod(4) \ \text{ (Euler's criterion)}^9.
\end{aligned}$$

□

Proof of the theorem of two squares

Let $n = n_1 n_2$ be the decomposition of n as a product of

- its prime factors belonging to Σ, which gives n_1;
- its prime factors not belonging to Σ, which gives n_2.

Since Σ is a monoid, $n_1 \in \Sigma$, so if n_2 is a square, then $n \in \Sigma$. Conversely, suppose that $n \in \Sigma$ and n_2 is not a square; we will show that this leads to a contradiction.

Let $n = a^2 + b^2$.

If q is a prime number congruent to -1 modulo 4, which divides both a and b, then q^2 divides n. Applying the same remark to n/q^2, we can suppose that none of the common prime divisors of a and b is congruent to -1 modulo 4.

[9] L. Euler 1707–1783.

If n_2 is not a square, then it has a prime factor q congruent to -1 modulo 4 whose exponent is odd.

It follows that q does not disappear in the reduction described above, and we are reduced to the case

$$a^2 + b^2 \equiv 0 (\text{mod } q)$$

with $a \not\equiv 0$ or $b \not\equiv 0$ (mod q), which implies $a \not\equiv 0$ and $b \not\equiv 0$.

Thus, we have $ab \not\equiv 0$ mod q, and dividing by b, we find that

$$\left(\frac{a}{b}\right)^2 + 1 \equiv 0 \pmod{q},$$

which is impossible since q is not congruent to 1 modulo 4. □

1.6.2 "Fermat-style" Proof of the Crucial Theorem

Lemma 1.6.1 *Let $N = a^2 + b^2$ be an element of Σ, and let $\ell = x^2 + y^2$ be a prime divisor of N. Then not only is N/ℓ an element of Σ, but furthermore, there exist integers u, v such that*

$$N\ell^{-1} = u^2 + v^2$$
$$a = ux + vy, \qquad b = |uy - vx|.$$

Remark 1.6.1 Naturally, we assume that a, b, x, y are positive.

Proof. The idea of the proof is that the **division** of N by ℓ can be done with the help of an astute use of **multiplication**. Let us write

$$N\ell = (ax \pm by)^2 + (ay \mp bx)^2, \tag{1}$$

and let us show that the sign can be chosen so that each of the terms of the right-hand side is divisible by ℓ^2 (this is the "astute" multiplication).

If we start with the the last term, we must show that $(ay - bx)(ay + bx)$ is divisible by ℓ. Indeed, we have

$$\begin{aligned}(ay - bx)(ay + bx) &= a^2y^2 - b^2x^2 \\ &= (a^2 + b^2)y^2 - b^2(x^2 + y^2) \\ &= Ny^2 - b^2\ell \equiv 0 \bmod \ell.\end{aligned}$$

Once the sign is correctly chosen, relation (1) shows that the first term of the right-hand side is also divisible by ℓ, so we have

$$ax \pm by = \ell u, \qquad ay \mp bx = \ell v \tag{2}$$

with u and $v \in \mathbb{Z}$. Hence

$$N\ell^{-1} = u^2 + v^2.$$

But we can solve (2) for a (resp. for b) by multiplying the first relation by x (resp. y) and the second by y (resp. by x). After simplifying by ℓ, we find

$$a = ux + vy, \qquad b = \pm(uy - vx). \qquad \square$$

Theorem 1.6.3 (Crucial Theorem) *If a prime number $p > 0$ is congruent to 1 modulo 4, then it is a sum of two squares.*

Proof. This proof is a simple reconstitution which follows Fermat's (purely literary) text step by step.

(1) If $p \equiv 1$ modulo 4, then -1 is a square modulo p, so there exists $a \in \mathbb{N}$ such that

$$a^2 + 1 \equiv 0 \ (\text{mod}\ p);$$

we can even choose $a < p/2$. Thus we have

$$\begin{cases} a^2 + b^2 = N'p, & 0 < N' < p \\ (a, b) = 1 \end{cases}$$

where we simply take $b = 1$; it plays no role in what follows.

(2) We will now adopt Fermat's reasoning by the absurd.

If $p \notin \Sigma$, we apply the preceding lemma to $N = N'p$, **assuming** that all the prime factors ℓ of N' lie in Σ. We deduce that $N/\ell = (N'/\ell)p$ belongs to Σ, and we can keep doing this as many times as necessary in order to obtain $p \in \Sigma$, which is absurd.

Thus N' has a prime factor $\ell < p$ which is not in Σ.

In this way, we construct a strictly decreasing sequence of prime numbers which do not lie in Σ.

Now, we started by assuming that a and b were relatively prime, so all these prime numbers must be congruent to 1 modulo 4, and the smallest such prime is 5. But $5 \in \Sigma$, contradiction! $\square$

Remark 1.6.2 The last sentence is not actually necessary, since there cannot be an infinite decreasing sequence of prime numbers $< p$. However, we put it in because it was part of Fermat's original argument.

1.6.3 Representations as Sums of Two Squares

Let $r(n)$ denote the number of representations of n in the form $a^2 + b^2$. Even if two representations differ only by the order and the signs of a and b, they are considered to be **distinct**.

Example 1.6.1

$$\begin{aligned} 2 &= (\pm1)^2 + (\pm1)^2 & r(2) &= 4 \\ 1 &= (\pm1)^2 + 0^2 = 0^2 + (\pm1)^2 & r(1) &= 4 \\ 5 &= (\pm2)^2 + (\pm1)^2 = (\pm1)^2 + (\pm2)^2 & r(5) &= 8. \end{aligned}$$

Using the fact that $\mathbb{Z}[i]$ is a principal ideal domain, we can prove the following result, which we leave as an exercise.

Theorem 1.6.4 *Let $n \in \mathbb{N}^*$ and let*

$$n = 2^{\alpha} \prod p^{v_p(n)} \prod q^{v_q(n)}$$

with $p \equiv 1 \pmod 4$ and $q \equiv -1 \pmod 4$ be the decomposition of n into prime factors. Then we have:

(i) $r(n) = 0$ *if all the $v_q(n)$ are not even.*
(ii) $r(n) = 4\prod_p (v_p(n) + 1)$ *if all the $v_q(n)$ are even.*

However, our main task is to count the **proper representations** of the odd numbers which admit no prime factors congruent to -1 modulo 4. By this, we mean decompositions of the type

$$\begin{cases} n = a^2 + b^2 \\ (a, b) = 1, \end{cases}$$

up to exchanging a and b, or changing their signs.

Example 1.6.2
Let $s(n)$ denote the number of proper representations of n. We have

$$s(1) = 0 \qquad s(5) = 1.$$

If p is prime and $p \equiv 1 \pmod 4$, then $s(p) = 1$ (see theorem 1.6.5).

Lemma 1.6.2 *Let $n \in \mathbb{N}^*$ and let $p = x^2 + y^2$ be an* **odd** *prime divisor of n.*
Assume that $n/p > 2$ and n/p admits the proper representation

$$n/p = u^2 + v^2.$$

Let

$$\begin{aligned} a &= ux + vy, & b &= uy - vx \\ a' &= ux - vy, & b' &= uy + vx. \end{aligned}$$

Then the representations $n = a^2 + b^2$ and $n = a'^2 + b'^2$ are distinct, and moreover:

(i) *If p does not divide n/p, the two representations $n = a^2 + b^2$ and $n = a'^2 + b'^2$ are proper.*

(ii) *If p divides n/p, then one is proper and the other is not.*

Proof. **(1)** We check that $a \neq \pm a'$, $b \neq \pm b'$, $a \neq \pm b'$ and $b \neq \pm a'$.

(2) Let us show that if p does not divide n/p , the two representations are proper. Indeed, solving for u and v, we have

$$\begin{cases} ax + by = up, \\ ay - bx = vp, \end{cases} \qquad \begin{cases} a'x + b'y = up, \\ a'y - b'x = -vp, \end{cases}$$

so (a, b) divides p (resp. (a', b') divides p). But if $(a, b) = p$ (resp. $(a', b') = p$), we see that p^2 divides n, which is absurd. Thus $(a, b) = (a', b') = 1$.

(3) Finally if p divides n/p, we will show that p divides bb'. Indeed, we have

$$bb' = u^2y^2 - v^2x^2 = \frac{n}{p}\,y^2 - v^2p.$$

Now, if p divides b (resp. b'), then p divides a (resp. a').

But if p divides a (resp. a'), we have

$$ux \equiv -vy \bmod p \quad (\text{resp. } ux \equiv vy \bmod p)$$

so

$$ux \not\equiv vy \bmod p \quad (\text{resp. } ux \not\equiv -vy \bmod p)$$

and p does not divide a' (resp. a).

Thus one and only one of the two representations is proper. □

Theorem 1.6.5 *Let $n = p_1^{v_{p_1}(n)} \ldots p_m^{v_{p_m}(n)}$ be a number in Σ admitting only prime factors congruent to 1 modulo 4. Then*

$$s(n) = 2^{m-1}.$$

Proof.

(1) We know that $s(p) = 1$, because

$$p = \pi\bar{\pi},$$

where π is unique modulo the units of $\mathbb{Z}[i]$.

By lemma 1.6.2, we also have

$$s(p^e) = s(p) = 1.$$

(2) Now, if $(p, n) = 1$ with $n > 2$, then lemma 1.6.2 shows that

$$s(p^e n) = 2s(n).$$

(3) Thus we obtain the result by induction on m. □

Corollary 1.6.1 *Let $n = p_1^{v_{p_1}(n)} \dots p_m^{v_{p_m}(n)}$ be as in theorem 1.6.5, and let $e \geq 1$. If $n^e = a^2 + b^2$ is a proper representation of $n^e \in \Sigma$, then n has a proper representation $u^2 + v^2$ such that*

$$a + ib = (u + iv)^e.$$

Proof. Set

$$(u + iv)^e = P_e(u, v) + iQ_e(u, v).$$

Then we want to show that by the theorem, the map

$$(u, v) \longmapsto (P_e(u, v), Q_e(u, v))$$

induces a bijection between the proper representations.

Decomposing n in $\mathbb{Z}[i]$, we have

$$\begin{aligned} n &= (\pi_1 \bar{\pi}_1)^{v_{p_1}(n)} \cdots (\pi_m \bar{\pi}_m)^{v_{p_m}(n)} \\ &= N_{\mathbb{Q}(i)/\mathbb{Q}}(\alpha) \end{aligned}$$

for every $\alpha \in \mathbb{Z}[i]$ of the form

$$\alpha = \pi_1^{\lambda_1} \bar{\pi}_1^{\mu_1} \cdots \pi_m^{\lambda_m} \bar{\pi}_m^{\mu_m}$$

with $\lambda_1 + \mu_1 = v_{p_1}(n), \dots, \lambda_m + \mu_m = v_{p_m}(n)$.

It is clear that if there exists an index ν such that $\lambda_\nu \mu_\nu \neq 0$, the representation of n in the form $X^2 + Y^2$ obtained using α (i.e. $n = \alpha\bar{\alpha}$) is not proper.

Moreover, α and $\overline{\alpha}$ give the same representation, so we obtain at most 2^{m-1} proper representations. By the theorem we obtain **only** proper representations in this way.

The same criterion applies to n^e, hence the result. □

1.7 EULER-STYLE PROOF OF FERMAT'S LAST THEOREM FOR $n = 3$

The proof we give here, which could easily have been Fermat's very own, was actually discovered by Euler[10] in 1753; it makes use only of the theory of the quadratic form

[10] L. Euler 1707–1783.

$X^2 + 3Y^2$, which is close to that of $X^2 + Y^2$. Later, in 1770, Euler gave a similar proof, in the language of the arithmetic of $\mathbb{Z}[\sqrt{-3}]$.

Consider the equation

$$x^3 + y^3 = z^3. \tag{1}$$

We may assume that x, y, z are pairwise relatively prime and z is even. Setting

$$x = p + q, \qquad y = p - q$$

we obtain

$$\frac{p}{4}(p^2 + 3q^2) = \left(\frac{z}{2}\right)^3, \quad pq \neq 0, \tag{2}$$

and since x and y are odd, $p^2 + 3q^2$ must be odd, which implies that $p/4 \in \mathbb{Z}$. We will distinguish two cases according to whether or not z is a multiple of 3.

(1) If 3 does not divide z

Then the two factors of the first term of (2) are relatively prime, so they are cubes:

$$p = 4r^3 \qquad p^2 + 3q^2 = s^3.$$

By corollary 1.6.1 (adapted to $\mathbb{Z}[\sqrt{-3}]$), there exist u and $v \in \mathbb{Z}$ such that

$$\begin{cases} p = P_3(u, v) = u(u + 3v)(u - 3v) \\ q = Q_3(u, v) = 3v(u^2 - v^2) \end{cases}$$

with $(u, v) = 1$.

Since q is odd, v is odd and u is even.

Since $p/4 \in \mathbb{Z}$, we see that $u/4 \in \mathbb{Z}$ and since $p/4 = r^3$ and 3 does not divide p, we see that $u/4$, $u + 3v$ and $u - 3v$ are cubes:

$$u = 4a^3, \qquad u + 3v = b^3, \qquad u - 3v = c^3.$$

Thus we obtain

$$(-2a)^3 + b^3 + c^3 = 0 \tag{3}$$

which is a second solution of (1), but in smaller integers (exercise).

(2) If z is a multiple of 3

Then p is divisible by 3, and as p and q are relatively prime, we see that q is not divisible by 3. We can write

$$\frac{p}{4}\left[q^2 + 3\left(\frac{p}{3}\right)^2\right] = 9\left(\frac{z}{6}\right)^3,$$

so $p/4$ is a multiple of 9, thus

$$\frac{p}{36}\left[q^2 + 3\left(\frac{p}{3}\right)^2\right] = \left(\frac{z}{6}\right)^3$$

and the two factors of the left-hand side are relatively prime cubes.

Corollary 1.6.1 implies the existence of relatively prime u and v such that

$$p = 36r^3, \qquad q = P_3(u, v), \qquad \frac{p}{3} = Q_3(u, v),$$

hence

$$-4r^3 = v(v + u)(v - u). \tag{4}$$

Furthermore, we find that u must be odd and v even since q is odd and

$$q = P_3(u, v) = u(u + 3v)(u - 3v).$$

Multiplying the relation (4) by 2, we see that $2v$, $v + u$, $v - u$ must be cubes, hence we also get a non-trivial solution of (3) in smaller integers (exercise). □

Remark 1.7.1 We will see in Exercise 1.7 that the equation

$$x^3 + y^3 = 2z^3 \tag{G_3}$$

can be treated in the same way, and its only non-trivial primitive solutions are $\pm(1, 1, 1)$.

1.8 KUMMER, 1847

Euler also introduced the ring $\mathbb{Z}[j]$, where $j = e^{2\pi i/3}$ is a primitive third root of unity, in order to study the Fermat equation of degree 3; he accepted the fact that the fundamental theorem of arithmetic extends to $\mathbb{Z}[j]$ (fortunately for him, this is actually the case, although it is not the case for the ring $\mathbb{Z}[\sqrt{-3}]$).

After him, other mathematicians (Lagrange, Gauss, Cauchy, Jacobi, Liouville) considered the possibility of generalising this idea to the rings $\mathbb{Z}[\zeta]$, where ζ denotes a root of unity of odd prime order, and extending the fundamental theorem of arithmetic to these fields.

In 1847, the Academy of Sciences in Paris was in a state of ferment over this question (see [Ed] p. 76–80), when Liouville announced that Kummer had proved three years earlier that the fundamental theorem of arithmetic did not extend to $\mathbb{Z}[\zeta]$, but that the theory of prime factorisation could be saved by the introduction of "ideal numbers".

This fundamental theory, published by Kummer[11] in 1846 ([Kum]), is essentially equivalent to the theory of ideals in the ring $\mathbb{Z}[\zeta]$.

The following paragraph is devoted to a brief sketch of the latter.

1.8.1 The Ring of Integers of $\mathbb{Q}(\zeta)$

For reasons of simplicity, we will assume that the primitive p-th root of unity ζ lies in $\mathbb{C}$; we can, if we wish, represent ζ by the number $e^{2\pi i/p}$, but the other primitive roots of unity are just as legitimate.

[11] E.E. Kummer 1810–1893.

The minimal polynomial of ζ over $\mathbb{Q}$ divides the cyclotomic polynomial

$$\Phi_p(X) = \frac{X^p - 1}{X - 1} = X^{p-1} + \cdots + 1,$$

so, if we prove that $\Phi_p(X)$ is irreducible in $\mathbb{Q}[X]$, then we will have proved that $\Phi_p(X)$ is equal to the minimal polynomial of ζ. But the irreducibility of $\Phi_p(X)$ follows easily from Eisenstein's criterion[12] ([VW1] p. 74). Indeed, the irreducibility of $\Phi_p(X)$ is equivalent to that of $\Phi_p(X+1)$, but we have

$$\Phi_p(X+1) = \frac{(X+1)^p - 1}{X} = X^{p-1} + pX^{p-2} + \cdots + \binom{p}{i} X^{p-i-1} + \cdots + p.$$

As all of the coefficients of $\Phi_p(X+1)$ except for the first one are divisible by p, and the last one is not divisible by p^2, Eisenstein's criterion implies that $\Phi_p(X+1)$ is irreducible, and thus, $\Phi_p(X)$ is irreducible, and it is the minimal polynomial of ζ.

It follows that the map

$$\begin{cases} \mathbb{Q}[X] \overset{\varphi}{\twoheadrightarrow} \mathbb{Q}(\zeta) \\ P(X) \longmapsto P(\zeta) \end{cases}$$

defines a homomorphism of $\mathbb{Q}[X]$ onto a subfield of $\mathbb{C}$ isomorphic to the quotient ring

$$\mathbb{Q}[X]/\operatorname{Ker}\varphi = \mathbb{Q}[X]/(\Phi_p(X)),$$

so $\mathbb{Q}(\zeta)$ is an extension of $\mathbb{Q}$ of degree $p-1$. We call it the field of p-th roots of unity.

Remark 1.8.1 Note that $\mathbb{Q}(\zeta)/\mathbb{Q}$ is a Galois extension, as all of the roots of $\Phi_p(X)$ are powers of ζ, so they belong to $\mathbb{Q}(\zeta)$; thus, every automorphism of $\mathbb{C}$ (continuous or not) sends $\mathbb{Q}(\zeta)$ onto itself (see Chapter 3).

Let $K/\mathbb{Q}$ be a field extension and let $\alpha \in K$. The map

$$\lambda_\alpha \begin{cases} K \longrightarrow K \\ x \longmapsto \alpha x \end{cases}$$

is an endomorphism of the $\mathbb{Q}$-vector space K.

Definition 1.8.1 *For every $\alpha \in K$, set*

$$\begin{cases} \mathrm{Norm}_{K/\mathbb{Q}}(\alpha) = N_{K/\mathbb{Q}}(\alpha) = \det(\lambda_\alpha) \\ \mathrm{Trace}_{K/\mathbb{Q}}(\alpha) = Tr_{K/\mathbb{Q}}(\alpha) = \mathrm{trace}(\lambda_\alpha) \end{cases}$$

[12] F.G. Eisenstein 1823–1852.

Example 1.8.1

Take $\alpha = \zeta$, and consider the following basis for $\mathbb{Q}(\zeta)$ as a vector space:

$$1, \zeta, \ldots, \zeta^{p-2}.$$

The matrix of λ_α in this basis is given by

$$\begin{pmatrix} 0 & 0 & & & -1 \\ 1 & 0 & & & -1 \\ 0 & 1 & & & -1 \\ \vdots & & \ddots & & \vdots \\ 0 & 0 & & 1 & -1 \end{pmatrix}$$

since $\lambda_\alpha(\zeta) = \zeta^{p-1} = -1 - \zeta - \cdots - \zeta^{p-2}$. If p is odd, we then have

$$\begin{cases} N_{\mathbb{Q}(\zeta)/\mathbb{Q}}(\zeta) = 1 \\ Tr_{\mathbb{Q}(\zeta)/\mathbb{Q}}(\zeta) = -1. \end{cases}$$

Proposition 1.8.1 *If $K/\mathbb{Q}(\alpha)$ is an extension of degree s, then*

$$\begin{cases} N_{K/\mathbb{Q}}(\alpha) = (\alpha_1\alpha_2 \cdots \alpha_r)^s \\ Tr_{K/\mathbb{Q}}(\alpha) = s(\alpha_1 + \cdots + \alpha_r), \end{cases}$$

where $\alpha_1, \ldots, \alpha_r$ denote the **conjugates** *of α over $\mathbb{Q}$.*

Remark 1.8.2

(1) The conjugates of α are the roots in $\mathbb{C}$ of the irreducible polynomial of α over $\mathbb{Q}$.
(2) For a proof of this proposition, see [Ca 1] appendix B.
(3) $N_{K/\mathbb{Q}}$ is a homomorphism from K^* to $\mathbb{Q}^*$, and $Tr_{K/\mathbb{Q}}$ is a homomorphism from $(K, +)$ to $(\mathbb{Q}, +)$.

Definition 1.8.2 *Let $x \in \mathbb{C}$. We say that x is* **integral** *if there exists a* **monic** *polynomial $F(X) \in \mathbb{Z}[X]$ such that $F(x) = 0$.*

Remark 1.8.3 A polynomial is called "monic" when the coefficient of the highest degree term is equal to 1.

Example 1.8.2

(1) Since ζ is a p-th root of unity, we see that ζ is a root of $F(X) = X^p - 1$, so ζ is integral.
(2) If x is integral and if $\sigma \in \mathrm{Aut}(\mathbb{C})$ is a (continuous or discontinuous) automorphism of $\mathbb{C}$, then $\sigma(x)$ is integral.
(3) If $x \in \mathbb{Q}$ is integral, then $x \in \mathbb{Z}$.

(4) If $x \in \mathbb{Q}(\zeta)$ is integral, then $N_{\mathbb{Q}(\zeta)/\mathbb{Q}}(x)$ and $Tr_{\mathbb{Q}(\zeta)/\mathbb{Q}}(x)$ are rational integers, so they lie in $\mathbb{Z}$.

In this text, we admit the following well-known theorem (cf [Sa] p. 35).

Theorem/Definition 1.8.1 *Let $K/\mathbb{Q}$ be a field extension. The set of integral elements of K forms a subring of K which we call the* **integral closure** *of $\mathbb{Z}$ in K, or the* ring of integers *of K.*

Example 1.8.3
The ring $\mathbb{Z}[\zeta]$ is the smallest subring (of $\mathbb{C}$) containing $\mathbb{Z}$ and ζ. As ζ is integral, $\mathbb{Z}[\zeta]$ is contained in the ring of integers of $\mathbb{Q}(\zeta)$. Let us now prove that $\mathbb{Z}[\zeta]$ is the integral closure of $\mathbb{Z}$ in $\mathbb{Q}(\zeta)$.

Lemma 1.8.1 *Let p be an odd prime. Then the following families form $\mathbb{Z}$-bases of $\mathbb{Z}[\zeta]$:*

(i) $1, \zeta, \ldots, \zeta^{p-2}$
(ii) $\zeta, \zeta^2, \ldots, \zeta^{p-1}$
(iii) $1, \pi, \ldots, \pi^{p-2}$ *with* $\pi = \zeta - 1$.

Proof. To start with, note that $1, \zeta, \ldots, \zeta^{p-1}$ forms a system of generators of $\mathbb{Z}[\zeta]$.

(i) Since $\zeta^{p-1} = -1 - \zeta \cdots - \zeta^{p-2}$, we see that the family given in the statement is a system of generators of $\mathbb{Z}[\zeta]$.

If we have

$$a_0 + a_1\zeta + \cdots + a_{p-2}\zeta^{p-2} = 0,$$

we deduce that $\Phi_p(X)$ divides the polynomial $a_0 + a_1X + \cdots + a_{p-2}X^{p-2}$, which shows that all the a_i are zero.

(ii) The matrix of the new family in the old basis is given by

$$\begin{pmatrix} 0 & 0 & & -1 \\ 1 & & & -1 \\ 0 & 1 & & -1 \\ \vdots & & \ddots & \vdots \\ 0 & 0 & 1 & -1 \end{pmatrix}.$$

Since this matrix has determinant 1, the new family is also a basis.

(iii) The same reasoning holds for the family $1, \pi, \ldots, \pi^{p-2}$, by computing the matrix with respect to $1, \zeta, \ldots, \zeta^{p-2}$. □

Lemma 1.8.2 *Set $\pi = \zeta - 1$. Then*

(i) $N_{\mathbb{Q}(\zeta)/\mathbb{Q}}(\pi) = p$.
(ii) *All the conjugates of π are associated to π (i.e. products of π and a unit).*

(iii) *p is associated to π^{p-1}.*
(iv) *If $n \in \mathbb{Z}$, then π divides n if and only if p divides n.*
(v) *The ideal generated by π is maximal in $\mathbb{Z}[\zeta]$.*

Proof.

(i) As π is a root of the polynomial $\Phi_p(X+1)$, it suffices to note that the constant term of this polynomial is equal to p.
(ii) The conjugates of π are

$$\zeta - 1, \zeta^2 - 1, \ldots, \zeta^{p-1} - 1,$$

and we see that

$$\zeta^2 - 1 = u_2\pi, \ldots, \zeta^{p-1} - 1 = u_{p-1}\pi,$$

where $u_2, \ldots, u_{p-1}$ are elements of $\mathbb{Z}[\zeta]$. Furthermore, it is clear that the norms of $u_2, \ldots, u_{p-1}$ are all equal to 1, so these elements are units.
(iii) Since $\Phi_p(X+1)$ is irreducible, its constant term is the product of the conjugates of π.
(iv) It suffices to show that if π divides n, then p divides n. To see this, it suffices to take the norms of the numbers n and $\pi\alpha$ to see that in fact p divides n^{p-1}, so p divides n.
(v) By (iii) of lemma 1.8.1, we see that

$$\mathbb{Z}[\zeta]/(\pi) \cong \mathbb{Z}/(\pi) \cap \mathbb{Z},$$

but by (iv) we have $(\pi) \cap \mathbb{Z} = (p)$, so $\mathbb{Z}[\zeta]/(\pi) \cong \mathbb{F}_p$, which is a field. □

Remark 1.8.4 Recall that $x \in \mathbb{C}^*$ is a unit if and only if both x and x^{-1} are integral. It follows that the product of the conjugates of a unit x must be an integer in $\mathbb{Z}$ which divides 1; i.e. if $x \in K$, then

$$N_{K/\mathbb{Q}}(x) = \pm 1.$$

Conversely, if $x \in K$ is integral of norm ± 1, then x is a unit since x^{-1} is a product of integral numbers. If x is a unit, all of its conjugates are units. The units of $\mathbb{Z}[\zeta]$ form a group under multiplication; it is the largest multiplicative group contained in $\mathbb{Z}[\zeta]$.

Theorem 1.8.2 *The integral closure of $\mathbb{Z}$ in $\mathbb{Q}(\zeta)$ is $\mathbb{Z}[\zeta]$.*

Proof. Let A be the integral closure of $\mathbb{Z}$ in $\mathbb{Q}(\zeta)$. We saw that $\mathbb{Z}[\zeta] \subset A$, so it suffices to show that $A \subset \mathbb{Z}[\zeta]$.
(i) Let $x \in A$; we can write

$$x = a_0 + a_1\zeta + \cdots + a_{p-2}\zeta^{p-2}$$

with $a_i \in \mathbb{Q}$. It follows that

$$(1 - \zeta)x = a_0 + (a_1 - a_0)\zeta + \cdots + (a_{p-2} - a_{p-3})\zeta^{p-2} - a_{p-2}\zeta^{p-1}.$$

Since A is a ring and $x \in A$, we have $(1 - \zeta)x \in A$, so

$$Tr_{\mathbb{Q}(\zeta)/\mathbb{Q}}(1 - \zeta)x = (p - 1)a_0 + (a_0 - a_1) + \cdots + (a_{p-3} - a_{p-2}) + a_{p-2} \in \mathbb{Z},$$

which simplifies to $p\, a_0 \in \mathbb{Z}$. Replacing x by $\zeta^{-1}x, \zeta^{-2}x$, etc., all of which lie in A, we obtain $p\, a_1 \in \mathbb{Z}, p\, a_2 \in \mathbb{Z}$, etc.

(ii) To show that the a_i actually lie in $\mathbb{Z}$, we use a trick which goes back at least to Kummer, and use the third basis of lemma 1.8.1 to write

$$px = b_0 + b_1\pi + \cdots + b_{p-2}\pi^{p-2}. \tag{1}$$

Since $p \sim \pi^{p-1}$, we see (successively) that π divides $b_0, b_1, \ldots, b_{p-2}$, so by (iv) of lemma 1.8.2, p divides $b_0, b_1, \ldots, b_{p-2}$. □

1.8.2 A Lemma of Kummer on the Units of $\mathbb{Z}[\zeta]$

The goal of this paragraph is to prove the following lemma, in which p is assumed to be an odd prime and ζ is assumed to lie in the field of complex numbers $\mathbb{C}$.

Lemma 1.8.3 *Let e be a unit contained in $\mathbb{Z}[\zeta]$, and let $\bar{e}$ be the complex conjugate unit. Then we have*

$$e/\bar{e} = \zeta^r$$

with $0 \leq r \leq p - 1$.

Proof. Kummer's proof can be separated into two parts. First, one shows that

$$e/\bar{e} = \pm\zeta^r,$$

and then that the sign is $+$.

(i) Kummer sets

$$e/\bar{e} = a_0 + a_1\zeta + \cdots + a_{p-1}\zeta^{p-1} = E(\zeta)$$

with $E(x) = a_0 + a_1X + \cdots + a_{p-1}X^{p-1}$. Naturally, this expression is not unique!

It follows that we have

$$1 = \bar{e}/\bar{\bar{e}}.\, e/\bar{e} = E(\zeta)E(\zeta^{p-1}) = R(\zeta)$$

where $R(X) = A_0 + A_1X + \cdots + A_{p-1}X^{p-1}$ is the remainder after dividing $E(X)E(X^{p-1})$ by $X^p - 1$.

Replacing X by 1 in the identity of the division, we obtain

$$A_0 + \cdots + A_{p-1} = (a_0 + \cdots + a_{p-1})^2 = 1. \tag{1}$$

Replacing X by ζ in the same identity, we obtain

$$A_0 + A_1\zeta + \cdots + A_{p-1}\zeta^{p-1} = 1, \tag{2}$$

which gives

$$A_0 - 1 = A_1 = \cdots = A_{p-1}.$$

Let k denote this integer. Replacing $A_0, \ldots, A_{p-1}$ by their values as functions of k in (1), we have

$$(a_0 + \cdots + a_{p-1})^2 = 1 + pk \equiv 1 \bmod p,$$

so

$$a_0 + \cdots + a_{p-1} \equiv \pm 1 \bmod p,$$

and, as the a_i are only unique up to translation, we may assume that

$$a_0 + \cdots + a_{p-1} = \pm 1, \tag{3}$$

which implies that for this particular choice of the a_i, we have $k = 0$, so

$$A_0 = 1 \quad \text{and} \quad A_1 = \cdots = A_{p-1} = 0.$$

But if we actually compute $R(X)$, we find that

$$A_0 = a_0^2 + a_1^2 + \cdots + a_{p-1}^2,$$

so one of the a_i^2 is equal to 1 and the others are zero. Thus

$$e/\bar{e} = \pm\zeta^r, \quad 0 \le r \le p-1.$$

(ii) To determine the sign, we write $u \in \mathbb{Z}[\zeta]$ in the basis $\zeta, \ldots, \zeta^{p-1}$ as

$$u = F(\zeta) = c_1\zeta + \cdots + c_{p-1}\zeta^{p-1}.$$

We have

$$u - \bar{u} = c_1(\zeta - \zeta^{-1}) + \cdots + c_{p-1}(\zeta^{p-1} - \zeta^{-p+1}),$$

so we see that $\zeta - \zeta^{-1}$ divides $u - \bar{u}$, so π divides $u - \bar{u}$.
Now if $e/\bar{e} = -\zeta^r = -\zeta^{r+p}$, we can assume that the exponent of ζ is equal to $2s$.
If we set $u = e\zeta^{-s}$, we see that $u = -\bar{u} \in \mathbb{Z}[\zeta]$.
It follows that $u - \bar{u} = 2u$, so p divides the norm of $2u$.
As this is equal to $\pm 2^{p-1}$ we obtain a contradiction. □

Remark 1.8.5 A more modern but less efficient method for proving the same result consists in noting that every unit of $\mathbb{Z}[\zeta]$ is of the form $\rho\zeta^r$ with $\rho \in \mathbb{Z}[\zeta] \cap \mathbb{R}$; for more detail see [St.T] p. 208.

1.8.3 The Ideals of $\mathbb{Z}[\zeta]$

The ring $\mathbb{Z}[\zeta]$ is not always factorial; in particular, it is not factorial when $p = 23$. However, the theory of factorisation in this ring is not completely lost, since $\mathbb{Z}[\zeta]$ is what is known as a *Dedekind domain*[13].

Definition 1.8.3 *Let A be a commutative integral ring. We say that A is a* **Dedekind domain** *if:*

(i) *A is Noetherian.*
(ii) *A is integrally closed.*
(iii) *Every non-zero prime ideal of A is maximal.*

Explanations

(i) We say that A is Noetherian if every ideal of A can be generated by a finite number of elements.
(ii) We say that A is integrally closed if any integral element of the field of fractions of A is contained in A (here, we can assume that A contains $\mathbb{Z}$).
(iii) We say that an ideal I of A is maximal when the only ideal of A which strictly contains I is A itself.

The set of non-zero ideals of a Dedekind domain A can be equipped with a multiplication which makes it into a commutative, simplifiable monoid, whose identity element is A itself. Curiously, the only thing we will need here is the definition of the multiplication of principal ideals, which is exactly as in the case of a principal ideal domain, i.e.

$$(\alpha)(\beta) := (\alpha\beta).$$

In the general case, the product of two ideals I and J is the ideal generated by the products ij of elements $i \in I$ and $j \in J$ (see [Sa] p. 57).

The essential theorem concerning $\mathbb{Z}[\zeta]$ is a special case of the following theorem, the proof of which can be found in [Sa] p. 60–61.

Fundamental Theorem 1.8.3 *In a Dedekind domain, every non-zero ideal can be written as a product of prime ideals, in a unique way up to reordering the factors.*

Now, we need to show that $\mathbb{Z}[\zeta]$ is a Dedekind domain.

Proposition 1.8.2 *Let A be a Dedekind domain with field of fractions K. Let L be a separable extension of finite degree of K, and let B be the integral closure of A in L (i.e. B is the set of $x \in L$ which are roots of some monic polynomial with coefficients in A). Then B is also a Dedekind domain.*

Once we know this criterion, we can take $A = \mathbb{Z}$ (which is principal, so a Dedekind domain), $K = \mathbb{Q}$ and $L = \mathbb{Q}(\zeta)$. Then we see that $B = \mathbb{Z}[\zeta]$ is a Dedekind domain.

[13] R. Dedekind 1831–1916.

Remark 1.8.6

(1) In fact, one can show that a Dedekind domain is much more than just a Noetherian ring; for instance an ideal of a Dedekind domain is generated by at most two elements [Hec1] p. 86.

(2) One essential articulation point of the theory is the equivalence of the two following assertions:

(i) the (non-zero) ideal I divides the principal ideal (α)
(ii) $\alpha \in I$ (we say that I "divides" α).

(3) To Kummer, ideals were "abstract numbers".

In fact, to each non-zero ideal $\mathfrak{b}$ of A, we can associate an integral number $b \in \mathbb{C}$ such that $\mathfrak{b}$ is the set of $x \in A$ which are divisible by b (cf. [Hec 1], p. 107, 108).

This remark illustrates the banal fact that when speaking of a mathematical object defined up to isomorphism, the properties of the object are more fundamental than its particular nature.

1.8.4 Kummer's Proof (1847)

To begin with, note that if $x \in \mathbb{Z}[\zeta]$, i.e. if

$$x = a_0 + a_1\zeta + \cdots + a_{p-2}\zeta^{p-2}$$

with $a_i \in \mathbb{Z}$, then

$$\begin{aligned} x^p &\equiv a_0^p + (a_1\zeta)^p + \cdots + (a_{p-2}\zeta^{p-2})^p \\ &\equiv a_0 + a_1 + \cdots + a_{p-2} \bmod p\mathbb{Z}[\zeta], \end{aligned} \tag{1}$$

i.e. x^p is congruent modulo p to an element of $\mathbb{Z}$.

Kummer needed the converse of this proposition when x is a unit, and he also needed to know that if I is an non-zero ideal such that I^p is principal, then I is principal.

Unfortunately, these assertions do not always hold, and Kummer eventually realised this. He then gave the following definition.

Definition 1.8.4 *The prime number p is said to be* **regular** *if the following condition (A) relative to the ring of integers* $\mathbb{Z}[\zeta]$ *of the field of the p-th roots of unity is satisfied:*

(A) *For every non-zero prime ideal I of* $\mathbb{Z}[\zeta]$, *we have*

$$(I^p \text{ is principal}) \implies (I \text{ is principal}).$$

Kummer showed that condition (A) implies the following assertion (B):

(B) *Every unit of* $\mathbb{Z}[\zeta]$ *congruent modulo p to an integer of* $\mathbb{Z}$ *is a p-th power in* $\mathbb{Z}[\zeta]$.

Theorem 1.8.4 (Kummer's Theorem) *If p is a* **regular** *prime > 3, then Fermat's assertion for the exponent p holds.*

Proof. The proof is divided into two cases.

(i) First case: assume that

$$x^p + y^p + z^p = 0$$

with x, y, z relatively prime and

$$xyz \not\equiv 0 \bmod p.$$

In 1847, everyone was considering the possibility of writing

$$(x+y)(x+\zeta y)\cdots(x+\zeta^{p-1}y) = -z^p, \tag{2}$$

and Kummer was able to establish that the factors of the left-hand side were pairwise relatively prime. Indeed, if two of them, say the first two, had a common "ideal" divisor D in $\mathbb{Z}[\zeta]$, then we would have

$$\begin{cases} (\zeta - 1)x \in D \\ (\zeta - 1)y \in D, \end{cases}$$

and as x and y are relatively prime, there exist u and $v \in \mathbb{Z}$ such that $ux + vy = 1$ by the theorem of Bachet (usually known as Bézout's theorem). It follows that $\pi \in D$, so p would divide the first term of (2), so it would divide z, which is impossible.

Since $\mathbb{Z}[\zeta]$ is a Dedekind domain, each of the principal ideals

$$(x+y), (x+\zeta y), \ldots, (x+\zeta^{p-1}y)$$

is the p-th power of a ideal. In particular, $(x+\zeta y) = I^p$, where I is an ideal of $\mathbb{Z}[\zeta]$, and by condition (A), the ideal I must be principal, so

$$x + \zeta y = e\, t^p \tag{3}$$

where $e, t \in \mathbb{Z}[\zeta]$ and e is a unit.

Complex conjugation transforms (3) into

$$x + \bar{\zeta} y = \bar{e}\, \bar{t}^p.$$

Now, we saw in (1) above that

$$t^p \equiv \bar{t}^p \bmod p\mathbb{Z}[\zeta]$$

and furthermore, we know by lemma 1.8.2 that $e/\bar{e} = \zeta^r$. Thus, we have

$$x + \bar{\zeta} y \equiv \zeta^{-r} e\, t^p \equiv \zeta^{-r}(x + \zeta y) \bmod p.$$

Multiplying by ζ^r we have

$$\zeta^r(x+\zeta^{-1}y) \equiv x+\zeta y \bmod p. \tag{4}$$

Note that $r=0$ is impossible, since it implies that

$$(\zeta^2-1)y \equiv 0 \bmod p.$$

So all the coordinates of the left-hand side in the basis (i) of lemma 1.8.1 must be divisible by p, which implies that $y \equiv 0 \bmod p$, contradiction!

If $r=1$, we have

$$\zeta x+y \equiv x+\zeta y \bmod p,$$

which implies that p divides $x-y$. If $r=2$, we have

$$\zeta(\zeta x+y) \equiv x+\zeta y \bmod p.$$

This gives $x \equiv 0 \bmod p$, contradiction!

Similarly, we see that the other values of r up to $p-2$ are impossible. Only $r=p-1$ remains; multiplying by ζ^2 we have the congruence

$$\zeta x+y \equiv \zeta^2+\zeta^3 y \pmod p,$$

which again gives $x \equiv 0 \bmod p$ by lemma 1.8.1. Thus, we must have $x \equiv y \bmod p$, and by symmetry,

$$x \equiv y \equiv z \bmod p.$$

It follows that

$$3x^p \equiv 0 \bmod p,$$

and as $p>3$, we obtain a contradiction.

(ii) Second case: assume that z is divisible by p.
Then all the factors of the left-hand side of (2) are divisible by π, since we have

$$x + y \equiv x^p + y^p \equiv 0 \bmod p.$$

Hence, for example

$$x + \zeta y = (x + y) + \pi y \equiv 0 \bmod \pi.$$

Moreover, we see that $x + \zeta y$ (for example) is divisible only once by π. Up to this difference, we can apply the reasoning of (i), and we obtain

$$\begin{cases} x + \zeta y = \pi \; e_1 \; t_1^p \\ x + \zeta^{-1} y = \pi \; e_{-1} \; t_{-1}^p \\ x + y = \pi \; e_0 \pi^{pk} \; t_0^p \end{cases}$$

where e_1, e_{-1}, e_0 are units and t_1, t_{-1}, t_0 are pairwise relatively prime integers of $\mathbb{Z}[\zeta]$ not divisible by π, and where $k = n(p-1) - 1$ with $n \geq 1$. Eliminating x and y by using the three linear equations, we obtain

$$e_1 t_1^p - e_0(1 + \zeta)\pi^{pk} t_0^p + \zeta e_{-1} t_{-1}^p = 0.$$

It is easy to see that $u = 1 + \zeta$ is a unit since $\zeta^2 - 1 = u\pi$, so we have

$$t_1^p + \zeta(e_{-1}/e_1) \, t_{-1}^p = e \, \pi^{pk} t_0^p$$

where e is a unit. But modulo p, we have

$$t_1^p + \zeta(e_{-1}/e_1) \, t_{-1}^p \equiv 0 \bmod p,$$

and it follows from the considerations at the beginning of 1.8.3 that $\zeta(e_{-1}/e_1)$ is congruent to a rational integer modulo p. By condition (B) above, it is then the p-th power of a unit of $\mathbb{Z}[\zeta]$, so we obtain an equation of the type

$$x^p + y^p = e \, \pi^{pk} z^p, \quad k \in \mathbb{N}^* \tag{5}$$

with $x, y, z \in \mathbb{Z}[\zeta]$ pairwise relatively prime, and such that

$$xyz \not\equiv 0 \bmod \pi.$$

Equation (5) is a generalisation of the second case of Fermat's equation, and can be treated in the same way, replacing y by $\zeta^j y$ with $j \in \{0, 1, \ldots, p-1\}$.

Indeed, we have

$$\begin{cases} x \equiv a_0 + a_1\pi \bmod \pi^2 \\ y \equiv b_0 + b_1\pi \bmod \pi^2 \end{cases}$$

with a_0, a_1, b_0, b_1 in $\mathbb{Z}$. Hence

$$\begin{aligned} x + \zeta^j y &\equiv (a_0 + a_1\pi) + (1+\pi)^j(b_0 + b_1\pi) \bmod \pi^2 \\ &\equiv (a_0 + b_0) + [a_1 + b_1 + jb_0]\pi \bmod \pi^2. \end{aligned}$$

As $b_0 \not\equiv 0 \bmod \pi$, there exists an integer $j \in \{0, 1, \ldots, p-1\}$ such that $a_1 + b_1 + jb_0 \equiv 0 \bmod p$. As π divides

$$(x+y)(x+\zeta y)\cdots(x+\zeta^{p-1}y) \equiv (a_0+b_0)^p \bmod \pi,$$

we see that $a_0 + b_0 \equiv 0 \bmod p$. Thus, if we replace y by $\zeta^j y$, we have

$$\begin{cases} x + \zeta y = \pi e_1\, t_1^p \\ x + y = \pi^{p(k-1)+1} e_0\, t_0^p, \text{ with } p(k-1)+1 \geq 2 \\ x + \zeta^{-1} y = \pi e_{-1}\, t_{-1}^p, \end{cases}$$

which implies that

$$k > 1. \tag{6}$$

As above, we find a new equation of the type

$$x^p + y^p = e\,\pi^{pk_1} z^p$$

with $xyz \not\equiv 0 \bmod \pi$ and $k_1 = k - 1$. After a certain number of steps, we arrive at

$$x^p + y^p = e\,\pi^{pk_n} z^p$$

with $k_n = 1$, which is is impossible by (6). □

Note that only the second part of the proof (**second case** of Fermat's equation) uses **infinite descent**.

Remark 1.8.7 It is more difficult to apply Kummer's methods to the equation

$$x^p + y^p + 2z^p = 0. \tag{G_p}$$

However, Dénes [Den] established a general theorem for a family of odd regular p which implies that (G_p) admits only the primitive non-trivial solution $\pm(1, 1, -1)$ when $p < 31$ (see Chapter 6, Section 6.7).

1.8.5 Regular Primes

Beautiful as it is, Kummer's proof would serve no purpose if there were no regular primes. Thus, Kummer immediately set to searching for a practical criterion to determine if a given odd prime number p is regular.

Kummer's criterion
The prime number $p \geq 5$ is regular if and only if p does not divide any of the numerators of the Bernoulli numbers $B_2, B_4, \ldots, B_{p-3}$.

A proof of this criterion lies beyond the scope of this book (see [Kum], [Ed] Chapter 6 or [B-C]); we restrict ourselves to noting that among the primes less than 100, only 37, 59 and 67 are irregular.

Kummer thought that there existed an infinite number of regular primes, but no one has yet been able to prove this conjecture. However, it is quite easy to prove that there exists an infinite number of irregular primes [I-R], p. 241.

It is now time to recall the definition of the Bernoulli numbers, numbers which first appeared in a treatise on probability entitled "Ars Conjectandi", which was published in 1713, eight years after the death of its author, Jakob Bernoulli[14].

In Jakob Bernoulli's treatise, the Bernoulli numbers are the universal coefficients $A, B, C, \ldots$ of a formula which must hold for every $c \in \mathbb{N}$:

$$\begin{aligned}1^c + \cdots + n^c = {} & \frac{1}{c+1} n^{c+1} + \frac{1}{2} n^c + \frac{c}{2} A\, n^{c-1} + \frac{c(c-1)(c-2)}{2 \cdot 3 \cdot 4} B\, n^{c-3} \\ & + \frac{c(c-1)(c-2)(c-3)(c-4)}{2 \cdot 3 \cdot 4 \cdot 5 \cdot 6} C\, n^{c-5} + \cdots\end{aligned}$$

A simple computation shows that

$$A = \frac{1}{6}, \quad B = -\frac{1}{30}, \quad C = \frac{1}{42}, \ldots.$$

Modern notation leads to the following definition.

Definition 1.8.5 *The sequence $B_0, B_1, B_2, \ldots$ of Bernoulli numbers is defined by $B_0 = 1$ and*

$$\begin{cases} 1 + 2B_1 = 0 \\ 1 + 3B_1 + 3B_2 = 0 \\ 1 + 4B_1 + 6B_2 + 4B_3 = 0 \\ 1 + 5B_1 + 10B_2 + 10B_3 + 5B_4 = 0 \\ \ldots \end{cases}$$

where the coefficients which appear in the left-hand side are the binomial coefficients.

[14] J. Bernoulli 1654–1705.

Proposition 1.8.3 *We have*

$$\frac{t}{e^t - 1} = \sum_{m=0}^{\infty} B_m \frac{t^m}{m!}.$$

Proof. If we multiply the two terms of the equation by $e^t - 1$, we obtain the induction relation of the definition. □

Theorem 1.8.5 (Bernoulli's Theorem) *For $m \geq 1$, set*

$$S_m(n) = 1^m + \cdots + (n-1)^m.$$

Then we have

$$(m+1)S_m(n) = \sum_{k=0}^{m} \binom{m+1}{k} B_k\, n^{m+1-k}.$$

Proof. We start from the identity

$$e^{kt} = \sum_{m=0}^{\infty} k^m \frac{t^m}{m!}, \qquad k = 0, 1, 2, \ldots, m-1,$$

and add to obtain

$$1 + e^t + e^{2t} + \cdots + e^{(n-1)t} = \sum_{m=0}^{\infty} S_m(n) \frac{t^m}{m!}.$$

We next remark that the left-hand side is

$$\frac{e^{nt} - 1}{e^t - 1} = \frac{e^{nt} - 1}{t} . \frac{t}{e^t - 1} = \left(\sum_{h=1}^{\infty} n^h \frac{t^{h-1}}{h!} \right) \left(\sum_{k=0}^{\infty} B_k \frac{t^k}{k!} \right),$$

and taking the product, we find the result. □

Remark 1.8.8 **(1)** We easily show that for odd $k > 1$, we have $B_k = 0$. Also, the number $(-1)^{m+1} B_{2m}$ is positive and $|B_{2m}| > 2(m/\pi e)^{2m}$; this implies that the numerators of the Bernoulli numbers of even index become very large as m tends to infinity.

(2) Certain identities concerning the Bernoulli numbers are much easier to memorise by using the so-called "symbolic" notation.

If we suppose that $f(X) = a_0 + a_1 X + \cdots + a_n X^n \in \mathbb{C}[X]$, then by definition we write

$$f(B) := a_0 B_0 + a_1 B_1 + \cdots + a_n B_n.$$

With this notation, Bernoulli's theorem becomes

$$S_m(n) = \frac{1}{m+1} \left[(n+B)^{m+1} - (B)^{m+1} \right],$$

i.e.

$$\sum_{h=0}^{n-1} h^m = \left[\frac{(X+B)^{m+1}}{m+1}\right]_0^n.$$

More generally, if $g(X)$ denotes an integral of the polynomial $f(X)$, we have the summation formula

$$\sum_{h=0}^{n-1} f(h) = [g(X+B)]_0^n.$$

(3) A famous formula of Euler[15] gives the values of $\zeta(2n)$ in terms of π^{2n} and B_{2n}. Indeed, recall that the **Riemann**[16] ζ **function** is defined by

$$\zeta(s) = \sum_{n\geq 1} \frac{1}{n^s} \quad \text{when } Re(s) > 1.$$

Furthermore, a classical identity of Euler ([Va] p. 43) gives

$$\pi \operatorname{cotg} \pi z = \frac{1}{z} + \sum_{m=1}^{\infty} \frac{2z}{z^2 - m^2}. \tag{7}$$

It follows that

$$\pi z \operatorname{cotg} \pi z = i\pi z + \frac{2i\pi z}{e^{2i\pi z} - 1} = 1 + 2\sum_{m=1}^{\infty} \frac{z^2}{z^2 - m^2},$$

i.e.

$$i\pi z + \sum_{m=0}^{\infty} (2i\pi)^m B_m \frac{z^m}{m!} = 1 - 2\sum_{n=0}^{\infty} \zeta(2n) z^{2n},$$

which gives **Euler's famous formula**

$$\zeta(2n) = -\frac{(2i\pi)^{2n}}{2(2n)!} B_{2n} = (-1)^{n+1} \frac{(2\pi)^{2n}}{2(2n)!} B_{2n}. \tag{8}$$

1.9 THE CURRENT APPROACH

Let us give a point-by-point summary of the mathematical history covered in this Chapter.

(a) In 1.5, we explained how Fermat, using infinite descent, was able to prove his famous assertion for $n = 4$, while remaining within the framework of the arithmetic of the ring of rational integers.

[15] L. Euler 1707–1783.

[16] B. Riemann 1826–1866.

(b) The case where $n = 3$ is more difficult (although Fermat did assert several times that he had solved it). The proof we gave above in 1.7 (and which could easily have been Fermat's own) obliges us to extend the narrow framework of the rational integers to

- either the representation of the positive integers by the quadratic form $X^2 + 3Y^2$
- or the arithmetic of the ring $\mathbb{Z}[\sqrt{-3}]$,

and, as Euler understood, these two approaches are actually closely related.

(c) When n is a relatively large prime number p, the first approach, i.e. the one using the representation of the rational integers by quadratic forms, does not give rise to any conclusions, but Kummer managed to develop the arithmetic of $\mathbb{Z}[\zeta]$ deeply enough to obtain a general proof of Fermat's assertion for a large number of values of p (as we saw in 1.8), though not for **all** values of p.

(d) The current approach (which made its first appearance in 1969, at the Journées Arithmétiques in Bordeaux, see the Appendix) consists in extending the framework for studying Fermat's assertion once again, by associating to every solution in relatively prime integers a, b, c of the Fermat equation

$$a^p + b^p + c^p = 0$$

a certain elliptic curve $E_{A,B,C}$ of equation

$$\begin{cases} Y^2 = X(X - a^p)(X + b^p) & \text{or} \\ Y^2 = X(X + a^p)(X - b^p) \end{cases}$$

(one of these curves is always a little better than the other).

In this way, one passes from the **finite** extension $\mathbb{Z}[\zeta]$ of $\mathbb{Z}$ to the **infinite** (and transcendental) extension $\mathbb{Z}[X, \sqrt{X(X \mp a^p)(X \pm b^p)}]$.

In 1969, it was not possible to develop a deep theory of the curves $E_{A,B,C}$, because the theory of elliptic curves was itself not developed enough to conclude (although it was already quite possible to perceive that the curve $E_{A,B,C}$ was "very beautiful"). But in 1986, K. Ribet showed that if a (relatively) old conjecture concerning certain elliptic curves was known to hold, then applying that conjecture to the curve $E_{A,B,C}$ gave a proof of Fermat's assertion.

This last point shows us the direction we will next take: an introduction to elliptic curves.

COMMENTARY

This Chapter is essentially historical in nature; it follows the various attempts to approach Fermat's last theorem, by methods which can be described by the following list.

The pedestrian route

This route was taken by Fermat himself for the exponent 4, then by Lamé and Lebesgue for the exponent 7.

The procedure consists in reducing the problem to the impossibility of finding integral solutions to the equations $f^2 = r^4 + 4s^4$ when $n = 4$, and $f^2 = r^4 - (3/4)\, r^2s^2 + (1/7)\, s^4$ when $n = 7$, using the method of infinite descent.

Quadratic forms
Perhaps, for odd p, Fermat associated the quadratic form $X^2+(-1)^{(p+1)/2}pY^2$ to his equation. In any case, it is certain that Euler did this for $p = 3$, and then Legendre and Dirichlet did it for $p = 5$.

Cyclotomic extensions
From De Moivre[17] in the 17th century, we know how to factor $x^p + y^p$ into a product of first-degree factors in x and y, which makes it natural to consider the set of numbers deduced from $\zeta_p = e^{2i\pi/p}$ by addition, subtraction and multiplication, i.e. the ring of cyclotomic integers. Using this, Kummer succeeded in proving Fermat's assertion for all "regular" primes. Unfortunately, it is still not known whether there exists an infinity of regular primes.

The elliptic approach
This approach is much more recent than the ones described above; it dates back to 1969, and the goal of this book is to describe it.

For details concerning "the pedestrian route" and "quadratic forms", the reader should study the book by A. Weil [Wei], the article by Christian Houzel [Hou 1], and the book by Catherine Goldstein [Go]. I would also like to indicate the intriguing theories developed by Van der Waerden *et al.* (see [F-G] and [VW1]), for those readers with prehistorical interests.

"Kummer's monument" is described with talent in the book by Edwards, [Ed]; this book provides a real introduction to algebraic number theory. We also refer to the book by Samuel [Sa] for theoretical notions about Dedekind domains.

The elliptic approach is in fact the main object of the present text. A great many different historical points of view on elliptic curves can be found in Chapter 4 and in the exercises and problems. For a more systematic presentation of the contributions of Fermat and Euler to this theory, we refer to Weil's book. We should also draw attention to the Appendix at the end of this volume, which reproduces the text of a lecture given in Cambridge, on November 28, 1995, explaining the circumstances which gave rise to the construction of elliptic curves linked to hypothetical non-trivial solutions of Fermat's equation.

Exercises and Problems for Chapter I

1.1 Give an exhaustive solution in rational numbers to the problem of Diophantus' which Fermat commented in the famous margin (see the introduction to Chapter 1).

1.2 Solve the following problem of Diophantus:

Find two (rational) numbers such that the square of each of them, augmented by the sum of these two numbers, forms a square.

Comment on the following solution, proposed by Diophantus:

We assume that the smallest number is x and the largest x + 1, so that $x^2 + (2x + 1) = \square$. *But we also need* $(x + 1)^2 + (2x + 1) = \square$.

We set:

$$x^2 + 4x + 2 = (x - 2)^2$$

and we find $(1/4, 5/4)$.

[17] A. De Moivre 1667–1754.

1.3 (A systematised version of Diophantus)
Consider the conic

$$Y^2 = aX^2 + bX + c.$$

Find a rational point on this conic if $a = \square$, $c = \square$ or $a + b + c = \square$.
Parametrise the conic.

1.4 (A systematised version of Diophantus)
Consider the biquadratic curve

$$\begin{cases} Y^2 = aX^2 + bX + c := P(X) \\ Z^2 = a'X^2 + b'X + c' := Q(X) \end{cases}$$

If P and Q have a common root on the projective line, this curve is of "genus 0" (parametrisable). Otherwise it is of "genus 1" (non-parametrisable).
Show that if (as in Diophantus) $P - Q$ has two rational solutions on the projective line, then the quadric

$$Y^2 - Z^2 = P - Q$$

has a rational parametrisation over $\mathbb{Q}$. Show that if a and a' (or c and c') are both squares, the biquadratic equation has two rational points (even in genus 1) in projective space.

1.5 We propose to show that the surface of equation

$$x^4 + y^4 = z^2$$

has no non-trivial integral points.

(i) Show that we can assume that $(x, y) = 1$.
(ii) Assuming that $(x, y) = 1$ and y is even, show that there exists a and $b \in \mathbb{Z}$ such that

$$x^2 = a^2 - b^2, \qquad y^2 = 2ab.$$

(iii) Deduce that there exists c and $d \in \mathbb{Z}$ such that

$$y^2 = 4cd(c^2 + d^2)$$

and that $(c, d) = 1$, $cd \equiv 0 \bmod 2$.
(iv) Deduce the existence of $u, v, w \in \mathbb{Z}$ such that

$$c = u^2, \qquad d = v^2, \qquad c^2 + d^2 = w^2.$$

(v) Conclude by using infinite descent.

1.6 Consider the equation in $\mathbb{Z}^3$

$$x^4 - y^4 = z^2 \qquad (E)$$

(i) Show that (E) has no **primitive** non-trivial solutions for which y is even (this can be deduced from Fermat's theorem on the area of Pythagorean triangles).

(ii) We propose to show that the equation (E) has no non-trivial solutions.

Show that if (a, b, c) is a **primitive** solution of (E) for which $|a|$ is minimal and b is odd, the equation (E) has one other solution of the same type for which $0 < |x| < |a|$. Conclude.

(iii) Using paragraph 1.5, show that all the primitive solutions of

$$x^4 + y^4 = 2z^2 \qquad (E')$$

satisfy $x^2 = y^2 = z^2 = 1$.

1.7 We propose to show that the equation

$$x^3 + y^3 = 2z^3 \qquad (G_3)$$

admits only the primitive solutions $\pm(1, -1, 0)$ and $\pm(1, 1, 1)$.

(i) Proceeding as in 1.7, show that

$$x = p + q, \qquad y = p - q$$

with p and q relatively prime of different parity, such that

$$p(p^2 + 3q^2) = z^3, \quad pq \neq 0$$

(ii) Show that if 3 does not divide z, we have

$$p = r^3, \qquad p^2 + 3q^2 = s^3.$$

Deduce that there exist relatively prime u and $v \in \mathbb{Z}$ such that

$$r^3 = p = u(u + 3v)(u - 3v), \qquad q = 3v(u^2 - v^2).$$

Deduce a second solution of (G_3) for which

$$a^3 = u + 3v, \qquad b^3 = u - 3v, \qquad c^3 = u$$

and show that $|c| < |z|$ if the solution (x, y, z) is not $\pm(1, 1, 1)$.
Show that $|c| > 1$ and conclude.

(iii) Now assume that 3 divides z.
Show that 3^2 divides p, but not q. Set

$$p = 3^2 p', \qquad z = 3z'.$$

Show that

$$p'(27p'^2 + q^2) = z'^3$$

and that

$$p' = r^3 \qquad 27p'^2 + q^2 = s^3.$$

Deduce that there exist relatively prime u and $v \in \mathbb{Z}$ such that

$$q = u(u+3v)(u-3v), \qquad r^3 = p' = v(u^2 - v^2).$$

Deduce a second solution of (G_3) for which

$$a^3 = v + u, \qquad b^3 = v - u, \qquad c^3 = v$$

and show that $|c| < |z|$.
Show that $|c| > 1$ and conclude.

1.8 Show that it is impossible to find four squares in an arithmetic progression.

1.9 Show that every divisor of $x^2 + 2y^2$ is of this form.

1.10 Show that every odd divisor of $x^2 + 3y^2$ is of this form.

1.11 Show that every divisor of $x^2 - 2y^2$ is of this form.

1.12 Show that if $N^3 = a^2 + Ab^2$, with $(a, b) = 1$ and $A = 2$, then $N = u^2 + Av^2$ with (u, v) such that

$$a = u^3 - 3Auv^2, \qquad b = 3u^2v - Av^3.$$

Solve the same problem for $A = 3$ and N odd.

1.13 **Mason's Theorem** (anticipated by Liouville, Korkine and Stothers).
For every non-zero $P(t) \in \mathbb{C}[t]$, set

$$\text{rad}\, P(t) = \prod_{\alpha \in \text{Roots}\,(P)} (t - \alpha) \in \mathbb{C}[t].$$

Note that if $P \in \mathbb{C}^*$, then $\text{rad}\, P = 1$.

(i) Consider three non-zero pairwise relatively prime polynomials A, B, C in $\mathbb{C}(t)$, such that

$$A + B + C = 0.$$

Show that we have

$$\begin{vmatrix} A' & B' \\ A & B \end{vmatrix} = \begin{vmatrix} B' & C' \\ B & C \end{vmatrix} = \begin{vmatrix} C' & A' \\ C & A \end{vmatrix}.$$

(ii) Deduce that if $AB \notin \mathbb{C}$, then

$$\deg\left(\frac{ABC}{\text{rad}(ABC)}\right) < \deg(AB).$$

(iii) Show that if $\deg(ABC) > 0$, then

$$\sup(\deg A, \deg B, \deg C) < \deg \text{rad}(ABC).$$

(iv) Deduce that if $n \geq 3$, the curve $X^n + Y^n + Z^n = 0$ cannot be parametrised by elements of $\mathbb{C}(t)$.

1.14 Use Mason's theorem to show that the biquadratic curve

$$\begin{cases} Y^2 = X^2 + T^2 \\ Z^2 = X^2 - T^2 \end{cases}$$

cannot be parametrised by elements of $\mathbb{C}(t)$ (giving a non-constant parametrisation).

1.15 Show that the area of a Pythagorean triangle with integral sides is always divisible by 6. Find a Pythagorean triangle with **integral** sides whose area is equal to 6.

Let n be a square-free integer. We say that n is "congruent" if it is the area of a Pythagorean triangle with rational coefficients.

What is the value of n if the sides of the right angle are $3/2$ and $20/3$? Can one find a triangle with integral sides having the same area?

1.16 Set $L = \mathbb{Q}(\sqrt{-5}) \subset \mathbb{C}$.

(i) Show that the integral closure of $\mathbb{Z}$ in L is $\mathbb{Z}[\sqrt{-5}]$.
(ii) Show that the units of $\mathbb{Z}[\sqrt{-5}]$ are the numbers ± 1.
(iii) Show that 2 and 3 remain irreducible in $\mathbb{Z}[\sqrt{-5}]$.
(iv) Show that $1 + \sqrt{-5}$ is irreducible in $\mathbb{Z}[\sqrt{-5}]$. What can one say of $1 - \sqrt{-5}$?
(v) Using the relation

$$2 \cdot 3 = (1 + \sqrt{-5})(1 - \sqrt{-5}),$$

show that $\mathbb{Z}[\sqrt{-5}]$ is not a factorial ring.
(vi) Knowing that $\mathbb{Z}[\sqrt{-5}]$ is a Dedekind domain, give a decomposition of the principal ideal (6) as a product of prime ideals.

1.17 Let us prove the following theorem of Bachet (1621): *Every positive integer is a sum of four (possibly zero) squares.*

(i) Prove Euler's identity:

$$(a^2 + b^2 + c^2 + d^2)(\alpha^2 + \beta^2 + \gamma^2 + \delta^2) = A^2 + B^2 + C^2 + D^2,$$

with

$$\begin{cases} A = a\alpha + b\beta + c\gamma + d\delta \\ B = a\beta - b\alpha + c\delta - d\gamma \\ C = a\gamma - c\alpha - b\delta + d\beta \\ D = a\delta - d\alpha + b\gamma - c\beta \end{cases}$$

(ii) Deduce that this suffices to prove the crucial theorem

Every prime number is a sum of 4 squares.

(iii) Show that if p is prime there exists $n \in \mathbb{N}$ and four integers a, b, c, d such that

$$\begin{cases} a^2 + b^2 + c^2 + d^2 = np \quad 0 < n < p \\ (a, b, c, d) = 1. \end{cases} \tag{1}$$

Let Σ denote the set of integers of $\mathbb{N}$ which are sums of two squares.
When $p \in \Sigma$, we will show that we can take $c = d = 0$.

When $p \notin \Sigma$, we denote by

$$1,\ r_2,\ r_3,\ldots,r_{(p-1)/2}$$

the set of quadratic residues of p (recall that -1 is not one). We set:

$$E = \{1, 1+1, 1+r_2, \ldots, 1+r_{(p-1)/2}\}.$$

Show that $\#E = (p+1)/2$, and deduce that E contains at least one non-residue. Deduce a relation of the desired type with $d = 0$.

(iv) Divide a, b, c, d by n so that

$$\begin{cases} a = xn + \alpha, & |\alpha| \leq \dfrac{n}{2} \\ b = yn + \beta, & |\beta| \leq \dfrac{n}{2} \\ e = zn + \gamma, & |\gamma| \leq \dfrac{n}{2} \\ d = tn + \delta, & |\delta| \leq \dfrac{n}{2}. \end{cases}$$

Show that

$$\alpha^2 + \beta^2 + \gamma^2 + \delta^2 = nr \quad \text{with } r \leq n. \tag{2}$$

(v) Show that in the case where $r = n$, we have $n = 2$, hence $a^2 + b^2 + c^2 + d^2 = 2p$. Deduce that p is a sum of 4 squares (compare with (vi)).

(vi) Assume now that $r < n$.

Multiplying (1) by (2), we obtain

$$A^2 + B^2 + C^2 + D^2 = prn^2;$$

check that A, B, C and D are divisible by n. Deduce that

$$A'^2 + B'^2 + C'^2 + D'^2 = pr.$$

(vii) Show that we can divide A', B', C', D' by their largest common divisor to obtain

$$a'^2 + b'^2 + c'^2 + d'^2 = ps \quad s \leq r < n < p$$
$$(a', b', c', d') = 1.$$

(viii) Conclude using infinite descent.

1.18 The goal of this exercise is to study the ring

$$A := \mathbb{Z}[\sqrt{-3}] = \{x + y\sqrt{-3};\ x \in \mathbb{Z} \text{ and } y \in \mathbb{Z}\} \subset \mathbb{C}.$$

As in the main text, set

$$N(x + y\sqrt{-3}) = x^2 + 3y^2.$$

(i) Show that N is a multiplicative homomorphism from the monoid $(A^*, \times)$ to $(\mathbb{N}^*, \times)$.

(ii) Determine the units of A.

(iii) Consider the relation

$$2 \times 2 = (1 + \sqrt{-3})(1 - \sqrt{-3}).$$

Show that 2, $1 + \sqrt{-3}$ are irreducible and not associated. Is the ring A a principal ideal domain?

(iv) Is the ring A Noetherian? (recall that every quotient of a Noetherian ring is Noetherian, and that if R is Noetherian then so is $R[X]$).

(v) Is the ring A integrally closed in its field of fractions?

(vi) Determine the integral closure $\mathcal{O}$ of $\mathbb{Z}$ in $\mathbb{Q}(\sqrt{-3})$ (set $j = (-1 + \sqrt{-3})/2$).

(vii) How can one interpret the relation in question (iii) in the Dedekind domain $\mathcal{O}$?

(viii) Show that the set $2A + (1 + \sqrt{-3})A$ is a maximal (so a prime) ideal of A.

(ix) Let $P = (2, 1 + \sqrt{-3})$ denote the ideal above. For every ideal Q, let PQ denote the ideal generated by the products of elements of P and Q. Show that

$$P^2 = (2) \cdot P$$

where (2) denotes the ideal $2A$.

(x) Deduce that A is not a Dedekind domain. Could we have foreseen this result?

1.19 We propose to establish the formula

$$\pi \operatorname{cotg} \pi z = \frac{1}{z} + \sum_{m=1}^{\infty} \frac{2z}{z^2 - m^2}$$

for $z \in \mathbb{C} \setminus \mathbb{Z}$.

(i) Show that the infinite product

$$f(z) = z \prod_{n=1}^{\infty} \left(1 - \frac{z}{n}\right) e^{z/n} \prod_{n=1}^{\infty} \left(1 - \frac{z}{n}\right) e^{-z/n}$$

converges uniformly on every compact subset of $\mathbb{C} \setminus \mathbb{Z}$.

(ii) Show that f is an odd integral function of period 1.

(iii) Set

$$g(z) := \frac{\pi}{\sin \pi z} - \frac{1}{f(z)}.$$

Show that g is bounded in $S := \{z \in \mathbb{C};\, 0 \leq Re(z) \leq 1,\ |\mathrm{Im}\, z| \geq 1\}$ and that $g(z)$ tends to zero when $|\mathrm{Im}(z)|$ tends to infinity.

(iv) Using Liouville's theorem, show that $g = 0$.

(v) Conclude by taking the logarithmic derivative of $(\sin \pi z)/\pi$.

Problem I

A. In the first part of this problem, we undertake to study the Diophantine equation

$$X^2 + 2Y^2 = Z^2. \tag{E}$$

(1) Give a rational parametrisation of the ellipse

$$x^2 + 2y^2 = 1.$$

(2) Deduce that every solution of (E) in $\mathbb{Z}^3$ can be written

$$X = \lambda(2t^2 - s^2), \qquad Y = 2\lambda ts, \qquad Z = \lambda(2t^2 + s^2),$$

with $t \in \mathbb{Z}$, $s \in \mathbb{N}$ and $\lambda \in \mathbb{Q}$.

(3) Let $M(\lambda, s, t)$ denote the solution described in question (2). Compare the solutions $M(\lambda, s, t)$,

$$M(-\lambda, s, t), M(\lambda, s, -t) \text{ and } M\left(\frac{\lambda}{2}, 2|t|, \frac{t}{|t|}s\right) \text{ when } st \neq 0.$$

(4) Show that every integral solution of (E) is proportional to a *primitive* solution (a, b, c), i.e. to a solution such that

$$(b, c) = (c, a) = (a, b) = 1.$$

(5) Show that every primitive solution of (E) in $\mathbb{Z}^3$ is of the type

$$X = \pm(2T^2 - S^2), \qquad Y = 2TS, \qquad |Z| = 2T^2 + S^2$$

with $S \in \mathbb{N}$, $T \in \mathbb{Z}$ and $(2T, S) = 1$.

(6) What can one say about the converse?

B. We now consider the Diophantine equation

$$2X^4 - Y^4 = Z^2. \tag{F}$$

(1) Show that this equation has a "non-trivial" solution (i.e. a solution such that $XYZ \neq 0$), and that for every solution $(x, y, z) \in \mathbb{Z}^3$, there exists $\lambda \in \mathbb{Z}$ and $(a, b, c) \in \mathbb{Z}^3$ such that

$$x = \lambda a, \qquad y = \lambda b, \qquad z = \lambda^2 c$$

and $(a, b) = 1$. Such a solution is called **primitive**.

(2) Show that the curve (F) is not unicursal (i.e. parametrisable in $\mathbb{C}(t)$). This can be done using Mason's theorem.

(3) Let (a, b, c) be a primitive solution of (F), show that $abc \not\equiv 0 \pmod 2$.

(4) Setting $b^2 + c = 2u$ and $b^2 - c = 2v$, show that there exists $(m, n) \in \mathbb{Z}^2$, with $(m, 2n) = 1$, such that

$$\begin{cases} a^2 = m^2 + n^2 \\ b^2 = m^2 - n^2 + 2mn = (m+n)^2 - 2n^2 \\ c = m^2 - n^2 - 2mn = (m-n)^2 - 2n^2. \end{cases}$$

(5) Deduce that there exist $(g, h) \in \mathbb{Z}^2$ and $(s, t) \in \mathbb{Z}^2$ such that

$$\rho gcd(g, h) = \rho gcd(2t, s) = 1 \quad gh \equiv 0 \bmod 2,$$

and

$$\begin{aligned} a^2 &= g^2 + h^2, & m &= g^2 - h^2, & n &= 2gh, \\ b &= \pm(2t^2 - s^2), & n &= 2st, & m + n &= 2t^2 + s^2. \end{aligned}$$

(6) Set $(g, s) = D,\ g = Da_1,\ s = Db_1$.
Deduce from question (5) that

$$h = Eb_1, \quad t = Ea_1 \qquad \text{with } (D, E) = 1$$

and

$$E^2(2a_1^2 + b_1^2) - 2EDa_1b_1 = D^2(a_1^2 - b_1^2).$$

(7) Deduce from question (6) that

$$2a_1^4 - b_1^4 = c_1^2$$

with

$$c_1 = E\frac{2a_1^2 + b_1^2}{D} - a_1b_1.$$

(8) Show that $|a_1| < |a|$ if and only if $(g, h) \neq (1, 0)$, and deduce, that by infinite descent every solution leads to the solution $(1, 1, 1)$.

(9) We now want to "run the computations backwards" to find the solution (a, b, c) from the solution (a_1, b_1, c_1).
Set

$$K_1 = \rho\text{gcd}(2a_1 + b_1^2, a_1b_1 + c_1)$$

and

$$K_2 = \rho\text{gcd}(2a_1^2 + b_1^2, a_1b_1 - c_1).$$

Show that we have either

$$D = \frac{2a_1^2 + b_1^2}{K_1}, \qquad E = \frac{a_1b_1 + c_1}{K_1}$$

or

$$D = \frac{2a_1^2 + b_1^2}{K_2}, \qquad E = \frac{a_1b_1 - c_1}{K_2}.$$

Then, choosing a system (D, E), show that

$$\begin{cases} |a| = (Da_1)^2 + E(b_1)^2 \\ b = \pm(2(Ea_1)^2 - (Db_1)^2) \\ c = ((Da_1)^2 - (Eb_1)^2 - 2DEa_1b_1)^2 - 8D^2E^2a_1^2b_1^2. \end{cases}$$

(10) Show that the solutions of (F) form a tree (each solution has one or two "children", ignoring the signs of a, b and c) whose "root" is $(1, 1, 1)$.
If $(a_1, b_1, c_1) = (1, 1, 1)$, then what is the value of $(|a|, |b|, |c|)$? Does one obtain an infinite number of solutions?

Problem 2

Principal Quadratic Imaginary Fields

Let $\mathbb{N}$ denote the set of natural integers, $\mathbb{Z}$ the ring of integers and $\mathbb{C}$ the field of complex numbers. If p is a prime number, let $\mathbb{F}_p$ denote the finite field $\mathbb{Z}/p\mathbb{Z}$.

Let S be a subring of $\mathbb{C}$, and denote by $M_n(S)$ the ring of $n \times n$ matrices with coefficients in S, and by $GL(n, S)$ the group of invertible elements of $M_n(S)$. For every M in $M_n(S)$, let M^* (resp. ${}^tM$) denote the adjoint matrix (resp. the transpose) of M.

We say that a Hermitian (resp. a real symmetric) matrix A is *positive definite* if the Hermitian form (resp. the symmetric bilinear form) associated to A is positive definite. We say that S is a *principal ideal domain* if every ideal of S can be generated by a single element, and *Euclidean* if there exists a map N from $S - \{0\}$ to $\mathbb{N}$ such that if a and b are two non-zero elements of S, there exists q and r belonging to S satisfying $a = bq + r$ and $r = 0$ or $N(r) < N(b)$.

I. Background

A. In this part, p denotes an odd prime.

(1) (a) Show that if u, v, w are three non-zero elements of $\mathbb{F}_p$, the equation

$$ux^2 + vy^2 = w$$

has a solution in $\mathbb{F}_p$. (Consider the cardinal of the set of elements of the form ux^2 (respectively of the form $w - vy^2$).)

(b) Let $n > 1$ be an integer such that p does not divide $4n - 1$. Show that there exist relative integers a, b and an integer $m \geq 1$ such that

$$a^2 + ab + nb^2 + 1 = mp.$$

(2) Assume that p is of the form $8k + 1$ or $8k + 3$, and let K be an extension of $\mathbb{F}_p$ obtained by adding a root of the polynomial $t^4 + 1$. Let b be a root of this polynomial in K, and set

$$x = b - b^{-1}.$$

(a) Show the following relations: $x^2 = -2$ and $x^p = x$. Deduce that x belongs to $\mathbb{F}_p$.

(b) Show that there exist integers a and m such that

$$2a^2 + 1 = (2m - 1)p,$$

and prove that the matrix

$$\begin{pmatrix} p & a & 0 \\ a & m & 1 \\ 0 & 1 & 2 \end{pmatrix}$$

is a symmetric positive definite matrix of determinant equal to 1.

Determine all the pairs (a, m) when $p = 17$.

B. Let $D \geq 1$ be a square-free integer. Set

$$\omega_D = \begin{cases} i\sqrt{D} & \text{if } D \equiv 1 \text{ or } 2 \pmod 4 \\ (1 + i\sqrt{D})/2 & \text{if } D \equiv 3 \pmod 4. \end{cases}$$

Let $\mathbb{Z}[\omega_D]$ denote the subring of $\mathbb{C}$ which is the set of elements of the form $\alpha + \beta\omega_D$ with $\alpha, \beta \in \mathbb{Z}$.

(1) Let p be a prime number not dividing D. Show that there exist integers a, b, m such that the matrix

$$\begin{pmatrix} p & a + b\omega_D \\ a + b\bar{\omega}_D & m \end{pmatrix}$$

is a positive definite Hermitian matrix with determinant equal to 1.

(2) In the Euclidean plane with respect to an orthonormal basis, let A, B, C denote the images of the numbers $0, 1, \omega_D$ respectively, and T the triangle which is the convex hull of the points A, B, C. Let R denote the radius of the circumscribed circle of T.

(a) Show that for every point M of T, we have

$$\inf(MA, MB, MC) \leq R.$$

(b) We set

$$k = \sup_{z \in \mathbb{C}} \Big(\inf_{u \in \mathbb{Z}[\omega_D]} |z - u|^2 \Big).$$

Prove the equality

$$k = \sup_{M \in T} (\inf(MA^2, MB^2, MC^2)).$$

(c) Deduce that we have

$$\begin{cases} k = \dfrac{D+1}{4} & \text{if } D \equiv 1 \text{ or } 2 \pmod 4 \\ k = \dfrac{(D+1)^2}{16D} & \text{if } D \equiv 3 \pmod 4. \end{cases}$$

(d) Let α, β be two elements of $\mathbb{Z}[\omega_D]$, with β non-zero. Show that there exists an element γ of $\mathbb{Z}[\omega_D]$ such that

$$|\alpha - \gamma\beta|^2 \leq k|\beta|^2.$$

Deduce that $\mathbb{Z}[\omega_D]$ is a Euclidean ring when D is equal to one of the following values: 1, 2, 3, 7, 11.

Application: determine γ when $D = 2$, $\alpha = 5 + 3\omega_2$, $\beta = -1 + 3\omega_2$.

II. Hermitian matrices of the form B^*B

In this part, S denotes the ring $\mathbb{Z}$ or one of the rings $\mathbb{Z}[\omega_D]$ for $D = 1, 2, 3, 7$ or 11. If $S = \mathbb{Z}$, set $k = \frac{1}{4}$, and if $S = \mathbb{Z}[\omega_D]$, then let k be the constant defined in I.B.2.b).

Two Hermitian matrices A and B of $M_n(S)$ are said to be *congruent* if there exists $U \in GL(n, S)$ such that $A = UBU^*$. The equivalence classes for this relation are called congruence classes.

To an element $x = (x_1, \ldots, x_n)$ of S^n, we associate a one-line matrix whose coefficients are the components of x; we use the same symbol x for this matrix. Let ${}^t x$ the transpose matrix, and x^* the matrix ${}^t\bar{x}$.

(1) Show that if A and B are two congruent Hermitian matrices, then $\det A = \det B$.

(2) **(a)** Let A be a positive definite Hermitian matrix belonging to $M_n(S)$. Show that there exists an integer $m(A) > 0$ and an element z belonging to S^n whose components are pairwise relatively prime, such that we have

$$m(A) = \inf_{x \in S^n \setminus \{0\}} xAx^* = zAz^*.$$

(b) Do we have $m(A) = m(B)$ whenever A and B are congruent?

(c) Determine $m(A)$ when $S = \mathbb{Z}$ and

$$A = \begin{pmatrix} 2 & 7 \\ 7 & 25 \end{pmatrix}.$$

A. The case $n = 2$

Let A be a positive definite Hermitian matrix in $M_2(S)$, and let z be an element of S^2 such that $m(A) = zAz^*$.

(1) **(a)** Show that ${}^t z$ is a column vector of an invertible matrix U_0 of $GL(2, S)$, and deduce the existence of a Hermitian matrix $B = (b_{ij})$, $1 \le i, j \le 2$, where $b_{12} = m(A)$, such that A and B are congruent.

(b) Show that there exists $s \in S$ such that

$$|b_{11}s + b_{12}| \le k^{1/2} b_{11}$$

and deduce the existence of a matrix C

$$C = \begin{pmatrix} a & b \\ \bar{b} & c \end{pmatrix}$$

congruent to A which satisfies the following two conditions

(i) $a = m(A) = m(C)$

(ii) $k^{-1/2}|b| \le a < c$.

(c) Show that if $A \in M_2(S)$ is a positive definite Hermitian matrix of determinant equal to d, then

$$m(A) \le (1 - k)^{-1/2} \, d^{1/2}.$$

(d) Deduce that the set of congruence classes of positive definite Hermitian matrices of order 2 with coefficients in S and given determinant is a finite set.

(2) **(a)** Assume that d is equal to 1 and S is one of the following rings:

$$S = \mathbb{Z}, \ S = \mathbb{Z}[\omega_D] \quad \text{for } D = 1, 3, 7.$$

Show that $m(A) = 1$ and that there exists $B \in GL(2, S)$ such that $A = B^*B$.

(b) Deduce the following properties:

(i) Every prime number is a sum of four squares.
(ii) For every prime number p, there exist relative integers a, b, c, d such that

$$p = a^2 + ab + b^2 + c^2 + cd + d^2.$$

(iii) For every prime number p, there exist relative integers a, b, c, d such that

$$p = a^2 + ab + 2b^2 + c^2 + cd + 2d^2.$$

B. Symmetric matrices with integral coefficients

(1) (a) Let $f : \mathbb{Z}^n \to \mathbb{Z}$ be a surjective homomorphism of Abelian groups, and let $x \in \mathbb{Z}^n$ be such that $f(x) = 1$. Show that $\mathbb{Z}^n$ is the direct sum of the subgroup generated by x and the kernel of f.

(b) Let $x = (x_1, \ldots, x_n)$ be an element of $\mathbb{Z}^n$. Show that the following conditions are equivalent:

(i) x belongs to a basis of $\mathbb{Z}^n$.
(ii) There exists $M \in GL(n, \mathbb{Z})$ admitting ${}^t x$ as a column vector.
(iii) There exist relative integers a_i, $1 \leq i \leq n$ such that

$$\sum_{i=1}^{n} a_i x_i = 1.$$

(iv) There exists a surjective homomorphism of Abelian groups $f : \mathbb{Z}^n \to \mathbb{Z}$ such that $f(x) = 1$.

(2) Let A be a symmetric positive definite matrix of order $n > 1$ with coefficients in $\mathbb{Z}$. Show that there exists a matrix $B = (b_{ij})$, $1 \leq i, j \leq n$, congruent to A and such that $b_{11} = m(A)$.

(3) Let $A = (a_{ij})$, $1 \leq i, j \leq n$, be a positive definite symmetric matrix with coefficients in $\mathbb{Z}$ such that $m(A) = a_{11}$. If $x = (x_1, \ldots, x_n)$ is an element of $\mathbb{Z}^n$, we define the element $y = (y_1, \ldots, y_n)$ by the following relations:

$$y_1 = x_1 + \sum_{i=2}^{n} a_{1i} a_{11}^{-1} x_i,$$
$$y_i = x_i \quad \text{for } 2 \leq i \leq n.$$

We set

$$z = (x_2, \ldots, x_n), \qquad {}^t y = U\, {}^t x.$$

(a) Show that we have

$$xA\, {}^t x = a_{11} y_1^2 + a_{11}^{-1} z B\, {}^t z$$

where B is a symmetric positive definite matrix belonging to $M_{n-1}(\mathbb{Z})$ and satisfying the two relations

$$A = {}^tU \begin{pmatrix} a_{11} & 0 \\ 0 & a_{11}^{-1}B \end{pmatrix} U$$

$$\det B = (a_{11})^{n-2} \det A.$$

(b) Show that we have

$$m(A) \leq \left(\frac{4}{3}\right)^{(n-1)/2} (\det A)^{1/n}.$$

Let us choose x so that we have

$$|y_1| \leq \tfrac{1}{2}; \qquad zB\,{}^tz = m(B).$$

(4) **(a)** Assume $n \leq 5$ and let $A \in M_n(\mathbb{Z})$ be a symmetric positive definite matrix whose determinant is equal to 1. Show that $m(A) = 1$ and deduce that there exist $B \in M_n(\mathbb{Z})$ such that $A = {}^tBB$.

(b) Show that every prime number of the form $8n+1$ or $8n+3$ is a sum of three squares.

III. Ideal classes and principal rings

Recall that two elements A and B of $M_n(\mathbb{Z})$ are similar if there exists an element Q of $GL(n, \mathbb{Z})$ such that $A = QBQ^{-1}$; the equivalence classes for this relation are called similarity classes.

A. Let $P(X)$ be a monic polynomial of degree $n > 1$ with coefficients in $\mathbb{Z}$, and irreducible over $\mathbb{Q}[X]$. If θ is a complex root of $P(X)$, let $\mathbb{Z}[\theta]$ denote the subring of $\mathbb{C}$ which is the set of elements of the form

$$\sum_{i=0}^{n-1} a_i\theta^i \quad \text{where } a_i \in \mathbb{Z} \text{ for } i = 0, \ldots, n-1.$$

We say that two ideals I and J of $\mathbb{Z}[\theta]$ belong to the same class if there exist two non-zero elements a and b of $\mathbb{Z}[\theta]$ such that $aI = bJ$.

Let A denote an element of $M_n(\mathbb{Z})$ such that $P(A) = 0$.

(1) Show that every non-zero ideal of $\mathbb{Z}[\theta]$ is a free Abelian group of rank n.

(2) **(a)** Show that there exists an element $x = (x_1, \ldots, x_n)$ of $\mathbb{Z}[\theta]^n \setminus \{0\}$ such that $A^tx = \theta^tx$.

(b) Show that $\mathbb{Z}x_1 + \cdots + \mathbb{Z}x_n$ is an ideal of $\mathbb{Z}[\theta]$ whose class is independent of the chosen eigenvector ${}^tx$.

(c) Let Q be an element of $GL(n, \mathbb{Z})$. Show that

$$I_A = I_{QAQ^{-1}}.$$

(3) Let $J = \mathbb{Z}y_1 + \cdots + \mathbb{Z}y_n$ be an ideal of $\mathbb{Z}[\theta]$, and set

$$y = (y_1, \ldots, y_n).$$

Show that there exists a matrix B with integral coefficients such that

$$B^ty = \theta^ty, \qquad P(B) = 0.$$

(4) Show that there exists a bijection between the set of similarity classes of matrices A in $M_n(\mathbb{Z})$ such that $P(A) = 0$ and the set of ideal classes of $\mathbb{Z}[\theta]$.

(5) Show that the following conditions are equivalent:

(i) $\mathbb{Z}[\theta]$ is a principal ring

(ii) There exists a single similarity class in $M_n(\mathbb{Z})$ of matrices A of order n with integral coefficients such that $P(A) = 0$.

B. Let $D \geq 1$ denote a square-free integer; let $\mathbb{Z}[\omega_D]$ be the ring introduced in I.B.

(1) Assume that $D \equiv 1$ or $2 \pmod 4$, and let

$$A(\alpha, \beta, \gamma) = \begin{pmatrix} -\alpha & \beta \\ \gamma & \alpha \end{pmatrix}$$

denote a matrix with coefficients in $\mathbb{Z}$ whose characteristic polynomial is

$$P(X) = X^2 + D.$$

Considering the values $\alpha = 0$ and $\alpha = 1$, show that $\mathbb{Z}[\omega_D]$ is principal if and only if $D = 1$ or 2.

(2) Assume that $D \equiv 3 \pmod 4$, and set

$$K = \frac{D+1}{4}.$$

Let A be an element of $M_2(\mathbb{Z})$ whose characteristic polynomial is

$$P(X) = X^2 - X + K.$$

(a) Let

$$B = \begin{pmatrix} -a & -b \\ c & a+1 \end{pmatrix}$$

be a matrix similar to A such that $|a|$ is minimal.

Computing PAP^{-1} when P is one of the following matrices:

$$\begin{pmatrix} 1 & n \\ 0 & 1 \end{pmatrix} \quad \begin{pmatrix} 1 & 0 \\ n & 1 \end{pmatrix} \quad \begin{pmatrix} 0 & 1 \\ 1 & 0 \end{pmatrix} \quad \begin{pmatrix} 1 & 0 \\ 0 & -1 \end{pmatrix}$$

show that we can assume that the coefficients of B satisfy

$$a \geq 0, \qquad c \geq 2a+1, \qquad b \geq 2a+1, \qquad 3(a^2+a)+1 \leq K.$$

(b) Let α, β, γ be three integers such that

$$0 \leq \alpha < K-1, \qquad 1 < \beta \leq \gamma, \qquad \beta\gamma = K + \alpha^2 + \alpha.$$

Show that for any element (x, y) of $\mathbb{Z}^2 \setminus \{0\}$, we have

$$\beta x^2 + \gamma y^2 + (2\alpha + 1)xy > y^2.$$

Deduce that the matrices

$$A = \begin{pmatrix} 0 & -K \\ 1 & 0 \end{pmatrix} \qquad M = \begin{pmatrix} -\alpha & -\gamma \\ \beta & \alpha + 1 \end{pmatrix}$$

are not similar.

(c) Assume that $\mathbb{Z}[\omega_D]$ is a principal ideal domain. Show that either $K = 1$ or $K + a^2 + a$ is a prime number for every integer a such that $0 \leq a < K - 1$.

(d) Assume that $K = 1$ or that $K + a^2 + a$ is prime for every $a \geq 0$ such that $3(a^2 + a) + 1 \leq K$. Prove that $\mathbb{Z}[\omega_D]$ is a principal ideal domain.

(e) Assume that $D \leq 200$. Prove that $\mathbb{Z}[\omega_D]$ is principal if and only if $D = 3, 7, 11, 19, 43, 67, 163$.

(f) Assume that $D \leq 10^6$. Write a program checking that the values you find are the only ones for which the ring $\mathbb{Z}[\omega_D]$ is principal.

C. Let S denote one of the rings $\mathbb{Z}[\omega_D]$ for $D = 19, 43, 67, 163$, and assume that S is Euclidean for a map N from $S \setminus \{0\}$ to $\mathbb{N}$. Let a be a non-invertible element of $S \setminus \{0\}$ such that $N(a)$ is minimal.

(1) Show that S/aS is isomorphic to one of the fields $\mathbb{F}_2$ or $\mathbb{F}_3$.

(2) Deduce that for $D = 19, 43, 67, 163$, $\mathbb{Z}[\omega_D]$ is a non-Euclidean principal ideal domain.

Problem 3

I. Let us give a translation into English of the passage in Jakob Bernoulli's text *Ars Conjectandi* (1713) where the "Bernoulli numbers" first appear.

> *We give ourselves the sequence* $1, 2, \ldots, n$ *and ask to compute the sum of these numbers, of their squares, their cubes, etc.*
> *As, in the table of combinations, the general term of the second column is* $n - 1$ *and we know that the sum of these* $n - 1$ *terms, or* $\int \overline{n-1}$, *is*
>
> $$\frac{n(n-1)}{2} = \frac{nn - n}{2}$$
>
> *the sum* $\int \overline{n-1}$ *or*
>
> $$\int n - \int 1 = \frac{nn - n}{2}.$$
>
> *Thus we have*
>
> $$\int n = \frac{nn - n}{2} + \int 1.$$
>
> *But* $\int 1 = n$, *so the sum of all the n is*
>
> $$\int n = \frac{nn - n}{2} + n = \frac{1}{2} n \cdot n + \frac{1}{2} n.$$

The general term of the third column is

$$\frac{(n-1)(n-2)}{1\cdot 2}=\frac{n\cdot n-3n+2}{2}$$

and the sum of all the terms (i.e. of all the $(n\cdot n-3n+2)/2)$ *is*

$$\frac{n(n-1)(n-2)}{1\cdot 2\cdot 3}=\frac{n^3-3nn+2n}{6}.$$

Thus

$$\int\overline{\frac{n^2-3n+2}{2}}=\int\frac{1}{2}nn-\int\frac{3}{2}n+\int 1=\frac{n^3-3nn+2n}{6}$$

and

$$\int\frac{1}{2}nn=\frac{n^3-3nn+2n}{6}+\int\frac{3}{2}n-\int 1.$$

But

$$\int\frac{3}{2}n=\frac{3}{2}\int n=\frac{3}{4}nn+\frac{3}{4}n$$

and

$$\int 1=n.$$

Substituting, we have

$$\begin{aligned}\int\frac{1}{2}nn&=\frac{n^3-3nn+2n}{6}+\frac{3nn+3n}{4}-n\\&=\frac{1}{6}n^3+\frac{1}{4}nn+\frac{1}{12}n\end{aligned}$$

so the double $\int nn$ *(sum of all the squares)* $=(1/3)n^3+(1/2)nn+(1/6)n$.
The general term of the fourth column is:

$$\frac{(n-1)(n-2)(n-3)}{1.2.3}=\frac{n^3-6nn+11n-6}{6}$$

and the sum of all the terms is

$$\frac{n(n-1)(n-2)(n-3)}{1.\,2.\,3.\,4}=\frac{n^4-6n^3+11nn-6n}{24}.$$

Thus we have

$$\begin{aligned}\int\overline{\frac{n^3-6nn+11n-6}{6}}&=\int\frac{1}{6}n^3-\int nn+\int\frac{11}{6}n-\int 1\\&=\frac{n^4-6n^3+11nn-6n}{24},\end{aligned}$$

hence

$$\int \frac{1}{6}n^3 = \frac{n^4 - 6n^3 + 11nn - 6n}{24} + \int nn - \int \frac{11}{6}n + \int 1.$$

Earlier, we obtained $\int nn = (1/3)n^3 + (1/2)nn + (1/6)n$, $\int (11/6)n = (11/6)\int n = (11/12)nn + (11/12)n$, *and* $\int 1 = n$.

After the substitutions, we find

$$\begin{aligned}\int \frac{1}{6}n^3 &= \frac{n^4 - 6n^3 + 11nn - 6n}{24} + \frac{1}{3}n^3 + \frac{1}{2}nn + \frac{1}{6}n - \frac{11}{12}nn - \frac{11}{12}n + n \\ &= \frac{1}{24}n^4 + \frac{1}{12}n^3 + \frac{1}{24}nn\end{aligned}$$

or, multiplying by 6,

$$\int n^3 = \frac{1}{4}n^4 + \frac{1}{2}n^3 + \frac{1}{4}nn.$$

Thus we successively obtain sums of higher and higher powers, and with little effort we obtain the following table:

Sums of Powers

$$\begin{cases} \int n = \frac{1}{2}nn + \frac{1}{2}n \\ \int nn = \frac{1}{3}n^3 + \frac{1}{2}nn + \frac{1}{6}n \\ \int n^3 = \frac{1}{4}n^4 + \frac{1}{2}n^3 + \frac{1}{2}nn \\ \int n^4 = \frac{1}{5}n^5 + \frac{1}{2}n^4 + \frac{1}{3}n^3 * -\frac{1}{30}n \\ \int n^5 = \frac{1}{6}n^6 + \frac{1}{2}n^5 + \frac{5}{12}n^4 * -\frac{1}{12}nn \\ \int n^6 = \frac{1}{7}n^7 + \frac{1}{2}n^6 + \frac{1}{2}n^5 * -\frac{1}{6}n^3 * +\frac{1}{42}n \\ \int n^7 = \frac{1}{8} + \frac{1}{2}n^7 + \frac{7}{12}n^6 * -\frac{7}{24}n^4 * +\frac{1}{12}nn \\ \int n^8 = \frac{1}{9}n^9 + \frac{1}{2}n^8 + \frac{2}{3}n^7 * -\frac{7}{15}n^5 * +\frac{2}{9}n^3 * -\frac{1}{30}n \\ \int n^9 = \cdots \\ \int n^{10} = \cdots \end{cases}$$

Anyone who carefully observes the symmetry properties of this table will easily be able to continue and complete it (the symbol $*$ *means there is a term of coefficient zero).*

If we let c denote an arbitrary exponent, we have

$$\int n^c = \frac{1}{c+1} n^{c+1} + \frac{1}{2} n^c + \frac{c}{2} A n^{c-1} + \frac{c(c-1)(c-2)}{2 \cdot 3 \cdot 4} B n^{c-3}$$
$$+ \frac{c(c-1)(c-2)(c-3)(c-4)}{2 \cdot 3 \cdot 4 \cdot 5 \cdot 6} C n^{c-5}$$
$$+ \frac{c(c-1)(c-2)(c-3)(c-4)(c-5)(c-6)}{2 \cdot 3 \cdot 4 \cdot 5 \cdot 6 \cdot 7 \cdot 8} D n^{c-7} + \cdots$$

etc., the exponents of n decreasing by multiples of 2 *until n or nn is reached. The capitals* $A, B, C \ldots$ *denote, in order, the last terms in the expressions of* $\int nn$, $\int n^4$, $\int n^6$, $\int n^8$, *etc., namely* $A = (1/6)$, $B = -(1/30)$, $C = (1/42)$, $D = (-1/30)$, *etc.*

These coefficients are such that each of them completes the other within an expression of unity. Thus $D = -1/30$, *since*

$$\frac{1}{9} + \frac{1}{2} + \frac{2}{3} - \frac{7}{15} + \frac{2}{9} + D = 1$$

Using this table, it took me less than a quarter of an hour to compute the sum of the tenth powers of the 1000 *first integers; the result is*

$$91,409,924,241,424,243,424,241,924,242,500$$

This example shows the uselessness of the book "Arithmetica infinitorum" by Ismael Bullialdus, which is entirely devoted to a tremendously large computation of the sums of the six first powers – less than what I accomplished here in a single page.

In this problem, we ask you to:

- **translate this text into the language of modern mathematics**,
- **name the points which need proofs**,
- **prove them completely**.

To do this, you may make use of the following problem.

II. The Bernoulli numbers B_m ($m \geq 1$) are defined by the series expansion

$$\frac{t}{e^t - 1} = 1 + \sum_{m=1}^{\infty} \frac{B_m}{m!} t^m, \tag{1}$$

and we also set $B_0 = 1$.

(a) Multiplying the two terms of (1) by $e^t - 1$, prove the the relation

$$1 + \sum_{k=1}^{m-1} \binom{k}{m} B_k = 0 \tag{2}$$

for $m \geq 2$.

(b) We use the following symbolic notation. Whenever

$$f(x) = a_0 + a_1 x + a_2 x^2 + \cdots \in \mathbb{C}[[x]],$$

we (symbolically!) set

$$f(B) = a_0 + a_1 B_1 + a_2 B_2 + \cdots$$

Naturally, when $f(x) \in \mathbb{C}[x]$, we have $f(B) \in \mathbb{C}$.

Show that

$$(1 + B)^m - B^m = 0.$$

(c) Show that if $2m + 1 > 1$, we have $B_{2m+1} = 0$.

(d) Set

$$S_k(n) = 1^k + 2^k + \cdots + (n-1)^k, \tag{3}$$

Show that the sums $S_k(n)$ satisfy the formula

$$(k+1)S_k(n) = (n+B)^{k+1} - B^{k+1}, \quad k \geq 1, \tag{4}$$

and show that this formula is equivalent to

$$(k+1)S_k(n) = \sum_{h=0}^{k} \binom{k+1}{h} B_h n^{k+1-h}.$$

(e) Consider the set

$$\mathbb{Z}_{(p)} = \left\{ r = \frac{a}{b} \in \mathbb{Q}\,;\; a \in \mathbb{Z},\; b \in \mathbb{N},\; p \nmid b \right\}.$$

Show that $\mathbb{Z}_{(p)}$ is a subring of Q, called the ring of p-integers.
What are the units of $\mathbb{Z}_{(p)}$?
What are the irreducible elements of $\mathbb{Z}_{(p)}$?
Show that $\mathbb{Z}_{(p)}$ is factorial and principal.

(f) Let p be a prime number and $m \geq 1$.
Show that pB_m is p-integral.
One can use induction on m, using relation (4) with $n = p$ and also the relation

(α) If $k \geq 1$, then $p^k/(k+1)$ is p-integral.

(g) Show that if $m \geq 2$ is even, we have the congruence

$$p\,B_m \equiv S_m(p) \bmod p.$$

For this, use the relation

(β) If $k \geq 2$, then $p^k/(k+1) \equiv 0 \bmod p$.

What can one say of this congruence when m and p are odd?

(h) Prove the relations (α) and (β).

(i) Let p be a prime number. Show that:

* if $p - 1$ does not divide k, $S_k(p) \equiv 0 \bmod p$.
* if $p - 1$ divides k, $S_k(p) \equiv -1 \bmod p$.

To do this, consider $\mathbb{F}_p^*$ and note that it is a cyclic group.

(j) Let $m \geq 1$, and show that there exists $A_{2m} \in \mathbb{Z}$ such that $B_{2m} = A_{2m} - \sum_{p-1|2m} 1/p$.

(k) Define the Bernoulli polynomials $B_n(x)$ by the formula

$$\frac{z\,e^{xz}}{e^z - 1} = \sum_{n=0}^{\infty} B_n(x)\,\frac{z^n}{n!}.$$

Show that $B_n(x) = (x + B)^n$ in symbolic notation.

(l) Show that the Bernoulli polynomials are given by the induction relations

$$\begin{cases} B_1(x) = x - \frac{1}{2} \\ \dfrac{d}{dx}\,B_n(x) = n\,B_{n-1}(x), \quad B_n(0) = B_n. \end{cases}$$

(m) Show that

$$B_n(x+1) = B_n(x) + nx^{n-1}, \quad \text{if } n \geq 1.$$

Deduce that we have

$$S_k(n) = \frac{B_{k+1}(n) - B_{k+1}}{k+1}.$$

(n) Let $\bar{B}_n(x)$ denote the function (of period 1) defined by $\bar{B}_n(x) = B_n(x - [x])$. Show (by induction on $k < n$) that

$$\frac{d^k}{dx^k}\bar{B}_n(x) = n(n-1)\cdots(n-k+1)\bar{B}_{n-k}(x)$$

and

$$\frac{d^n}{dx^n}\bar{B}_n(x) = n!\left(1 - \sum_{k\in\mathbb{Z}} \delta(x-k)\right)$$

where δ is the "Dirac measure".

(o) Assume that a and b are in $\mathbb{Z}$, and prove, by induction on n, the Euler–MacLaurin formula

$$\sum_{r=a}^{b-1} f(r) = \int_a^b f(x)\,dx + \sum_{k=1}^{n} \frac{B_k}{k!}\left[f^{(k-1)}(b) - f^{(k-1)}(a)\right]$$

$$+ \frac{(-1)^{n-1}}{n!}\int_a^b \bar{B}_n(x)\,f^{(n)}(x)\,dx,$$

in which f is assumed to be sufficiently continuously differentiable on $[a, b]$.

(p) Deduce that if $f(x) \in \mathbb{C}[x]$ and if g is an integral of f, we have

$$\sum_{r=a}^{b-1} f(r) = g(b+B) - g(a+B)$$

in symbolic notation.

(q) Assume that $g \in \mathbb{N}$ is such that $\bar{g}$ is a generator of $\mathbb{F}_p^*$. Set

$$F(t) = \frac{gt}{e^{gt} - 1} - \frac{t}{e^t - 1} = \sum_{m=1}^{\infty} \frac{B_n(g^m - 1)}{m!} t^m$$

$$u = e^t - 1 \quad \text{and} \quad G(u) = \frac{F(t)}{t}.$$

Show that

$$G(u) = \sum_{k=0}^{\infty} c_k u^k,$$

where the c_k are p-integers.

(r) Deduce that

$$G(u) = G(e^t - 1) = \sum_{m=0}^{\infty} \frac{A_m}{m!} t^m,$$

where the A_m are p-integers such that

$$A_{m+h(p-1)} \equiv A_m \bmod p \quad \text{for } h \geq 1 \text{ and } m > 1.$$

(s) Show that we have

$$\frac{B_m}{m}(g^m - 1) = A_{m-1}.$$

(t) Deduce the following statements, known as the **Kummer congruences**, from the preceding questions .

Let m and m' be two even positive integers such that

$$m \equiv m' \bmod p - 1.$$

Show that if m and m' are not divisible by $p - 1$, then

$$\frac{B_m}{m} \equiv \frac{B_{m'}}{m'} \bmod p.$$

Problem 4

Extracts from some letters to Holmboë [Ab]

Read the following extract from letters written by Abel to Holmboë (dated August 3, 1823).

Copenhague, in the year $\sqrt[3]{6064321219}$
(counting the decimal part).

You recall the little memoir which speaks of inverse functions of elliptic transcendentals, and in which I proved an impossible thing; I begged M. Degen to look over it, but he could not find any defect in the conclusion nor understand where there could be an error. The devil if I know how to solve this.

I tried to prove that it was impossible to solve the equation

$$a^n = b^n + c^n$$

in integers, when n is greater than 2, but I only succeeded in proving the following rather curious theorems.

Theorem I

The equation $a^n = b^n + c^n$, where n is a prime number, is impossible, whenever one or more of the quantities $a, b, c, a+b, a+c, b-c, \sqrt[m]{a}, \sqrt[m]{b}, \sqrt[m]{c}$ are prime numbers.

Theorem II

If we have

$$a^n = b^n + c^n,$$

then each of the quantities a, b, c will always be factorisable into two relatively prime factors, in such a way that setting $a = \alpha \cdot a',\ b = \beta \cdot b',\ c = \gamma \cdot c'$, one of the following 5 cases will hold:

1. $a = \dfrac{a'^n + b'^n + c'^n}{2}$, $\quad b = \dfrac{a'^n + b'^n - c'^n}{2}$, $\quad c = \dfrac{a'^n + c'^n - b'^n}{2}$

2. $a = \dfrac{n^{n-1}a'^n + b'^n + c'^n}{2}$, $\quad b = \dfrac{n^{n-1}a'^n + b'^n - c'^n}{2}$, $\quad c = \dfrac{n^{n-1}a'^n + c'^n - b'^n}{2}$

3. $a = \dfrac{a'^n + n^{n-1}b'^n + c'^n}{2}$, $\quad b = \dfrac{a'^n + n^{n-1}b'^n - c'^n}{2}$, $\quad c = \dfrac{a'^n + c'^n - n^{n-1}b^n}{2}$

4. $a = \dfrac{n^{n-1}(a'^n + b'^n) + c'^n}{2}$, $\quad b = \dfrac{n^{n-1}(a'^n + b'^n) - c'^n}{2}$, $\quad c = \dfrac{n^{n-1}(a'^n - b'^n) + c'^n}{2}$

5. $a = \dfrac{a'^n + n^{n-1}(b'^n + c'^n)}{2}$, $\quad b = \dfrac{a'^n + n^{n-1}(b'^n - c'^n)}{2}$, $\quad c = \dfrac{a'^n - n^{n-1}(b'^n - c'^n)}{2}$

Theorem III

For the equation $a^n = b^n + c^n$ to be possible, it is necessary for a to have one of the following three forms:

1. $a = \dfrac{x^n + y^n + z^n}{2}$,

2. $a = \dfrac{x^n + y^n + n^{n-1}z^n}{2}$,

3. $a = \dfrac{x^n + n^{n-1}(y^n + z^n)}{2}$,

where x, y and z have no common factors.

Theorem IV

The quantity a cannot be less than $(9^n + 5^n + 4^n)/2$, and the smallest of the quantities a, b, c cannot be less than $(9^n - 5^n + 4^n)/2$.

Answer the following questions concerning Abel's letter.

(1) Prove the first assertion of theorem II under the assumptions that n is prime, $(a, b, c) = 1$ and $abc \not\equiv 0 \bmod n$.
(2) Can you prove the same assertion when $(a, b, c) \neq 1$? What can you conclude?
(3) From now on, assume that n is a prime > 2, $(a, b, c) = 1$ and $abc \not\equiv 0 \bmod n$. Furthermore, set $x = -a,\ y = b,\ z = c$, so that we have

$$x^n + y^n + z^n = 0.$$

Show that

$$\begin{cases} y + z = a'^n & y^{n-1} - y^{n-2}z + \cdots + z^{n-1} = \alpha^n \\ z + x = b'^n & z^{n-1} - z^{n-2}x + \cdots + x^{n-1} = \beta^n \\ x + y = c'^n & x^{n-1} - x^{n-2}y + \cdots + y^{n-1} = \gamma^n \end{cases}$$

with new numbers $a', \alpha, b', \beta, c', \gamma$, which are pairwise relatively prime and such that

$$x = -a'\alpha, \qquad y = -b'\beta, \qquad z = -c'\gamma.$$

(4) Assume the existence of a prime number $p \neq n$ such that

(i) $x^n + y^n + z^n \equiv 0 \bmod p$ implies $xyz \equiv 0 \bmod p$.

Show that we have $a'b'c' \equiv 0 \bmod p$.
(5) Show that if $x \equiv 0 \bmod p$, then

$$a' \equiv 0 \mod p.$$

To do this, use the relations of question (3).
(6) Deduce from (3) and (5) that $\alpha^n \equiv ny^{n-1} \bmod p$.
(7) Deduce from (3) and (6) that n is an n-th power modulo p.
(8) From now on, assume that

(ii) n is not an n-th power modulo p.

Show that conditions (i) and (ii) imply that $abc \equiv 0 \bmod n$.
(9) Assume that $n = 3$, and give a value of p satisfying (i) and (ii).
(10) Show that if $2n + 1$ is prime and $p > 3$, then conditions (i) and (ii) are satisfied. Conclusion?

Remark: This result is due to Sophie Germain[18].

[18] Sophie Germain 1776–1831.

Problem 5

Recall that if $a(t) = (t - r_1)^{\nu_1} \dots (t - r_s)^{\nu_s} \in \mathbb{C}[t]$, we set

$$\operatorname{rad} a(t) = (t - r_1) \cdots (t - r_s).$$

Consider a matrix

$$\Gamma = \begin{pmatrix} \gamma_{1,1} & \gamma_{1,2} \\ \gamma_{2,1} & \gamma_{2,2} \\ \vdots & \vdots \\ \gamma_{n,1} & \gamma_{n,2} \end{pmatrix} \in M_{n,2}(\mathbb{C}),$$

and assume that the 2×2 minors of the matrix

$$M = \begin{pmatrix} 1 & 0 \\ 0 & 1 \\ \gamma_{1,1} & \gamma_{1,2} \\ \vdots & \vdots \\ \gamma_{n,1} & \gamma_{n,2} \end{pmatrix}$$

are all non-zero.

(1) Show that if a_1 and a_2 are two relatively prime elements of $\mathbb{C}[t]$, then any two elements of the column of $n+2$ polynomials

$$M \begin{pmatrix} a_1 \\ a_2 \end{pmatrix}$$

are relatively prime.

(2) Set

$$M \begin{pmatrix} a_1 \\ a_2 \end{pmatrix} = \begin{pmatrix} a_1 \\ a_2 \\ \varphi_1(v) \\ \vdots \\ \varphi_n(v) \end{pmatrix} = \begin{pmatrix} w_1 \\ w_2 \\ w_3 \\ \vdots \\ w_{n+2} \end{pmatrix}$$

where v denotes the vector $\begin{pmatrix} a_1 \\ a_2 \end{pmatrix}$.

Show that if a_1/a_2 is not a constant (i.e. is not in $\mathbb{C}$), then none of the determinants

$$\begin{vmatrix} w_i & w_j \\ w_i' & w_j' \end{vmatrix} \quad 1 \le i < j \le n+2$$

is equal to zero.

(3) Show that these determinants are all asssociated in $\mathbb{C}[t]$ (i.e. all equal up to a non-zero multiplicative constant). Let $\Delta(t)$ denote the monic element which represents them all.

(4) Set

$$P(X_1, X_2) = X_1X_2\varphi_1(X_1, X_2)\dots\varphi_n(X_1, X_2),$$

and show that the polynomial $P(x, 1) = x\varphi_1(x, 1)\dots\varphi_n(x, 1)$ is a polynomial of degree $n + 1$ with no multiple roots.

(5) Set $m(t) = P(a_1(t), a_2(t))$. Show that if $r \in \mathbb{C}$ is a root of $m(t)$ of order $\rho \geq 1$, then r is a root $\Delta(t)$ of order $\rho - 1$.

(6) Show that $\deg\ \Delta(t) \leq \deg\ (a_1a_2) - 1$.

(7) Deduce that

$$\deg\ m(t) - (\deg\ (a_1a_2) - 1) \leq \deg\ \text{rad}\ m(t).$$

(8) Prove that if $\deg\ a_1 \neq \deg\ a_2$, then

$$n\ \sup\ (\deg\ a_1,\ \deg a_2)\ \leq\ \deg\ \text{rad}\ m(t) - 1.$$

(9) Prove the same inequality in the general case.

If $\deg\ \varphi_i(v) < \sup\ (\deg\ a_1,\ \deg\ a_2)$ for a certain $i \in \{1, \dots, n\}$, we can replace a_2 by $\varphi_i(v)$ as a basis element, and reduce to (8).

(10) Show that this result generalises Mason's theorem, whose statement is as follows:

If $(a_1, a_2) = 1$ and $a_3 = a_1 + a_2$ (with $a_2/a_1 \notin \mathbb{C}$), then

$$\deg\ \text{rad}\ (a_1a_2a_3) > \sup\ (\deg\ a_1,\ \deg\ a_2).$$

(11) Let n be an integer ≥ 1 and let $x_1, y_1, x_2, y_2, x_3, y_3$ be elements of $\mathbb{C}[t]$ such that x_1y_1, x_2y_2 and x_3y_3 are relatively prime and not all in $\mathbb{C}$.

Show that the relations

$$\begin{cases} x_1y_1^n + x_2y_2^n + x_3y_3^n = 0 \\ x_1^ny_1 + x_2^ny_2 + x_3^ny_3 = 0 \end{cases}$$

are impossible whenever $n \geq 5$.

(12) Take a_1 and a_2 as above and assume that t does not divide a_1 and $1 - t$ does not divide $a_1 + a_2$.

Applying Mason's theorem from Exercise 1.13 above to suitable relations linking a_i, a_j, a_k for all the possible choices of the indices, show that

$$\deg\ \text{rad}\ (a_1a_2a_3a_4) \geq \frac{4}{3}\ \sup\ (\deg\ a_1,\ \deg\ a_2) + C^{st}$$

when $a_3 = a_1 + a_2,\ \ a_4 = a_1 + ta_2$.

Give a value of the constant.

Problem 6

Notation and Goals of the Problem

Equip the Euclidean affine oriented plane $\mathcal{P}$ with an orthonormal basis $\mathcal{R}$ with origin O. To every point M of coordinates (x, y) in $\mathcal{R}$, we associate the complex number $z = x + iy$; in this way we identify the plane with the set $\mathbb{C}$ of complex numbers.

An **integral point** of the plane is a point whose coordinates are integers. The set of all integral points forms a **lattice**. The lattice can be identified with the subset $\mathbb{Z}[i] = a + ib;\ (a, b) \in \mathbb{Z} \times \mathbb{Z}$ of $\mathbb{C}$.

The general goal of this problem is to find and study planar configurations subject to conditions involving integers:

Theme A: find regular polygons whose vertices belong to the lattice.
Theme B: find and study subsets of the plane such that the distances between any two points are integers.
Theme C: find and study configurations containing a fixed number of points of the lattice.

The notation and goals specific to each theme are given in their headings. The themes B and C are independent and can be attacked in either order. Both, however, depend on the first question of theme A.

Theme A: Regular Polygons with Integral Vertices

We propose to show that the only convex regular polygons with integral vertices are squares. For this, we first establish a preliminary result which will be used again in themes B and C. Throughout theme A, the coordinates of the points are defined in $\mathcal{R}$.

A.I. Preliminary question

A.I.1. Let θ be a real number, and n an integer greater than or equal to 1. Show that

$$\cos(n+1)\theta = 2\cos\theta \cdot \cos n\theta - \cos(n-1)\theta.$$

A.I.2. Deduce that there exists a sequence $(P_n)_{n\geq 1}$ of polynomials such that for every $n \in \mathbb{N}^*$, P_n satisfies the following properties:

- P_n is a polynomial of degree n with integral coefficients, and P_n is a monic polynomial (i.e. the coefficient of X^n is equal to 1).
- for every real number θ, $P_n(2\cos\theta) = 2\cos n\theta$.

A.I.3. Let θ be a real number such that θ/π is rational. Show that $2\cos\theta$ is a solution of an equation of the form

$$X^n + a_{n-1}X^{n-1} + \cdots + a_0 = 0, \tag{1}$$

where $n \in \mathbb{N}^*$ and $a_i \in \mathbb{Z}$ for $0 \leq i \leq n-1$.

A.I.4. Let θ be a real number, and assume that θ/π and $\cos\theta$ are rational numbers.
Show that $\cos\theta \in \{-1, -1/2, 0, 1/2, 1\}$.
(Begin by showing that every rational solution of equation (1) is a relative integer.)

A.II. Application to regular polygons with integral vertices

In this part, n denotes an integer greater than or equal to 3. Recall that a sequence $(A_1, \ldots, A_n)$ of n distinct points of the plane defines a convex regular polygon P whose vertices are exactly the n points, if there exists a rotation r of angle $2\pi/n$ or $-(2\pi/n)$ such that $r(A_i) = A_{i+1}$ for $1 \leq i \leq n-1$, and $r(A_n) = A_1$. We know that such a rotation must be unique. Let us write $P = (A_1, \ldots, A_n)$. The centre Ω of the rotation r is called the centre of P.

A.II.1. Let $P = (A_1, \ldots, A_n)$ be a convex regular polygon whose n vertices are integral points. Let Ω be its centre.

(a) Show that Ω is the isobarycentre of the set of vertices of P, and deduce that Ω has rational coordinates.
(b) Let ω denote the complex number associated to Ω; recall the analytic representation of the rotation r of centre Ω and angle $2\pi/n$ or $-(2\pi/n)$, using associated complexes.
(c) Write $r(A_1) = A_2$, and show that $\cos(2\pi/n)$ and $\sin(2\pi/n)$ are rationals. Deduce using A.I.4 that $n = 4$, i.e. that P is a square.

A.II.2. Let A_1 and A_2 be two distinct integral points. Show that the two squares $C = (A_1, A_2, A_3, A_4)$ and $C' = (A_1, A_2, A'_3, A'_4)$ admitting A_1 and A_2 as consecutive vertices have only integral vertices. Give the coordinates in $\mathcal{R}$ of A_3, A_4, A'_3, A'_4 in terms of the coordinates (x_1, y_1) of A_1 and (x_2, y_2) of A_2.

Theme B: Sets with Integral Distances

A non-empty subset E of points of the plane is said to be a **set with integral distances**, if for any points A and B belonging to E, the distance AB is an integer. Part B.I studies some examples. Part B.II establishes that an infinite set with integral distances must be contained in a line. However, part B.III shows that for every integer ($n \geq 3$), there exists a set with integral distances, consisting of n points such that any three of them cannot lie on a line.

B.I. Some examples

B.I.1. Can the vertices of a square, a rectangle, or a diamond form a set with integral distances?

B.I.2. Let ABC be an equilateral triangle of side 112.

(a) Justify the existence and uniqueness of the point D defined by the following conditions: $AD = 73$, $BD = 57$, D and C are on the same side of the line (AB).
(b) Compute the coordinates x and y of D in the basis $\mathcal{R}' = (O', \vec{i}, \vec{j})$ where O' is the midpoint of $[AB]$, and we have

$$\vec{i} = \frac{\overrightarrow{O'A}}{\| \overrightarrow{O'A} \|}, \qquad \vec{j} = \frac{\overrightarrow{O'C}}{\| \overrightarrow{O'C} \|}.$$

Observe that x is rational and write it as an irreducible fraction. Observe also that $y = y_1\sqrt{3}$ where y_1 is a rational number which we also express as an irreducible fraction.
(c) Show that $E = \{A, B, C, D\}$ is a set with integral distances.

B.II. Infinite sets with integral distances

B.II.1. Let H be a hyperbola and $\mathcal{R}''$ a Cartesian basis of the plane in which H has equation

$$xy = 1.$$

Let Γ be a curve in the plane, whose equation in $\mathcal{R}''$ is given by

$$ax^2 + bxy + cy^2 + dx + ey + f = 0,$$

where a, b, c, d, e are not all zero.

Show that if $\Gamma \cap H$ is infinite, then $\Gamma = H$, and give an upper bound for the number of points in $\Gamma \cap H$ when $\Gamma \neq H$.

B.II.2. Let E be a set with integral distances containing three non-collinear points A, B, C. Set $p = AB$, $q = AC$, and for $j \in \{0, \ldots, p\}$ and $k \in \{0, \ldots, q\}$, set

$$U_j = \{M \in \mathcal{P}; \mid MA - MB \mid = j\} \quad \text{and} \quad V_k = \{M \in \mathcal{P}; \mid MA - MC \mid = k\}.$$

(a) Describe the geometric nature of the set U_j and V_k for $j \in \{0, \ldots, p\}$ and $k \in \{0, \ldots, q\}$. Distinguish the case $j = 0$ and $j = p$ (resp. $k = 0$ and $k = q$) from the case $0 < j < p$ (resp. $0 < k < q$).

(b) Deduce from B.II.1 that for every $j \in \{0, \ldots, p\}$ and $k \in \{0, \ldots, q\}$, $U_j \cap V_k$ is a finite (possibly empty) subset of the plane.

(c) Prove that $E \subset \bigcup_{0 \leq j \leq p} U_j$ and $E \subset \bigcup_{0 \leq k \leq q} V_k$, and deduce that E is finite.

B.II.3. Given a point A and a vector $\vec{v}$, let $E_{A,\vec{v}}$ denote the set of all the points M of the plane such that $\overrightarrow{AM} = x\vec{v}$ with $x \in \mathbb{Z}$.

Let E be an infinite subset of the plane. Show that the following properties are equivalent:

(*) E is a set with integral distances;

(**) There exists a point A and a vector $\vec{v}$ of norm 1 such that $E \subset E_{A,\vec{v}}$.

B.III. Finite sets with integral distances

Let ϕ be the real number defined by $\cos\phi = 4/5$, $0 < \phi < \pi$. For every natural integer p, let M_p be the point whose associated complex number is $e^{2ip\phi}$.

B.III.1. Show that the points M_p with p in $\mathbb{N}$ are pairwise distinct.

B.III.2. Let p and q be two natural integers. Prove that the distance M_pM_q is equal to the quantity $2 \mid \sin(p-q)\phi \mid$. Deduce that M_pM_q is a rational number.

B.III.3. Let n be an integer greater than or equal to 3. Show that there exists a set with integral distances, consisting of n points, contained in a circle of centre O.

Theme C: Configurations Containing a Fixed Number of Points of the Lattice

After studying some examples (part C.I), we propose to establish that for every integer $n \in \mathbb{N}^*$, there exists

- a circle containing exactly n points of the lattice in its interior (part C.II);
- a square containing exactly n points of the lattice in its interior (part C.III);
- a circle passing through exactly n points of the lattice (part C.IV).

Throughout theme C we assume that coordinates are defined in the basis $\mathcal{R}$, unless we explicitly assume otherwise.

C.I. Some examples

C.I.1. Without justification, construct four circles C_1, C_2, C_3 and C_4 such that for each $j \in \{1, 2, 3, 4\}$, exactly j points of the lattice lie in the interior of C_j; give their radii and the coordinates of their centres.

C.I.2. Let n belong to $\mathbb{N}^*$. Without justification, give the coordinates of the vertices of a square containing n^2 points of the lattice in its interior.

C.II. A circle containing *n* points of the lattice in its interior

C.II.1. Classification of the points of the lattice

(a) Let B be a bounded subset of the plane. Show that B contains only a finite number of points of the lattice.

(b) Let A be the point of coordinates $(\sqrt{2}, 1/3)$. Show that there cannot exist two points of the lattice at the same distance from A.

Deduce that we can classify the points of the lattice as a sequence $(M_n)_{n \geq 1}$ such that $AM_n < AM_{n+1}$ for every $n \in \mathbb{N}^*$.

C.II.2. Application

Given an integer $n \in \mathbb{N}^*$, deduce from C.II.1 b that there exists a circle which contains exactly n points of the lattice in its interior.

C.III. A square containing *n* points of lattice in its interior

C.III.1. Definition of a function on the lattice

Let D_1 be the line of equation $x+y\sqrt{3}-\frac{1}{3}=0$ and D_2 the line of equation $x\sqrt{3}-y-(1/\sqrt{3})=0$.

(a) Show that D_1 and D_2 are perpendicular, and give the coordinates of their intersection point Ω. Draw these two lines.

(b) Set

$$X = (1/2)\left(x\sqrt{3} - y - 1/\sqrt{3}\right), \qquad Y = (1/2)\left(x + y\sqrt{3} - 1/3\right).$$

Show that this defines a change of basis of the orthonormal basis, such that the new axes lie respectively along D_1 and D_2.

Let M be a point in the plane, and let (x, y) denote its coordinates in $\mathcal{R}$ and (X, Y) its coordinates in the new basis. Set

$$f(M) = \mid X \mid + \mid Y \mid = (1/2)\left|x\sqrt{3} - y - 1/\sqrt{3}\right| + 1/2\left|x + y\sqrt{3} - 1/3\right|.$$

C.III.2. Injectivity of the function *f*

Consider two points M_1 and M_2 of the lattice, of coordinates respectively (x_1, y_1) and (x_2, y_2) in $\mathcal{R}$, such that $f(M_1) = f(M_2)$.

(a) Show that there exist four real numbers $\alpha, \beta, \gamma, \delta$ satisfying $\alpha^2 = \beta^2 = \gamma^2 = \delta^2 = 1$ and such that $\alpha x_1 + \beta y_1 - \gamma x_2 - \delta y_2 + (\gamma - \alpha)/3 = 0,\ \ \beta x_1 + \alpha y_1 - \delta x_2 - \gamma y_2 + (\delta - \beta)/3 = 0$. (Observe that for every real number x, we have $\mid x \mid = \lambda x$ with $\lambda^2 = 1$.)

(b) Deduce that $M_1 = M_2$. (Begin by showing that $\gamma - \alpha = \delta - \beta = 0$.)

C.III.3. A new classification of the points of the lattice

Using the same method as in C.II.1, show that we can classify the points of the lattice as a sequence $(N_n)_{n\geq 1}$ such that $f(N_n) < f(N_{n+1})$ for every n in $\mathbb{N}^*$.

C.III.4. Let a be a strictly positive real number. Show that the set of points of the plane whose coordinates (X, Y) satisfy $|X| + |Y| < a$ is the interior of a square C_a; give the vertices of the square.

Deduce that for every integer n in $\mathbb{N}^*$, there exists a square C_a whose interior contains exactly n points of the lattice.

C.IV. A circle passing through *n* points of the lattice

C.IV.1. Number of integral solutions of the equation $x^2 + y^2 = 5^n$

Let n be an integer in $\mathbb{N}$; and associate to it the following two sets:

$$\mathcal{E}_n = \{(x, y) \in \mathbb{Z} \times \mathbb{Z};\, x^2 + y^2 = 5^n\}$$
$$E_n = \{z \in \mathbb{Z}[i];\, |z|^2 = 5^n\}.$$

(a) Show that $\mathcal{E}_n$ and E_n are finite sets of the same cardinal.

(b) Determine E_0.

For every element ω in E_0 and every integer p such that $0 \leq p \leq n$, set

$$Z_{\omega,p} = \omega(2 + i)^p(2 - i)^{n-p}.$$

(c) Prove that $Z_{\omega,p}$ belongs to E_n and the map $(\omega, p) \to Z_{\omega,p}$, from $E_0 \times \{0, \dots, n\}$ to E_n, is injective. (One can show that if $Z_{\omega,p} = Z_{\omega',q}$ with ω and ω' elements of E_0 and p and q integers less than or equal to n, then $((2 + i)/(2 - i))^{4(p-q)} = 1$, and use (A.I.4.).)

(d) Let $z = x + iy$ be an element of E_n, with $n \geq 1$. Show that (x, y) satisfies one of the following systems of relations:

$$(1)\ \begin{cases} 2x - y \equiv 0 \bmod 5 \\ x + 2y \equiv 0 \bmod 5 \end{cases} \quad \text{or} \quad (2)\ \begin{cases} 2x + y \equiv 0 \bmod 5 \\ -x + 2y \equiv 0 \bmod 5. \end{cases}$$

Deduce that one of the two numbers $z/(2 + i)$ or $z/(2 - i)$ belongs to E_{n-1}.

(e) Prove that the map $(\omega, p) \to Z_{\omega,p}$ from $E_0 \times \{0, \dots, n\}$ to E_n, is bijective. Deduce the number of elements of $\mathcal{E}_n$.

C.IV.2. A circle passing through an even number of points of the lattice

(a) For every integer $n \in \mathbb{N}^*$, set

$$A_n = \{(x, y) \in \mathcal{E}_n;\, x \text{ even and } y \text{ odd}\}$$

and

$$B_n = \{(x, y) \in \mathcal{E}_n;\, x \text{ odd and } y \text{ even}\}.$$

Show that A_n and B_n have the same cardinal and that $\mathcal{E}_n = A_n \cup B_n$ and $A_n \cap B_n = \emptyset$.

(b) Let $k \in \mathbb{N}^*$. Determine the number of points of the lattice belonging to the circle whose centre is the point of coordinates $(1/2, 0)$ and whose radius is $(1/2)(5^{(k-1)/2})$.

C.IV.3. A circle passing through an odd number of points of the lattice

Let $k \in \mathbb{N}^*$ and Γ_k be the circle whose centre is the point of coordinates $(1/3, 0)$ and whose radius is $(1/3)5^k$.

(a) Show that the number of points of the lattice belonging to Γ_k is equal to the cardinal of the set F_k defined by

$$F_k = \{z = x + iy;\, z \in E_{2k},\ x \equiv -1 \bmod 3,\ y \equiv 0 \bmod 3\}.$$

The goal of questions (b), (c) and (d) is to compute the cardinal of F_k.

(b) Show that for any $\omega \in E_0$ and $z \in E_{2k}$, ωz and $\omega\bar{z}$ belong to E_{2k}. Prove that the relation (R), defined over E_{2k} by

"For z and z' in E_{2k}, we have $z(R)z'$ if and only if there exists $\omega \in E_0$ such that $z' = \omega z$ or such that $z' = \omega\bar{z}$", is an equivalence relation on E_{2k}.

Denote by $(R)(z)$ the equivalence class of an element z of E_{2k}.

(c) Let $z = x + iy$ be an element of E_{2k}.

- Assume $xy \neq 0$. Give the elements of $(R)(z)$ in terms of x and y and show that $(R)(z) \cap F_k$ contains two elements.
- Assume $xy = 0$. Give the elements of $(R)(z)$ and show that $(R)(z) \cap F_k$ contains one element.

(d) Deduce that F_k contains $2k + 1$ elements.

Problem 7

Legendre's Theorem

Let a, b and c be three non-zero integers in $\mathbb{Z}$, and consider the Diophantine equation in x, y, z given by

$$ax^2 + by^2 + cz^2 = 0. \qquad (L_{a,b,c})$$

We will establish necessary and sufficient conditions for this equation to admit a solution in $\mathbb{Z}^3 \setminus \{(0, 0, 0)\}$. Obviously we may assume that the greatest common divisor of a, b and c is 1 and that they are square-free.

Show that we may also assume that a, b and c are pairwise relatively prime. We make this hypothesis from now on.

I Necessary conditions

(1) Show that a, b, c cannot be all three positive or all three negative.

(2) Show that $-bc$ must be a square modulo a, that $-ca$ must be a square modulo b and that $-ab$ must be a square modulo c.

II We want to show that conditions (1) and (2) of I are sufficient. Assume that

$$|a| \leq |b| \leq |c|.$$

Then we have

$$|ab| \leq |ac| \leq |bc|,$$

and we say that $|ac$ is the index I of the equation $L_{a,b,c}$. Let us reason by induction on I, assuming that conditions (1) and (2) hold.

(1) Show that if $I = 1$, the equation $L_{a,b,c}$ admits an infinite number of primitive solutions (i.e. solutions such that x, y and z are relatively prime) in $\mathbb{Z}^3$.

(2) Assume that $I \geq 2$, and that the theorem holds for all equations of index $< I$. Show that there exists integers Q and r, with $|r| \leq (1/2)|c|$, such that:

$$ar^2 + b = cQ$$

and $|Q| < I$.
(3) Show that if $Q = 0$, the theorem holds for $L_{a,b,c}$.
(4) If $Q \neq 0$, let A denote the greatest common divisor of ar^2, b and cQ. Show that A divides r^2 and Q, and that A is square-free. Deduce that A divides r.
(5) Set

$$\alpha = \frac{r}{A}, \quad \beta = \frac{b}{A}, \quad q = \frac{Q}{A} = C\gamma^2,$$

where C is the square-free part of Q, and set $B = \alpha\beta$.

Show that the equation $L_{a,b,c}$ satisfies all the necessary conditions of the statement.
(6) Show that

$$\begin{cases} |AB| = |ab| < I \\ |AC| \leq |AC|\gamma^2 = |Q| < I. \end{cases}$$

(7) Let (X, Y, Z) be a primitive solution of $L_{a,b,c}$. Set

$$\begin{cases} x = A\alpha X - \beta Y \\ y = X + a\alpha Y \\ z = C\gamma Z. \end{cases}$$

Show that:

$$ax^2 + by^2 + cz^2 = cC\gamma^2(AX^2 + BY^2 + CZ^2),$$

and deduce that (x, y, z) is a non-zero solution of the equation $L_{a,b,c}$.

2

ELLIPTIC FUNCTIONS

Like many other mathematical objects, elliptic functions were born twice: the first time in the 18th century and then a second time during the romantic period of the early 19th century. We will try to give an idea of the circumstances of this birth and rebirth, and then in Sections 2.4–2.8 we will use the theory of functions of one complex variable, a theory developed around the time of the second birth of elliptic functions, to develop the standard construction due to Weierstrass, which is very fashionable today.

From this point of view the essential results are the Liouville theorems (Section 2.4), the theorem given in Section 2.5, those of Section 2.7 and Abel's theorem (Section 2.8). Finally, in Sections 2.9–2.11, we backtrack and give a second construction, which becomes equivalent to the standard construction over the field of complex numbers. This construction was developed in particular by Rausenberger, and has the advantage of passing meaningfully to the ultrametric framework (see the exercises of Chapter 5). It leads to the construction of a "universal elliptic curve" known as the Tate curve E_q.

The Tate curve is a powerful tool in the modern theory of elliptic curves and we will use it in a very typical manner in Chapter 5. From this point of view the essential results are the theorem given in Section 2.10 and the formulae (27) and (28).

2.1 ELLIPTIC INTEGRALS

Around the time when Kepler's laws[19] became known, and integral calculus was in the air, the natural desire arose to compute the path of the orbit of a planet or a comet.

The computation of an arc of an ellipse was attempted by Wallis[20] in 1655, and series expansions were given by Newton[21] and Euler.

We will give this computation for the most general (non-decomposable) conic, given by the equation

$$y^2 = 2px + (e^2 - 1)x^2, \quad p \neq 0,$$

[19] J. Kepler 1571–1630.
[20] J. Wallis 1616–1703.
[21] I. Newton 1642–1727.

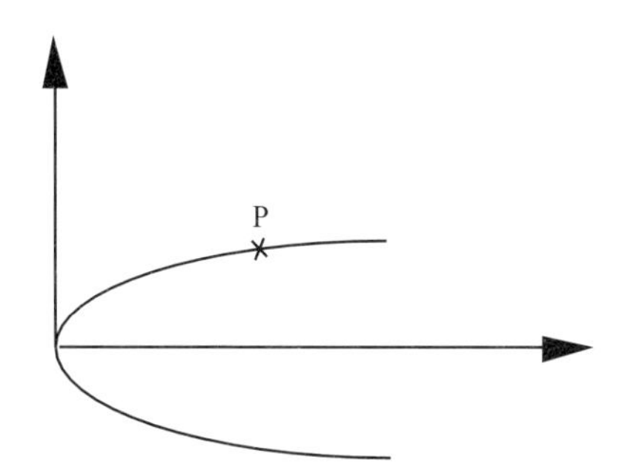

where e is the eccentricity ($e = c/a$) of this conic and $p = b^2/a$ is its parameter (following the tradition that a is the large semi-axis and b the small semi-axis of the conic when it has a centre).

We know that this conic can be parametrised by taking $t = y/x$. We find

$$\begin{cases} x = \dfrac{2p}{t^2 + (1 - e^2)} \\ y = \dfrac{2p\,t}{t^2 + (1 - e^2)}. \end{cases}$$

From now on, we set

$$D := D(t) = t^2 + (1 - e^2).$$

The arc element of the conic is given by

$$\begin{aligned} ds^2 &= dx^2 + dy^2 \\ &= (2p)^2 \frac{[t^2 + (1+e)^2]\,[t^2 + (1-e)^2]}{D^4}\, dt^2. \end{aligned}$$

Thus, we have

$$\overset{\frown}{\mathrm{OP}} = 2p \int_{t(P)}^{\infty} \frac{\sqrt{(t^2 + (1+e)^2)\,(t^2 + (1-e)^2)}}{D^2}\, dt. \tag{1}$$

It is an Abelian integral attached to the curve of equation

$$u^2 = [t^2 + (1+e)^2][t^2 + (1-e)^2]. \tag{2}$$

We assume that $e \geq 0$, so we see that the second term of the equation of this curve has double roots only if $e \in \{0, 1\}$.

Special case: **(1)** If the eccentricity $e = 0$, the conic is a circle of radius $a = r$, and we have

$$\overset{\frown}{\mathrm{OP}} = 2r \int_{t(P)}^{\infty} \frac{dt}{1 + t^2} = 2r\left(\frac{\pi}{2} - \mathrm{Arctg}\,\frac{y}{x}\right).$$

The curve (2) decomposes into two parabolas

$$u = \pm(t^2 + 1).$$

(2) If the eccentricity $e = 1$, the conic is a parabola, and we have

$$\widehat{\mathrm{OP}} = 2p \int_{t(P)}^{\infty} \frac{\sqrt{t^2+4}}{t^3}\, dt.$$

This integral can be computed using "elementary transcendentals", since the curve (2) is unicursal; it is given by

$$u^2 = t^2(t^2 + 4).$$

Remark 2.1.1 **(1)** One can check that if $e \notin \{0, 1\}$, the arc of the conic is not computable using elementary transcendentals. Check that the curve (2) is not unicursal, for example by using Mason's theorem (see Exercise 1.13 of Chapter 1).

(2) The change of variables

$$t = \frac{b}{a}\sqrt{\frac{1+\theta}{1-\theta}}, \qquad \theta \in]-1, 1[$$

makes it possible to express the arclength in a more classical form in the case of the ellipse.

A simple computation gives

$$dt = \frac{b}{a}\left[\frac{d\theta}{(1-\theta)\sqrt{1-\theta^2}}\right]$$
$$D(t) = \frac{b^2}{a^2}\left(\frac{1+\theta}{1-\theta}\right) + (1 - e^2).$$

If we are dealing with an ellipse, then

$$1 - e^2 = \frac{b^2}{a^2}$$

and we have

$$D(t) = \frac{2b^2}{a^2}\,\frac{1}{1-\theta}.$$

It remains to compute the numerator.

We set

$$N(t) = \sqrt{[t^2 + (1+e)^2][t^2 + (1-e)^2]}.$$

Note that

$$t^2 + (1+e)^2 = \frac{b^2}{a^2}\Big(\frac{1+\theta}{1-\theta}\Big) + (1+e)^2 = \frac{1}{a^2}\,\frac{b^2(1+\theta) + a^2(1+e)^2(1-\theta)}{1-\theta},$$

whose numerator is equal to

$$\theta(b^2 - a^2(1+e)^2) + b^2 + a^2(1+e)^2.$$

As we have

$$\begin{cases} b^2 - a^2(1+e)^2 = a^2(1-e^2) - a^2(1+e)^2 = -2a^2e(1+e) \\ b^2 + a^2(1+e)^2 = a^2(1-e^2) + a^2(1+e)^2 = 2a^2(1+e), \end{cases}$$

we see that

$$(1-\theta)N(t) = 2\sqrt{(1+e)(1-e\theta)(1-e)(1+e\theta)}.$$

It follows that

$$\widehat{\mathrm{OP}} = a\int_{\theta(P)}^{1} \frac{1 - e^2\theta^2}{\sqrt{(1-\theta^2)(1-e^2\theta^2)}}\,d\theta.$$

In the case of the hyperbola, we would obtain an analogous result.

(3) Because hyperbolic functions were introduced by J.-H. Lambert[22], these integrals could not be called "hyperbolic", so they were called elliptic!

2.2 THE DISCOVERY OF ELLIPTIC FUNCTIONS IN 1718

Elliptic functions made their very first appearance in a remarkable work by Fagnano[23], dating from 1718, on the rectification of the **Bernoulli lemniscate** (introduced by Jakob Bernoulli in 1694).

Given two points F_1 and F_2 of the plane, at a distance of $2c$ from each other, the Bernoulli lemniscate is defined to be the locus of the points M in the plane such that $MF_1 \cdot MF_2 = c^2$.

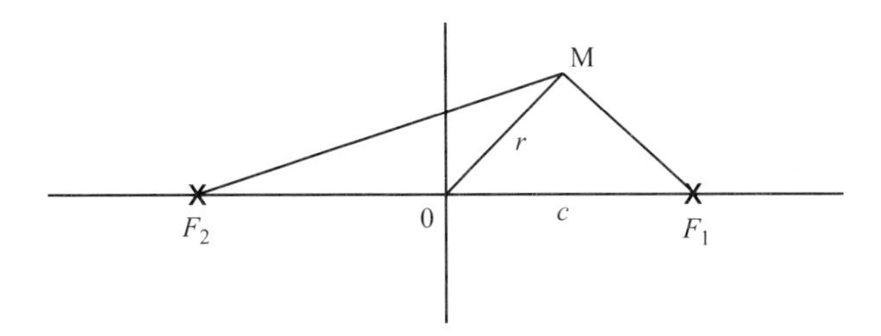

If we set

$$r_1 = MF_1, \quad r_2 = MF_2, \quad r = M0,$$

[22] J.H. Lambert 1728–1777.

[23] G.C. Fagnano 1682–1766.

the point M lies on the lemniscate if and only if $r_1^2 r_2^2 = c^4$. Now, we have

$$\begin{cases} r_1^2 &= r^2 - 2cx + c^2 \\ r_2^2 &= r^2 + 2cx + c^2, \end{cases}$$

hence

$$r^4 + 2c^2r^2 - 4c^2x^2 = 0.$$

To simplify, set

$$2c^2 = 1.$$

Then the equation of the lemniscate becomes

$$r^4 + r^2 - 2x^2 = 0. \tag{3}$$

Remark 2.2.1 In **homogeneous Cartesian coordinates**, equation (3) becomes

$$(X^2 + Y^2)^2 + (Y^2 - X^2)Z^2 = 0,$$

so we see that the lemniscate is a circular quartic; it passes through the cyclic points.

We can also note that it is the inverse of an equilateral hyperbola with respect to its centre.

Here is its shape:

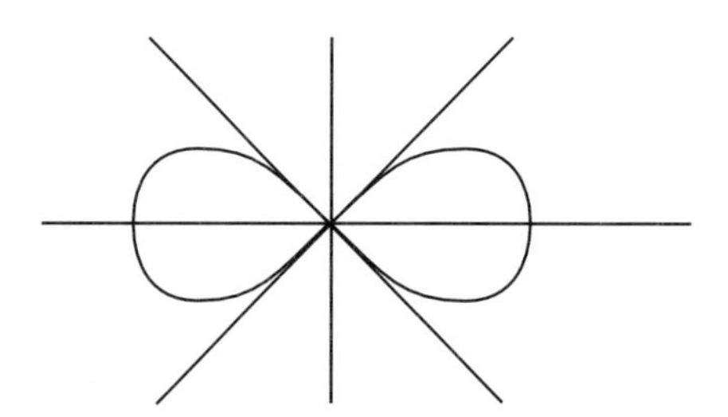

From now on we parametrise this curve using r, via

$$\begin{cases} 2x^2 = r^2 + r^4 \\ 2y^2 = r^2 - r^4. \end{cases}$$

We wish to express arclengths in terms of r. We have

$$\begin{cases} 2x\,\dfrac{dx}{dr} = r + 2r^3 \\[2ex] 2y\,\dfrac{dy}{dr} = r - 2r^3, \end{cases}$$

therefore

$$\left(\frac{ds}{dr}\right)^2 = \frac{1}{1 - r^4}$$

and

$$s(r) = \int_0^r \frac{dr}{\sqrt{1 - r^4}}. \tag{4}$$

Let us compare formula (4) with the classical integral

$$\operatorname{Arcsin} r = \int_0^r \frac{dr}{\sqrt{1 - r^2}}$$

and recall that the circle

$$1 - r^2 = \sigma^2$$

can be parametrised by

$$r = \frac{2t}{1 + t^2}, \qquad \sigma = \frac{1 - t^2}{1 + t^2}.$$

Thus, for this circle, we have

$$\frac{dr}{dt} = \frac{2(1 - t^2)}{(1 + t^2)^2} = \frac{2\sigma}{1 + t^2}$$

and

$$\int \frac{dr}{\sigma} = \int \frac{2\,dt}{1 + t^2}.$$

□

Apparently Fagnano was inspired by this method, when he set

$$r^2 = \frac{2t^2}{1 + t^4}, \qquad \sigma = \sqrt{1 - r^4} = \frac{1 - t^4}{1 + t^4}.$$

This gives

$$r\frac{dr}{dt} = \frac{2t(1 - t^4)}{(1 + t^4)^2}$$

and

$$\frac{dr}{\sigma} = \sqrt{2}\,\frac{dt}{\sqrt{1 + t^4}},$$

which gives the formula

$$\int_0^r \frac{dr}{\sqrt{1-r^4}} = \sqrt{2}\int_0^{t(r)} \frac{dt}{\sqrt{1+t^4}}.$$

If we now consider this last integral, we are led to set

$$t^2 = \frac{2u^2}{1-u^4}, \qquad s = \sqrt{1+t^4} = \frac{1+u^4}{1-u^4}.$$

It follows that

$$t\frac{dt}{du} = \frac{2u(1+u^4)}{(1-u^4)^2}$$

and

$$\frac{dt}{s} = \sqrt{2}\,\frac{du}{\sqrt{1-u^4}},$$

hence

$$\int_0^r \frac{dr}{\sqrt{1-r^4}} = \sqrt{2}\int_0^{t(r)} \frac{dt}{\sqrt{1+t^4}} = 2\int_0^{u(r)} \frac{du}{\sqrt{1-u^4}}.$$

Noting that r can be computed from u by the formula

$$r^2 = \frac{4u^2(1-u^4)}{(1+u^4)^2},$$

we obtain the marvellous formula which enabled Fagnano to duplicate the arc of a lemniscate using a ruler and compass construction.

Remark 2.2.2 To the above curve

$$1 - r^4 = \sigma^2, \tag{5}$$

we can associate the Diophantine equation

$$x^4 - y^4 = z^2 \tag{5'}$$

which has no non-trivial relatively prime solutions (see Exercise 1.6 of Chapter 1).

To see this, we show that if (5′) admits a non-trivial solution, then it admits a non-trivial solution for which y is even (same exercise).

We will show that the Fermat descent which we are led to apply to (5′) is analogous to the procedure used to duplicate the arc of a lemniscate.

Since $x^4 = z^2 + y^4$, we can set

$$x^2 = a^2 + b^2, \qquad y^2 = 2ab, \qquad z = a^2 - b^2$$

which gives (choosing b even)

$$a = u^2, \qquad b = 2v^2.$$

We deduce the equation

$$x^2 = a^2 + b^2 = u^4 + 4v^4$$

which recalls the differential element $dt/\sqrt{1+t^4} = (1/\sqrt{2})(dr/\sigma)$ since $y/x = (2/\sqrt{u^4/v^4+4})(u/v)$.
We now set

$$u^2 = a'^2 - b'^2, \quad 2v^2 = 2a'b';$$

b' is necessarily even, so

$$a' = x'^2 \quad \text{and} \quad b' = y'^2$$

and we recover the form

$$u^2 = x'^4 - y'^4, \tag{5''}$$

which recalls the differential element $du/\sqrt{1-u^4} = (1/\sqrt{2})(dt/s)$ since $v/u = (1/\sqrt{1-(y'^4/x'^4)}).\ y'/x'$.

2.3 EULER'S CONTRIBUTION (1753)

Euler delved deeply into Fagnano's work in 1751, and he realised that Fagnano's formula was the analogue of the relation

$$\int_0^{2u\sqrt{1-u^2}} \frac{du}{\sqrt{1-u^2}} = 2\int_0^u \frac{du}{\sqrt{1-u^2}}.$$

This formula conceals a trigonometric identity which generalises to

$$\begin{cases} \int_0^r \dfrac{dr}{\sqrt{1-r^2}} = \int_0^u \dfrac{du}{\sqrt{1-u^2}} + \int_0^v \dfrac{dv}{\sqrt{1-v^2}} \\ \text{when } r = u\sqrt{1-v^2} + v\sqrt{1-u^2}. \end{cases}$$

What could the analogous identity for elliptic integrals be? More precisely, how should we choose r so that

$$\int_0^r \frac{dr}{\sqrt{1-r^4}} = \int_0^u \frac{du}{\sqrt{1-u^4}} + \int_0^v \frac{dv}{\sqrt{1-v^4}}? \tag{1}$$

We could try

$$r = u\sqrt{1-v^4} + v\sqrt{1-u^4},$$

but we find that for $u = v$, we have $r = 2u\sqrt{1-u^4}$, whereas we saw above that

$$r = 2u\,\frac{\sqrt{1-u^4}}{1+u^4}.$$

Thus we are led to correct the choice of r and try

$$r = \frac{u\sqrt{1-v^4} + v\sqrt{1-u^4}}{1+u^2v^2}. \tag{2}$$

In particular, we would like to show that the relation $du/\sqrt{1-u^4} = -dv/\sqrt{1-v^4}$ implies that r is constant.

Set

$$U = \sqrt{1-u^4}, \qquad V = \sqrt{1-v^4};$$

we want to show that if $du/U + dv/V = 0$, then $dr = 0$, i.e.

$$(V\,du + u\,dV + U\,dv + v\,dU)\,(1+u^2v^2) - (uV + vU)\,(2uv^2du + 2vu^2dv) = 0.$$

To obtain this result, we substitute

$$dV = -\frac{2v^3}{V}\,dv, \qquad dU = -\frac{2u^3}{U}\,du$$

and order the terms in du and dv. We find

$$\begin{aligned}
&\left[\left(V - \frac{2u^3v}{U}\right)(1+u^2v^2) - (uV+vU)2uv^2\right]du \\
&+\left[\left(U - \frac{2v^3u}{V}\right)(1+u^2v^2) - (uV+vU)2vu^2\right]dv \\
&= \left[(UV - 2u^3v)\,(1+u^2v^2) - 2uv^2(uUV + v(1-u^4))\right]\frac{du}{U} \\
&+\left[(UV - 2v^3u)\,(1+u^2v^2) - 2vu^2(vUV + u(1-v^4))\right]\frac{dv}{V},
\end{aligned}$$

and the two terms in square brackets are equal.

To conclude, we note that, at least for u and v near zero, $dr = 0$ is equivalent to $(du/U) + (dv/V) = 0$. Thus, if r is constant, we have

$$\int_0^u \frac{du}{U} + \int_r^v \frac{dv}{V} = 0,$$

i.e. $s(u) + s(v) = s(r)$ when (2) is satisfied.

The equivalence

$$(2) \iff s(u) + s(v) = s(r),$$

which holds for u and v sufficiently small, is known as the theorem of addition of lemniscatic integrals.

2.4 ELLIPTIC FUNCTIONS: STRUCTURE THEOREMS

The idea of inverting elliptic integrals can be found in the notebooks of Gauss[24] dating back to 1796 (and 1797 for lemniscatic integrals), which were published only after his death.

Abel[25] rediscovered this idea around 1823 and published it in 1827 in his *Recherches* [Ab] t.I, pp. 266–278. Then Jacobi[26], probably inspired by an article by Abel, used the idea in 1828 (complete works t.I, p. 43).

Since these functions are meromorphic and doubly periodic, only Cauchy's theory[27] was lacking in order to complete their study; this was done by Liouville[28] (1844) and Eisenstein (1847).

From the whole of Cauchy's theory, we only use two tools – but what powerful tools they are!

Liouville's Theorem 2.4.1 *Let f be a holomorphic function $\mathbb{C} \to \mathbb{C}$ which is bounded in module in a neighbourhood of infinity (i.e. bounded in module in all of $\mathbb{C}$).*

Then f is a constant function.

Remark 2.4.1 This theorem is in fact due to Cauchy[28], who published it in 1844 (Comptes Rendus, XIX, pp. 1377–1378).

Proof. Assume that $|f(z)| < K$ for every $z \in \mathbb{C}$, and let us show that if z and z' lie in $\mathbb{C}$, then $f(z) = f(z')$.

Let C be a circle of centre z and radius $r \geq 2|z - z'|$. By Cauchy's formula, we have

$$f(z') - f(z) = \frac{1}{2i\pi} \int_C \left[\frac{1}{\zeta - z'} - \frac{1}{\zeta - z} \right] f(\zeta)\, d\zeta.$$

Setting $\zeta = z + re^{i\theta}$, we obtain

$$|f(z') - f(z)| = \left| \frac{1}{2\pi} \int_0^{2\pi} \frac{z' - z}{\zeta - z'} f(\zeta)\, d\theta \right|.$$

Since $|\zeta - z'| \geq |\,|\zeta - z| - |z - z'|\,| \geq r/2$, we have

$$\begin{aligned} |f(z') - f(z)| &\leq \frac{1}{2\pi} \int_0^{2\pi} \frac{|z' - z|}{r/2} K\, d\theta \\ &= 2|z' - z| K r^{-1}. \end{aligned}$$

It follows that $|f(z') - f(z)|$ is smaller than any given positive real number. □

[24] C.F. Gauss 1777–1855.
[25] N.H. Abel 1802–1829.
[26] C. Jacobi 1804–1851.
[27] A.L. Cauchy 1789–1857.
[28] J. Liouville 1809–1882.

Theorem 2.4.2 *If a sequence of holomorphic functions converges uniformly on an open set, then the limit is holomorphic on this open set. Moreover, the order of differentiating and taking limits can be exchanged.*

Proof. See [W-W] p. 91. □

Let ω_1 and ω_2 be two $\mathbb{R}$-linearly independent complex numbers, and let Λ be a lattice

$$\Lambda = \mathbb{Z}\,\omega_1 + \mathbb{Z}\,\omega_2 = \mathbb{Z}\,\omega_1 \oplus \mathbb{Z}\,\omega_2.$$

Definition 2.4.1 *A function $f : \mathbb{C} \longrightarrow \mathbb{C} \cup \{\infty\}$ is called* **elliptic** *if f is meromorphic in $\mathbb{C}$ and if there exists a lattice Λ such that*

$$f(z + \omega) = f(z)$$

for every $z \in \mathbb{C}$ and every $\omega \in \Lambda$.

Example 2.4.1 Every constant function is an elliptic function.

Remark 2.4.2 It suffices in fact for f to be meromorphic and such that $f(z+\omega_1) = f(z) = f(z + \omega_2)$ for every $z \in \mathbb{C}$.

Notation 2.4.1 *Given ω_1 and ω_2, the* **fundamental parallelogram** Π_{ω_1,ω_2} *is the parallelogram $(0, \omega_1, \omega_1 + \omega_2, \omega_2)$ from which we remove ω_1, ω_2 and the sides adjacent to $\omega_1 + \omega_2$.*

A translation of $\Pi := \Pi_{\omega_1,\omega_2}$ is a parallelogram of type $\alpha + \Pi$.

Note that the elliptic functions with a given period lattice Λ form a field which contains the constant functions (the constant functions, of course, form a subfield isomorphic to $\mathbb{C}$).

Liouville's Theorem 2.4.3 *Every entire elliptic function is constant.*

Proof. Let f be such a function. The function f must be bounded on $\bar{\Pi}$ since this set is compact, so it is bounded on all of $\mathbb{C}$. Thus it is constant by Liouville's theorem. □

Corollary 2.4.1 *If two elliptic functions of the same lattice Λ have the same poles with the same principal parts in Π, then their difference is a constant.*

Now, note that since an elliptic function of lattice Λ has only a finite number of poles in a bounded domain of $\mathbb{C}$, we can choose $\alpha \in \mathbb{C}$ so that it admits no poles on the boundary of $\alpha + \Pi$.

Liouville's Theorem 2.4.4 *Let f be an elliptic function of lattice Λ which has no poles on the boundary of $\alpha + \Pi$; then the sum of the residues of f in $\alpha + \Pi$ is zero.*

Proof. This sum is equal to

$$\frac{1}{2i\pi}\int_{\partial C} f(z)\,dz$$

where $C = \alpha + \Pi$.

The integrals along the opposite sides of $\alpha + \Pi$ cancel out since dz changes sign on two opposite sides and $f(z)$ takes the same values, so the integral is zero. □

Corollary 2.4.2 *Let f be non-constant and elliptic of lattice Λ. Then for $\alpha + \Pi$ chosen as above, we can say that f admits either a multiple pole in $\alpha + \Pi$ or at least two simple poles of opposite residues.*

Liouville's Theorem 2.4.5 *Let f and $\alpha + \Pi$ be as in theorem 2.4.4. Let m_a (resp. n_b) denote the orders of the zeros (resp. poles) of f in $\alpha + \Pi$. Then we have*

$$\Sigma m_a = \Sigma n_b$$

and this number does not depend on α.

Proof. Note that

$$\Sigma m_a - \Sigma n_b = \frac{1}{2i\pi}\int_{\partial C} \frac{f'(z)}{f(z)}\,dz$$

and $g(z) = f'(z)/f(z)$ is still elliptic, of lattice Γ.

We now apply theorem 2.4.4 to g. □

Definition 2.4.2 *The number n of poles (or zeros) of an elliptic function f of lattice Λ which lie in a parallelogram $\alpha + \Pi$ is called the* **order** *of f.*

Example 2.4.2 If f is not constant, its order is greater than or equal to 2.

Liouville's Theorem 2.4.6 *Let f and $\alpha + \Pi$ be as in theorem 2.4.4, and let n be the order of f.*

If $a_1, \ldots, a_n$ (resp. $b_1, \ldots, b_n$) denote the suitably repeated zeros (resp. poles) of f lying in $C = \alpha + \Pi$, then

$$a_1 + \cdots + a_n \equiv b_1 + \cdots + b_n \bmod \Lambda.$$

Proof. We have

$$\sum_{\nu=1}^{n}(a_\nu - b_\nu) = \frac{1}{2i\pi}\int_{\partial C} z\frac{f'(z)}{f(z)}\,dz.$$

But on two opposite sides of ∂C, the values of $z(f'(z)/f(z))$ at corresponding points differ by $\pm\,\omega_j\,(f'(z)/f(z))$, $j \in \{1, 2\}$, so we have

$$\frac{1}{2i\pi}\int_{\partial C} z\frac{f'(z)}{f(z)}\,dz = \frac{1}{2i\pi}\left[\int_0^1 -\omega_2\frac{f'(\alpha + t\omega_1)}{f(\alpha + t\omega_1)}\omega_1\,dt + \int_0^1 \omega_1\frac{f'(\alpha + t\omega_2)}{f(\alpha + t\omega_2)}\omega_2\,dt\right]$$
$$= \frac{1}{2i\pi}\left\{-\omega_2[\log f(z)]_\alpha^{\alpha+\omega_1} + \omega_1[\log f(z)]_\alpha^{\alpha+\omega_2}\right\}.$$

As $f(\alpha) = f(\alpha + \omega_j)$ for $j \in \{1, 2\}$, we see that the variation of $\log f(z)$ corresponds to an integer multiple of $2i\pi$, which gives the result. □

Corollary 2.4.3 *Let f and $\alpha + \Pi$ be as in theorem 2.4.6 and let $c \in \mathbb{C} \cup \{\infty\}$. Then*

(i) *$f(z) = c$ has n solutions $z_1, \ldots, z_n$ in $\alpha + \Pi$ or in a neighbouring parallelogram.*
(ii) *The sum $z_1 + \cdots + z_n$, considered modulo Λ, does not depend on c or on α.*

Proof. Apply theorems 2.4.5 and 2.4.6 to $g(z) = f(z) - c$, and note that the poles of g are the same as those of f. □

Naturally, Liouville knew of the existence of non-constant elliptic functions, but we still need to prove it here; we give a constructive proof following the approach of Weierstrass[29].

Remark 2.4.3 The existence of non-constant elliptic functions was known to Abel in 1823 (letter to Holmboë).

2.5 WEIERSTRASS-STYLE ELLIPTIC FUNCTIONS

If we want to construct a meromorphic function of period lattice Λ, we may first think of

$$f(z) := \sum_{\omega\in\Lambda} \frac{1}{z - \omega}, \tag{1}$$

but this does not turn out to be a very good idea; f would be of order 1 and Liouville's theorem 2.4.5 shows that there exists no elliptic function of order 1. This shows that (1) does not make sense.

A better idea would be to consider the function

$$g(z) := \sum_{\omega\in\Lambda} \frac{1}{(z - \omega)^2}, \tag{2}$$

for which the above objection is no longer valid. Unfortunately, we see (taking $z = \omega_1/2$ for example) that the right-hand series does not converge!

[29] K. Weierstrass 1825–1897.

An excellent idea is to take

$$h(z) := \sum_{\omega \in \Lambda} \frac{1}{(z-\omega)^3}, \tag{3}$$

because now the series is uniformly convergent (starting from a certain rank) on every compact subset of $\mathbb{C}$. To see this, we need a lemma.

Lemma 2.5.1 *If the real number s is >2, the series*

$$\sum_{\omega \in \Lambda}{}' \frac{1}{|\omega|^s},$$

where $\sum'_{\omega \in \Lambda}$ *denotes* $\sum_{\omega \in \Lambda \setminus \{0\}}$, *is convergent.*

Proof. Identify $\mathbb{C}$ with $\mathbb{R}\omega_1 \oplus \mathbb{R}\omega_2$, and recall that the norm on $\mathbb{C}$ and $\| \ \|_\infty$ are two equivalent norms.

Thus we have

$$\sum_{\omega \in \Lambda}{}' \frac{1}{|\omega|^s} \leq C \sum_{(m,n) \in \mathbb{Z}^2}{}' \frac{1}{[\sup(m,n)]^s},$$

where C denotes a certain constant.

Now, if $\| \ \|$ denotes the norm $\| \ \|_\infty$, we have

$$\sum_{\|\omega\|=N} \frac{1}{\|\omega\|^s} = \frac{8N}{N^s} = \frac{8}{N^{s-1}},$$

so $\sum_{N=1}^{\infty} \left(\sum_{\|\omega\|=N} 1/\|\omega\|^s \right)$ converges if $s > 2$. □

Theorem 2.5.1 *The series which appears in the right-hand side of (3) is absolutely uniformly convergent on every compact subset of* $\mathbb{C}$, *and the function h is an odd elliptic function of lattice* Λ *and of order 3.*

Proof.

(i) The general term of the series is equivalent in module to $1/|\omega|^3$, and by the lemma, the series is absolutely convergent. It is even absolutely uniformly convergent if $|z|$ is bounded above.

(ii) Since the functions $1/(z-\omega)^3$ are meromorphic for every $\omega \in \Lambda$, it follows from theorem 2.4.4 above that h is meromorphic on every compact subset, so in all of $\mathbb{C}$.

(iii) Since the series is absolutely convergent, it is also commutatively convergent ([Di,1], ch V, §3), so h is periodic of lattice Λ and is also odd.

(iv) The only pole of h in $-((\omega_1+\omega_2)/2) + \Pi$ is the origin, and it is a triple pole, so h is of order three.

□

Corollary 2.5.1 *The roots of h in Π are the three points $\omega_1/2$, $\omega_2/2$, $(\omega_1+\omega_2)/2$.*

Proof.
If $z = \omega/2$, with $\omega \in \Lambda$, we have $-z = -\omega/2$ so $-z \equiv z \bmod \Lambda$.
If $z \notin \Lambda$, then $f(z) \neq \infty$, and by periodicity, we have

$$f(-z) = f(z).$$

But since f is odd, we also have

$$f(-z) = -f(z),$$

hence $2f(z) = 0$.
We thus find three roots: $\omega_1/2$, $\omega_2/2$, $(\omega_1+\omega_2)/2$ in Π. As there cannot be more than three, we have obtained them all. □

If we integrate $-2h(z) + 2/z^3$ term by term in the neighbourhood of zero, we obtain a function $H(z)$, holomorphic in this neighbourhood, which is the sum of the series of the integrals

$$H(z) = \sum_{\omega\in\Lambda}{}' \left[\frac{1}{(z-\omega)^2} - \frac{1}{\omega^2}\right] := \sum_{\omega\in\Lambda\setminus\{0\}} \left[\frac{1}{(z-\omega)^2} - \frac{1}{\omega^2}\right]. \tag{4}$$

By lemma 2.5.1, we know that this series is uniformly convergent on every compact subset. It follows that if we set

$$\wp_\Lambda(z) = \frac{1}{z^2} + H(z), \tag{5}$$

then $\wp_\Lambda$ is a meromorphic function on $\mathbb{C}$ and $\wp'_\Lambda(z) = -2h(z)$.

Definition 2.5.1 *The function $\wp_\Lambda$ is called the* **Weierstrass function** *of the lattice Λ.*

Remark 2.5.1 When Λ is evident from the context, we simply write $\wp$ instead of $\wp_\Lambda$.

Theorem 2.5.2
(i) *The function $\wp_\Lambda$ is an elliptic function of lattice Λ.*
(ii) *If $\lambda \in \mathbb{C}^*$, we have the homogeneity relation*

$$\wp_\Lambda(z) = \lambda^2 \wp_{\lambda\Lambda}(\lambda z).$$

(iii) *Moreover, $\wp_\Lambda$ is an even function of order 2.*

Proof. To simplify the notation, let us write $\wp := \wp_\Lambda$.
(1) We already know that $\wp$ is meromorphic on $\mathbb{C}$.
(2) The homogeneity relation is clear. Taking $\lambda = -1$ and noting that $-\Lambda = \Lambda$, we see that $\wp$ is even.

(3) Since $\wp'(z) = -2h(z)$, we have

$$\wp'(z+\omega) = \wp'(z)$$

for every $\omega \in \Lambda$.

Choose a pair of generators of Λ $\omega \in \{\omega_1, \omega_2\}$, and integrate the preceding relation; we obtain

$$\wp(z+\omega_j) = \wp(z) + C_j$$

for $j \in \{1, 2\}$.

Since the series on the right-hand side of (4) is uniformly convergent on every compact subset, we see that the only poles of $\wp$ are the points of Λ.

It follows that 0 is the only pole of $\wp$ in Π; as it is a double pole, $\wp$ is of order 2 if it is elliptic.

It also follows that $-\omega_1/2$ and $-\omega_2/2$ are not poles of $\wp$. If we set $z = -\omega_j/2$, we obtain

$$\wp\left(\frac{\omega_j}{2}\right) = \wp\left(-\frac{\omega_j}{2}\right) + C_j,$$

and since $\wp$ is even, $C_j = 0$.

We have shown that $\wp$ is periodic of lattice Λ. □

Corollary 2.5.2

(1) *For every $u \in \mathbb{C}$, the equation*

$$\wp(z) = u$$

has either two simple roots or one double root in Π.

(2) *A necessary and sufficient condition for the second case is that*

$$u \in \left\{\wp\left(\frac{\omega_1}{2}\right), \wp\left(\frac{\omega_1}{2}\right), \wp\left(\frac{\omega_1+\omega_2}{2}\right)\right\}.$$

Proof.

(1) The first point follows from Liouville's theorem 2.4.5 and from the fact that $\wp$ is of order 2.

(2) Saying that z_0 is a multiple root of $\wp(z) = u$ is the same as saying that

$$\wp(z_0) = u, \qquad \wp'(z_0) = 0,$$

and corollary 2.5.1 then gives the result. □

Let σ be the involution of $\mathbb{C}/\Lambda$ which associates to every element z its opposite $-z$. In Π, the involution σ can be represented by the map which associates to a point a of Π the representative of $-a$ modulo Λ which lies in Π.

It is clear that the group $\{\mathrm{id}, \sigma\}$ acts on $\mathbb{C}/\Lambda$ and that if $a \notin \{0, \omega_1/2, \omega_2/2, (\omega_1+\omega_2)/2\}$, the orbit of a contains two elements a and a'.

Lemma 2.5.2 *Let f be an even elliptic function of the lattice Λ, and let $a^* = \{a, a'\}$ (resp. $a^* = \{a\}$) be the orbit of a under the action of $\{\mathrm{id}, \sigma\}$.*

(1) *Then $f(a) = f(a')$ and f has the same order at a and a'.*
(2) *If $a = a'$, the order of f at a is even.*

Proof.
(1) The first assertion follows from the parity and the periodicity of f.
(2) The second assertion necessitates a local study.
If $f(a+z) = a_m z^m + \cdots$, then $f(-a-z) = a_m z^m + \cdots$ since f is even.
But $f(-a-z) = f(a'-z) = a'_{m'}(-z)^{m'} + \cdots$, so f has the same order at a and at a', i.e. $m = m'$.
(3) If $a = a'$, we have

$$a_m z^m + \cdots = f(a+z) = f(a-z) = a_m(-z)^m + \cdots,$$

so m is even since we assume that $a_m \neq 0$. □

Corollary 2.5.3 *Every even elliptic function of lattice Λ is a rational function of $\wp$.*

Proof. Let f be such an elliptic function and let a_μ (resp. b_ν) be representatives of the zeros (resp. poles) of f in $\alpha + \Pi \setminus \{0\}$ modulo $\{\mathrm{id}, \sigma\}$, where $|\alpha|$ is assumed to be very small and such that f has neither zeros nor poles on the boundary of $\alpha + \Pi$. We denote by r_μ (resp. s_ν) the order of multiplicity of a_μ (resp. b_ν) divided by the order of its isotropy group (i.e. by 1 or 2) in $\{\mathrm{id}, \sigma\}$.

More precisely, the divisor is equal to 1 if and only if a_μ (resp. b_ν) is not in $\{\omega_1/2, \omega_2/2, (\omega_1 + \omega_2)/2\}$, otherwise it is equal to 2. The lemma shows that r_μ (resp. s_ν) is an integer.

We then set

$$g(z) := \frac{\Pi[\rho(z) - \wp(a_\mu)]^{r_\mu}}{\Pi[\wp(z) - \wp(b_\nu)]^{s_\nu}},$$

which comes down to constructing an elliptic function of order n which imitates the behaviour of f at its zeros and its non-zero poles in $\alpha + \Pi$.

Since g is of order n, it follows from Liouville's theorem 2.4.5 that f and g have the same order at the origin.

Thus f/g is an elliptic function of lattice Λ, holomorphic in $\alpha + \Pi$. By Liouville's theorem 2.4.3, it is constant.

Thus if we set $f/g = C$ and

$$\varphi(X) = C\,\frac{\Pi[X - \wp(a_\mu)]^{r_\mu}}{\Pi[X - \wp(b_\nu)]^{s_\nu}},$$

we have $f = \varphi(\wp)$. □

Example 2.5.1 Consider $\wp'^2$.

Since $\wp'^2$ has no poles other than 0 in Π, φ is a polynomial.

Since $\wp'^2$ admits the double zeros $\omega_1/2$, $\omega_2/2$, $(\omega_1+\omega_2)/2$, we have

$$\wp'^2 = C\left[\wp - \wp\left(\frac{\omega_1}{2}\right)\right]\left[\wp - \wp\left(\frac{\omega_2}{2}\right)\right]\left[\wp - \wp\left(\frac{\omega_1+\omega_2}{2}\right)\right].$$

Finally, $\wp'(z)^2 = 4/z^6 + \cdots$ so $C = 4$ and

$$\varphi(X) = 4\left[X - \wp\left(\frac{\omega_1}{2}\right)\right]\left[X - \wp\left(\frac{\omega_2}{2}\right)\right]\left[X - \wp\left(\frac{\omega_1+\omega_2}{2}\right)\right].$$

Note that the polynomial $\varphi(X)$ has only simple roots, since otherwise $\omega_1/2$, $\omega_2/2$ or $(\omega_1+\omega_2)/2$ would be a zero of $\wp'^2$ of order at least equal to four.

Theorem 2.5.3

(1) *Every elliptic function f of lattice Λ can be written*

$$f = \varphi(\wp) + \wp'\psi(\wp),$$

where φ and ψ are two rational functions.

(2) *The field of elliptic functions of lattice Λ is $\mathbb{C}(\wp, \wp')$.*

Proof.

(1) The assertion follows from corollary 2.5.3 if f is even.

(2) Set

$$g := \frac{f(z)+f(-z)}{2}, \qquad h := \frac{f(z)-f(-z)}{2};$$

it is clear that g and h are two elliptic functions of lattice Λ and that g (resp. h) is even (resp. odd).

Thus $g = \varphi(\wp)$ and $h/\wp' = \psi(\wp)$. □

2.6 EISENSTEIN SERIES

Given a lattice Λ of $\mathbb{C}$, for each integer $m \geq 3$, we set

$$G_m(\Lambda) := \sum_{\omega\in\Lambda}{}' \frac{1}{\omega^m} = \sum_{\omega\in\Lambda\setminus\{0\}} \frac{1}{\omega^m}.$$

Note that $G_m(\Lambda) = 0$ if m is odd.

Notation 2.6.1 *G_{2k} is called the* **Eisenstein series of index $2k$** (or k, according to the author) *of the lattice Λ.*

The Eisenstein series can be used to express the Laurent[30] expansion of $\wp$ at the origin. Indeed, we have

$$\wp(z) - \frac{1}{z^2} = \sum{}' \left[\frac{1}{(z-\omega)^2} - \frac{1}{\omega^2}\right].$$

But

$$\frac{1}{(z-\omega)^2} = \frac{1}{\omega^2} + 2\,\frac{z}{\omega^3} + 3\,\frac{z^2}{\omega^4} + \cdots$$

Thus

$$\wp(z) = \frac{1}{z^2} + 3G_4 z^2 + 5G_6 z^4 + 7G_8 z^6 + \cdots \tag{6}$$

We saw in the preceding section that

$$\wp'^2 = 4\wp^3 + a\wp^2 + b\wp + c \tag{7}$$

with $a, b, c \in \mathbb{C}$. By Liouville's theorem 2.4.6, we have $a = 0$.

We now propose to compute a, b, c in terms of the Eisenstein series by formally identifying the Laurent expansions at the origin of the two sides of (7), using the expression (6) as well as the derived expression

$$\wp'(z) = -\frac{2}{z^3} + 6G_4 z + 20G_6 z^3 + 42G_8 z^5 + \cdots \tag{8}$$

Remark 2.6.1

(1) We can show (exercise) that $G_{2m} \not\equiv 0$ when $m \geq 2$, i.e. that there exists a lattice Λ such that $G_{2m}(\Lambda) \neq 0$.

(2) The coefficients a, b, c of the right-hand side of (7) are unique. This can be seen on the Laurent expansion of $\wp'^2 - 4\wp^3$.

Theorem 2.6.1 *Let G_4 and G_6 be the Eisenstein series of indices 4 and 6 associated to Λ, and let $\wp$ be the Weierstrass function of Λ.*

Then $\wp$ is a solution of the differential equation

$$Y'^2 = 4Y^3 - 60G_4 Y - 140G_6.$$

Remark 2.6.2 $60G_4$ and $140G_6$ are traditionally denoted g_2 and g_3.

Proof. **(1)** If we replace Λ by $\alpha\Lambda$ ($\alpha \in \mathbb{C}^*$) and z by αz, then $\wp$, G_4 and G_6 become $\alpha^{-2}\wp$, $\alpha^{-4}G_4$ and $\alpha^{-6}G_6$; we will say that $\wp$ is of weight 2, G_4 of weight 4 and G_6 of weight 6, etc.

[30] P.A. Laurent 1813–1854.

For relation (7) to be homogeneous, we need b to be of weight 4 and c of weight 6. We see that in fact we have

$$b = \lambda G_4 \quad \text{and} \quad c = \mu G_6.$$

(2) We can compute λ and μ by formal computation using (6) and (8).

Computing $\boldsymbol{\lambda}$: since G_6 has to disappear from the result, we identify the z^{-2} terms in the expressions

$$\left(-\frac{2}{z^3} + 6G_4 z\right)^2 \quad \text{and} \quad 4\left(\frac{1}{z^2} + 3G_4 z^2\right)^3 + \lambda\,\frac{G_4}{z^2} + \cdots,$$

which gives $-24 = 36 + \lambda$.

Computing $\boldsymbol{\mu}$: since G_4 must disappear from the result, we identify the constant terms in the expressions

$$\left(-\frac{2}{z^3} + 20G_6 z^3\right)^2 \qquad \text{and} \qquad 4\left(\frac{1}{z^2} + 5G_6 z^4\right)^3 + \mu G_6,$$

which gives $-80 = 60 + \mu$. □

Remark 2.6.3 The theorem implies the following relations

(i) $\wp(\omega_1/2) + \wp(\omega_2/2) + \wp(\omega_1 + \omega_2)/2 = 0.$
(ii) $\wp'' = 6\wp^2 - 30G_4.$

Relation (ii) makes it possible to express G_{2m} as a polynomial in G_4 and G_6 (see Exercise 2.3 and also [Se 1] p. 151).

2.7 THE WEIERSTRASS CUBIC

Keeping the definitions of Section 2.6, we say that the cubic

$$Y^2 = 4X^3 - g_2 X - g_3 \qquad (E)$$

is the *Weierstrass cubic associated to the lattice* Λ.

This cubic is a special case of cubics of the type

$$Y^2 = D(X) = aX^3 + bX^2 + cX + d,$$

where $D(X)$ is a polynomial of degree three without multiple roots; the homogeneous equation of this cubic is

$$Y^2 Z - (aX^3 + bX^2 Z + cXZ^2 + dZ^3) = 0. \qquad (\Gamma)$$

Theorem 2.7.1 *When the polynomial $D(X)$ has no multiple roots, the cubic (Γ) is a non-singular (or smooth) curve in the projective plane $\mathbb{P}_2(\mathbb{C})$.*

Proof. Let $F(X, Y, Z)$ denote the left-hand side of the equation (Γ). It suffices to prove that

$$F'_X = F'_Y = F'_Z = 0$$

implies $(X, Y, Z) = (0, 0, 0)$, which is an impossible choice for a system of homogeneous coordinates!

To simplify the notation, we again set

$$D(X, Z) = aX^3 + bX^2Z + cXZ^2 + dZ^3;$$

then the system becomes

$$-D'_X(X, Z) = 2YZ = Y^2 - D'_Z(X, Z) = 0.$$

If $Z \neq 0$, it follows that $Y = 0$, so

$$D'_X(X, Z) = D'_Z(X, Z) = 0,$$

which is impossible since $D(X, 1)$ has no multiple roots.

Thus $Z = 0$ and since $a \neq 0$, it follows that $X = 0$, and finally $Y = 0$! □

Since (E) has no multiple points, we do not know how to construct a Diophantus-type parametrisation for this cubic.

However, for $z \in \mathbb{C}/\Lambda$, set

$$z \longmapsto f(z) = \begin{cases} (\wp(z), \wp'(z), 1) & \text{if } z \notin \Lambda \\ (0, 1, 0) & \text{if } z \in \Lambda; \end{cases}$$

we then have the following result.

Theorem 2.7.2 *The map $f : \mathbb{C}/\Lambda \to \mathbb{P}_2(\mathbb{C})$ is a bijective map from $\mathbb{C}/\Lambda$ to the set $\bar{E}(\mathbb{C})$ of the points of the projective completion of E in the plane $\mathbb{P}_2(\mathbb{C})$.*

Moreover, this map is holomorphic.

Proof. **(1)** Let us show that f maps $\Pi\setminus\{0\}$ injectively to $E(\mathbb{C})$.

We already saw that $f(z) \in E(\mathbb{C})$ if $z \in \Pi\setminus\{0\}$. Assume that $f(z_1) = f(z_2)$ and $\wp'(z_1) \neq 0$. Then we have

$$\wp(z_1) = \wp(z_2), \qquad \wp'(z_1) = \wp'(z_2).$$

The first relation implies $z_2 \in \{z_1, z'_1\}$, and with the second, we obtain $z_2 = z_1$. (Recall that $z'_1 = \sigma(z_1)$).

If now $\wp'(z_1) = 0$, we have

$$z_1 \in \left\{\frac{\omega_1}{2}, \frac{\omega_2}{2}, \frac{\omega_1 + \omega_2}{2}\right\},$$

and as the numbers $\wp(\omega_1/2)$, $\wp(\omega_2/2)$, $\wp(\omega_1 + \omega_2)/2$ are distinct, we again have $z_2 = z_1$.

(2) Let us show that f maps $\Pi\backslash\{0\}$ onto $E(\mathbb{C})$.
Let $(a, b) \in \mathbb{C}^2$ be such that

$$b^2 = 4a^3 - g_2 a - g_3. \tag{9}$$

Since $\wp$ is of order 2, the equation

$$\wp(z) = a$$

has two solutions z_1 and z_1' in $\Pi\backslash\{0\}$. Relation (9) implies that

$$\wp'(z_1)^2 = \wp'(z_1')^2 = b^2.$$

Now, we know that $\wp'(z_1') = -\wp'(z_1)$, so we have $f(z_1) = (a, b, 1)$ or $f(z_1') = (a, b, 1)$.
(3) Clearly $z \mapsto f(z)$ is holomorphic on $\mathbb{C}\backslash\Lambda$.
Suppose now that $z_0 \in \Lambda$; is f still holomorphic in the neighbourhood of z_0?
Since $f(z_0) = (0, 1, 0)$, we need to represent $f(z)$ in another affine chart of $\mathbb{P}_2(\mathbb{C})$.
The equation of $\bar{E}$ is given by

$$Y^2 Z = 4X^3 - g_2\, X\, Z^2 - g_3 Z^3,$$

and since we want to study it in the neighbourhood of $(0, 1, 0)$, we set $Y = 1$, which gives the cubic

$$Z = 4X^3 - g_2 X Z^2 - g_3 Z^3.$$

In this system of coordinates, we have

$$f(z) = (X, Z) = \left(\frac{\wp(z)}{\wp'(z)}, \frac{1}{\wp'(z)}\right),$$

and these are indeed holomorphic functions in the neighbourhood of $z = 0$. □

2.8 ABEL'S THEOREM

We saw in Section 2.7 that the map f is a bijection from $\mathbb{C}/\Lambda$ onto $\bar{E}(\mathbb{C})$.

Since $\mathbb{C}/\Lambda$ is a group, we can transport the group structure and define a group law on $E(\mathbb{C})$. Precisely, if $P = f(u)$ and $Q = f(v)$, we set

$$P \oplus Q := f(u + v) = f(f^{-1}(P) + f^{-1}(Q)).$$

Now our problem is to characterise $P \oplus Q$ without making use of the bijection f!

Before considering the general case, we note that the identity element O of $\bar{E}(\mathbb{C})$ is the point $(0, 1, 0)$, and the opposite of the point (X, Y, Z) is the point $(X, -Y, Z)$.

Abel's Theorem 2.8.1 *A necessary and sufficient condition for three points P, Q, R of $\bar{E}(\mathbb{C})$ to be colinear is that $P \oplus Q \oplus R = O$.*

Proof. (Modern style)

(1) Assume that P, Q, R all lie on the line $aX + bY + cZ = 0$; then their elliptic parameters $u = f^{-1}(P)$, $v = f^{-1}(Q)$ and $w = f^{-1}(R)$ are the zeros of the elliptic function $g(z) = a\wp + b\wp' + c$.

If $b \neq 0$, this function admits a unique pole of order 3 at $z = 0 \in \mathbb{C}/\Lambda$, so by Liouville's theorem 2.4.6, $u + v + w = 0 \in \mathbb{C}/\Lambda$.

If $b = 0$ but $a \neq 0$, the line is $aX + c$ and it passes through the point $(0, 1, 0) \in \overline{E}(\mathbb{C})$.

Thus we can assume that $R = O$, of elliptic parameter $w = 0 \in \mathbb{C}/\Lambda$. As g admits a unique pole of order 2 in $\mathbb{C}/\Lambda$, we again have $u + v = 0$ by Liouville's theorem 2.4.6, so $u + v + w = 0$.

Only the case $a = b = 0$ and $c \neq 0$ remains.

But the intersection of the Weierstrass cubic

$$Y^2Z = 4X^3 - g_2XZ - g_3Z^3$$

with the line $Z = 0$ gives the point $(0, 1, 0) = 0$ counted three times ($Z = 0$ is an inflection tangent at O). Thus the result still holds.

(2) Conversely, if we have $u + v + w = 0$, we see that the line defined by $f(u)$ and $f(v)$ intersects $\bar{E}(\mathbb{C})$ at a point $f(w')$ such that (by the first part)

$$u + v + w' = 0.$$

It follows that $w' = w$, and the theorem is proved. □

Proof. (Abel's style) We will not attempt to give a complete and rigorous proof; the argument here should be taken as essentially "heuristic".

Let us change the notation and write $(x_1, y_1), (x_2, y_2), (x_3, y_3)$ for the points P, Q and R; for $j \in \{1, 2, 3\}$ we write

$$(x_j, y_j) = f(z_j), \quad z_j \in \mathbb{C}/\Lambda.$$

Assume that P, Q and R lie on a line of equation $y = mx + n$; then we want to show that when m and n vary, the sum $z_1 + z_2 + z_3$ remains constant. To see this, it suffices to prove that

$$dz_1 + dz_2 + dz_3 = 0.$$

Now, we have

$$\frac{dx}{y} = \frac{\wp'(z)}{\wp'(z)}\, dz = dz,$$

so it suffices to show that

$$\frac{dx_1}{y_1} + \frac{dx_2}{y_2} + \frac{dx_3}{y_3} = 0.$$

The elements x_1, x_2, x_3 are the roots of the polynomial

$$P(X) = 4X^3 - g_2X - g_3 - (mX + n)^2,$$

so since $P(x_j) = 0$, we have

$$P'(x_j)\, dx_j + \frac{\partial P}{\partial m}\, dm + \frac{\partial P}{\partial n}\, dn = 0,$$

i.e.

$$P'(x_j)\, dx_j = 2(mx_j + n)\,(x_j\, dm + dn).$$

If the roots x_1, x_2, x_3 are simple, this gives the relation

$$\sum_{j=1}^{3} \frac{dx_j}{y_j} = \sum_{j=1}^{3} \frac{dx_j}{mx_j + n} = 2 \sum_{j=1}^{3} \frac{x_j\, dm + dn}{P'(x_j)}. \tag{10}$$

Abel's trick was to consider the rational function $X(X\, dm + dn)/P(X)$, with the assumption that x_1, x_2 and x_3 are distinct, and to decompose it into elementary fractions. We have

$$\frac{X(X\, dm + dn)}{P(X)} = \sum_{j=1}^{3} \frac{x_j(x_j\, dm + dn)}{P'(x_j)\,(X - x_j)}.$$

Plugging $X = 0$ into this relation, we obtain the desired result, i.e. $dz_1 + dz_2 + dz_3 = 0$, hence:

$$z_1 + z_2 + z_3 = C^{te}. \tag{10'}$$

If we now let z_1, z_2 and z_3 tend simultaneously to zero while respecting the conditions listed above, we see that the constant in (10′) is zero. □

Special case

(1) If $P = Q$, then the line $aX + bY + cZ = 0$ is the tangent at P to the curve $\bar{E}(\mathbb{C})$. The parameter w of R is $-2u$.

If $2u = 0$, the tangent at $f(u)$ passes through the origin O, i.e. it is parallel to the axis of the ordinates.

(2) If $P = Q = R$ (which is possible, as we have already seen) then the line $aX+bY+cZ = 0$ will be an **inflection tangent** at P of the curve $\bar{E}(\mathbb{C})$.

Thus the inflection points of $\bar{E}(\mathbb{C})$ are the points whose parameters satisfy the equation $3u = 0 \in \mathbb{C}/\Lambda$.

We easily see that there are exactly nine of these inflection points and that they form collinear triples.

Geometric Construction

We represent the group law in the following diagram:

$$(P, Q) \longmapsto P \oplus Q$$

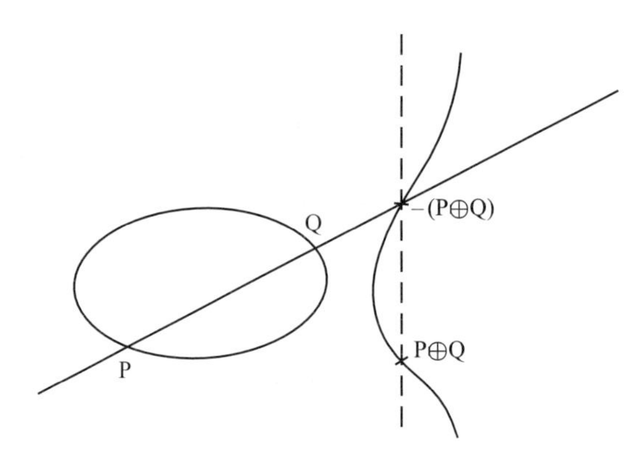

Remark 2.8.1

(1) Let K denote a subfield of $\mathbb{C}$. If $(g_2, g_3) \in K^2$, then the preceding construction puts an Abelian group structure on $\bar{E}(K)$, making it into a subgroup of $\bar{E}(\mathbb{C})$.

(2) In Chapter 5, Section 5.4, theorem 5.4.2, we will prove that every curve of equation

$$Y^2Z = 4X^3 - aXZ^2 - bZ^3$$

having no double points, i.e. such that $a^3 - 27b^2 \neq 0$, can be parametrised by elliptic functions.

2.9 LOXODROMIC FUNCTIONS

We now give a second construction of the elliptic functions, which is much simpler than the one given in Sections 2.4 and 2.5, and which furthermore has the advantage of generalising from $\mathbb{C}$ to the case of an arbitrary complete valuation field ("Tate functions").

The main idea is to note that every function $g(z)$ of the form $f(e^{2i\pi z/\omega_1})$ admits the group of periods $\mathbb{Z}\omega_1$. If moreover we have

$$f(q\zeta) = f(\zeta) \tag{11}$$

for every $\zeta \in \mathbb{C}^*$ and for a certain $q = e^{2i\pi\omega_2/\omega_1}$, with $\Im(\omega_2/\omega_1) > 0$, we see that g admits the period lattice $\Lambda = \mathbb{Z}\omega_1 + \mathbb{Z}\omega_2$. Finally if f is meromorphic in $\mathbb{C}^*$, then g is an elliptic function of lattice Λ.

Definition 2.9.1 *Let $q \in \mathbb{C}^*$ be a complex number of module < 1. We say that a map $f : \mathbb{C}^* \to \mathbb{C} \cup \{\infty\}$ is a* **loxodromic function of multiplicator** *q iff f is meromorphic and satisfies equation (11).*

It is clear that the loxodromic functions of multiplicator q form a field, which we call $\mathcal{L}_q$.

It is also clear that the loxodromic functions satisfy structure theorems analogous to the theorems concerning elliptic functions.

Theorem 2.9.1 *Every entire loxodromic function is constant.*

Proof. Indeed, let $f \in \mathcal{L}_q$.

Equation (11) shows that f is defined by its restriction to the annulus $\overline{C_q(R)} = \{z \in \mathbb{C};\ |q|R \leq |z| \leq R\}$. As $\overline{C_q(R)}$ is compact, f is bounded in $\overline{C_q(R)}$ so in $\mathbb{C}^*$. A theorem from the theory of functions of one complex variable ([Va] p. 389) states that f is holomorphic at the origin since f is holomorphic and bounded in the neighbourhood of the origin.

Thus, we can apply the Liouville theorems of Section 2.4. □

Theorem 2.9.2 *Assume that $f \in \mathcal{L}_q$ has neither zeros nor poles on the boundary of the annulus $C_q(R) = \{z \in \mathbb{C};\ |q|R \leq |z| < R\}$. Then the sum of the residues of the poles of $f(z)/z$ lying in $C_q(R)$ is zero.*

Proof. Let Γ and Γ' denote the circumferences bounding $C_q(R)$; we know that the sum is equal to

$$\frac{1}{2i\pi}\left(\int_{\Gamma'_+}\frac{f(\zeta)}{\zeta}\,d\zeta - \int_{\Gamma_+}\frac{f(z)}{z}\,dz\right).$$

Setting $z = q\zeta$ in the second integral, we recover the first one. □

Corollary 2.9.1 *Every non-constant loxodromic function of multiplicator q has at least two poles (and two zeros) in every annulus $C_q(R)$.*

Theorem 2.9.3 *Let $f \in \mathcal{L}_q$ and let $C_q(R)$ be an annulus chosen as in theorem 2.9.2. Let m_a (resp. n_b) denote the orders of the zeros (resp. poles) of f in $C_q(R)$. Then we have*

$$\Sigma m_a = \Sigma n_b.$$

Proof. Indeed, $\Sigma m_a - \Sigma n_b$ is equal to

$$\frac{1}{2i\pi}\left(\int_{\Gamma'+}\frac{f'(\zeta)}{f(\zeta)}\,d\zeta - \int_{\Gamma+}\frac{f'(z)}{f(z)}\,dz\right),$$

and we see as in theorem 2.9.2 that the two integrals are equal. □

Before proving theorem 2.9.4, we take a moment to give some examples of non-trivial loxodromic functions.

Example 2.9.1 For $z \in \mathbb{C}^*$, set

$$S(z) = \prod_0^\infty (1 - q^n z) \prod_1^\infty (1 - q^n z^{-1}).$$

We easily see that each infinite product is convergent, and that the map $S : \mathbb{C}^* \to \mathbb{C}$ is holomorphic. A simple computation (absolute convergence) shows that

$$S(qz) = -z^{-1}S(z), \qquad S\left(\frac{1}{z}\right) = -z^{-1}S(z). \tag{12}$$

Now consider $a_1, \ldots, a_m \in \mathbb{C}^*$, $b_1, \ldots, b_m \in \mathbb{C}^*$ (not necessarily distinct) such that $a_i \neq b_j$ and

$$a_1 \cdots a_m = b_1 \cdots b_m.$$

If for $z \in \mathbb{C}^*$ not equal to $b_1, \ldots, b_m$ modulo $\langle q \rangle$, we set

$$M(z) := \frac{S(z/a_1) \ldots S(z/a_m)}{S(z/b_1) \ldots S(z/b_m)},$$

then we see that M defines a meromorphic function on $\mathbb{C}^*$ whose poles are congruent to $b_1, \ldots, b_m \mod \langle q \rangle$. Furthermore, the relations (12) imply that

$$M(qz) = M(z), \qquad M\left(\frac{1}{z}\right) = \frac{S(a_1 z) \cdots S(a_m z)}{S(b_1 z) \cdots S(b_m z)}, \tag{13}$$

so that in particular we see that $M \in \mathcal{L}_q$.

Theorem 2.9.4 *For every R such that the boundary of $C_q(R)$ contains neither zeros nor poles of $f \in \mathcal{L}_q$, let λ denote the quotient $(a_1 \cdots a_m)/(b_1 \cdots b_m)$ of the product of the zeros of f in $C_q(R)$ by the product of its poles in $C_q(R)$. Then $\lambda \in \langle q \rangle$.*

Proof. Keep the notation of the preceding example without assuming that $a_1 \cdots a_m = b_1 \cdots b_m$. Then the first of the relations (13) must be replaced by

$$M(qz) = \lambda M(z).$$

Now consider the function

$$g(z) = \frac{f(z)}{M(z)}.$$

By the construction of M, g has no poles or zeros in $C_q(R)$, and since $f \in \mathcal{L}_q$, it follows that g is entire, as well as its inverse $1/g$.

Let $\sum_{-\infty}^{\infty} c_n z^n$ be the Laurent expansion of g in $\mathbb{C}^*$. As $\lambda g(qz) = g(z)$ and at least one of the c_n is different from zero, we see that $(\lambda q^n - 1)c_n = 0$, hence $\lambda = q^{-n} \in \langle q \rangle$. □

Remark 2.9.1 The preceding proof also shows that $c_\nu = 0$ if $\nu \neq n$, so we have the following result.

Theorem 2.9.5 *If $\lambda = q^n$, the loxodromic function f of theorem 2.9.4 is of the form*

$$C^{te}\frac{S(z/a_1)\cdots S(z/a_m)}{S(q^n z/b_1)S(z/b_2)\cdots S(z/b_m)}.$$

Let us now study a particular loxodromic function ρ.

2.10 THE FUNCTION ρ

The function we will consider now is neither more nor less than a somewhat disguised version of the Weierstrass $\wp$ function (or perhaps $\wp$ is a disguised version of ρ!)

Let $\mathcal{M}$ be the complex vector space of meromorphic functions on $\mathbb{C}^*$, and let λ be the map

$$\lambda\begin{cases}\mathcal{M} \longrightarrow \mathcal{M} \\ f(z) \longmapsto z\dfrac{f'(z)}{f(z)}.\end{cases}$$

The image of $\mathcal{L}_q$ under λ is contained in $\mathcal{L}_q$, which enables us to construct new loxodromic functions.

However, we will be particularly interested in the image of the set of functions h such that

$$h(qz) = -z^{-1}h(z). \tag{14}$$

This image is contained in the set $\mathcal{S}_q$ of solutions of Abel's equation:

$$v(qz) = v(z) - 1. \tag{15}$$

Lemma 2.10.1 *Set $\chi(z) := zS'(z)/S(z)$. Then*

(1) *$\mathcal{S}_q$ is the affine $\mathbb{C}$-space $\chi + \mathcal{L}_q$.*

(2) *The map $D : v(z) \mapsto zv'(z)$ maps $\mathcal{S}_q$ to $\mathcal{L}_q$.*

Proof. Indeed, S satisfies the relations (12). □

Remark 2.10.1

(1) A simple computation gives

$$\chi(z) = \sum_1^\infty \frac{q^n z^{-1}}{1 - q^n z^{-1}} - \sum_0^\infty \frac{q^n z}{1 - q^n z}. \tag{16}$$

(2) The second relation of (12) implies that

$$\chi(z) + \chi\left(\frac{1}{z}\right) = 1. \tag{17}$$

As it is particularly interesting to consider the function $D(\chi)$, we are led to the following definition.

Definition 2.10.1 *Denote by ρ the function $-D(\chi)$, i.e.*

$$\rho(z) = \sum_{-\infty}^{\infty} \frac{q^n z}{(1 - q^n z)^2}.$$

Remark 2.10.2 Relation (17) implies that

$$\rho(z) = \rho\left(\frac{1}{z}\right). \tag{18}$$

Expansion of ρ in the annulus $\Gamma=\{z \in \mathbb{C}; |q|<|z|<|q^{-1}|\}$

If we subtract from $\rho(z)$ the term corresponding to $n = 0$, we obtain a sum of rational functions whose poles lie outside of Γ, so for $z \in \Gamma$, we can write

$$\begin{cases} \text{If } n > 0: & \dfrac{q^n z}{(1 - q^n z)^2} = q^n z + \cdots + m q^{nm} z^m + \cdots \\ \text{If } n < 0: & \dfrac{q^n z}{(1 - q^n z)^2} = q^{-n}\left(\dfrac{1}{z}\right) + \cdots + m q^{-nm}\left(\dfrac{1}{z}\right)^m + \cdots \end{cases}$$

As these series are absolutely convergent in Γ, they form a summable family and we have

$$\rho(z) - \frac{z}{(1 - z)^2} = \sum_{m=1}^{\infty} \frac{m q^m}{1 - q^m}\left(z^m + \frac{1}{z^m}\right). \tag{19}$$

Note that the function $\rho(z) - z/(1 - z)^2$ is invariant under the symmetry $z \to 1/z$ (this follows from (18) or (19)), one of whose fixed points is $1 \in \Gamma$. Thus it is natural to consider the Taylor expansion of $\rho(z) - z/(1 - z)^2$ in the neighbourhood of the identity element of the group $\langle q \rangle$, i.e. of the point $z = 1$. Set $z = 1 + \zeta$. Then by (19) we have

$$\rho(z) - \frac{z}{(z - 1)^2} = 2\sum_{m=1}^{\infty} \frac{m q^m}{1 - q^m} + \sum_{2}^{\infty} \gamma_n \zeta^n \tag{20}$$

with $\gamma_2 = \sum_{m=1}^{\infty} (m^3 q^m)/(1 - q^m)$.

We now wish to follow a procedure analogous to the one in Section 2.6 which we used to find the differential equation of the Weierstrass $\wp$ function. Set $\rho_1(z) = \rho(z) + c$, where c is a constant which we will determine later in order to state a simple result. By (20), we have

$$\rho_1(1 + \zeta) = \zeta^{-2} + \zeta^{-1} + (c + \gamma_0) + \sum_{2}^{\infty} \gamma_n \zeta^n, \tag{21}$$

with $\gamma_0 = 2\sum_{m=1}^{\infty} (m q^m)/(1 - q^m)$.

If we apply the operator D to the two sides of (21), we find

$$(D\rho_1)(1+\zeta) = -2\zeta^{-3} - 3\zeta^{-2} - \zeta^{-1} + \sum_{2}^{\infty} n\gamma_n(\zeta+1)\zeta^{n-1}. \tag{22}$$

If we compute $[D\rho_1]^2 - 4\rho_1^3$ as in Section 2.6, we find

$$[D\rho_1]^2 - 4\rho_1^3 = A\zeta^{-4} + B\zeta^{-3} + C\zeta^{-2} + D\zeta^{-1} + \cdots$$

with

$$A = 1 - 12(c+\gamma_0), \quad B = 2A, \quad C = 1 - 20\gamma_2 - 12(c+\gamma_0) - 12(c+\gamma_0)^2.$$

If we determine c by the condition $A = 0$, we find $B = 0$ and $C = -(1/12) - 20\gamma_2$. Since the loxodromic function $[D\rho_1]^2 - 4\rho_1^3 - C\rho_1$ has at most one simple pole in Γ, it must be constant by the corollary to theorem 2.9.2 of Section 2.9. We have obtained the following result.

Theorem 2.10.1 *The loxodromic function of multiplicator q:*

$$\rho_1(z) = \rho(z) + \frac{1}{12} - 2\sum_{m=1}^{\infty} \frac{mq^m}{1-q^m}$$

satisfies the differential equation:

$$[z\rho_1'(z)]^2 = 4\rho_1^3 - g_4\rho_1 - g_6 \tag{23}$$

where

$$g_4 = \frac{1}{12} + 20\sum_{1}^{\infty} \frac{m^3q^m}{1-q^m} \quad \text{and} \quad g_6 = \frac{1}{216} - \frac{7}{3}\sum_{1}^{\infty} \frac{m^5q^m}{1-q^m}.$$

2.11 COMPUTATION OF THE DISCRIMINANT

We easily obtain the arguments z for which $\rho_1'(z)$ is zero, from the relation (18). Indeed, differentiating this relation, we have

$$z\rho'(z) = -\frac{1}{z}\rho'\left(\frac{1}{z}\right),$$

hence we deduce that $\rho'(-1) = 0$.

But since $\rho(qz) = \rho(z)$, we also have

$$qz\rho'(qz) = -\frac{1}{z}\rho'\left(\frac{1}{z}\right),$$

and taking $z = q^{-1/2}$, we also obtain $\rho'(q^{1/2}) = 0$. As $\sqrt{q^{-1}}$ has two determinations, we also have $\rho'(-q^{1/2}) = 0$. This shows that ρ' admits at least three zeros in the annulus

$\{z \in \mathbb{C};\ |q|(1+\varepsilon) \leq |z| < 1+\varepsilon\}$ with $\varepsilon > 0$ sufficiently small. As it admits a single pole of order three (at the point $z = 1$) in the same annulus, theorem 2.9.3 of Section 2.9 shows that these are the only zeros of ρ' in this annulus. Thus we see that the discriminant of the right-hand side of equation (23) is, up to a factor, equal to

$$\Delta = g_4^3 - 27g_6^2 = 16[\rho(-1) - \rho(\sqrt{q})]^2[\rho(-1) - \rho(-\sqrt{q})]^2[\rho(\sqrt{q}) - \rho(-\sqrt{q})]^2.$$

Our goal is to simplify this expression of Δ.

Consider z and z' in $\mathbb{C}^*$. Since the function $z \mapsto \rho(z) - \rho(z')$ is a loxodromic function in $\mathcal{L}_q$, Theorem 2.9.5 implies that

$$\rho(z) - \rho(z') = \mu(z')\frac{S(z/z')S(zz')}{S(z)^2}$$

with $\mu(z') \in \mathbb{C}^*$, constant with respect to z. For more symmetry we will write

$$\rho(z) - \rho(z') = \lambda(z')\frac{S(z/z')S(zz')}{S(z)^2S(z')^2} \tag{24}$$

with $\lambda(z') \in \mathbb{C}^*$, constant with respect to z. Setting the principal parts of the two sides of (24) at the pole $z = 1$ equal, we find

$$1 = \lambda(z')\frac{S(1/z')}{S(z')}\lim_{z=1}\left(\frac{z-1}{S(z)}\right)^2 = -\frac{\lambda(z')}{z'}\prod_1^\infty(1-q^n)^{-4},$$

hence

$$\lambda(z') = -z'\prod_1^\infty(1-q^n)^4. \tag{25}$$

Substituting these values of $\pm(\rho(z) - \rho(z'))$ into the expression of Δ given by the lemma, we have

$$\Delta = 16\left\{\lambda(-1)^2\lambda(-\sqrt{q})\frac{S(-\sqrt{q})^2}{S(\sqrt{q})^2S(-1)^2}\frac{S(\sqrt{q})^2}{S(-\sqrt{q})^2S(-1)^2}\frac{S(-1)S(-q)}{S(\sqrt{q})^2S(-\sqrt{q})^2}\right\}^2.$$

We easily deduce from (12) that

$$S(-q) = S(-1),$$

so

$$\Delta = 16\left\{\frac{\sqrt{q}\prod(1-q^n)^{12}}{S(-1)^2S(\sqrt{q})^2S(-\sqrt{q})^2}\right\}^2.$$

Now, we have

$$\begin{cases} S(-1) = 2\prod_1^{\infty}(1+q^n)^2 \\ S(\sqrt{q}) = \prod_0^{\infty}(1-q^{n+1/2})^2 \\ S(-\sqrt{q}) = \prod_0^{\infty}(1+q^{n+1/2})^2. \end{cases}$$

This gives

$$S(-1)S(\sqrt{q})S(-\sqrt{q}) = 2\prod_1^{\infty}(1+q^n)^2\prod_0^{\infty}(1-q^{2n+1})^2 = 2,$$

as we see by multiplying the right-hand term by $\prod_1^{\infty}(1-q^n)$.

Finally, we obtain

Lemma 2.11.1 *The discriminant* Δ *is given by the formula*

$$\Delta = q\prod_1^{\infty}(1-q^n)^{24}. \tag{26}$$

□

2.12 RELATION TO ELLIPTIC FUNCTIONS

We saw at the beginning of Section 2.9 that if f is a loxodromic function of multiplicator

$$q = e^{2i\pi\omega_2/\omega_1}, \quad \text{Im}(\omega_2/\omega_1) > 0,$$

then $f(e^{-2i\pi u/\omega_1})$ is an elliptic function of lattice $\Lambda = \mathbb{Z}\omega_1 + \mathbb{Z}\omega_2$.

Conversely, if g is an elliptic function of lattice $\Lambda = \mathbb{Z}\omega_1 + \mathbb{Z}\omega_2$, with Im $\omega_2/\omega_1 > 0$ and if $z \in \mathbb{C}^*$, the function

$$f(z) := g\Big(\frac{\omega_1}{2i\pi}\log z\Big)$$

is well-defined because $\log z \in \mathbb{C}/2i\pi\mathbb{Z}$ and g admits the period ω_1. And if we set

$$q = e^{2i\omega_2/\omega_1},$$

we have $f(qz) = g((\omega_1/2i\pi)\log(qz)) = f(z)$.

Moreover, it is clear that this correspondence preserves the analyticity of the functions.

Thus, we write

$$z = e^{2i\pi u/\omega_1} = 1 + \frac{2i\pi u}{\omega_1} - \frac{2\pi^2 u^2}{\omega_1^2} + \cdots.$$

With the notation of Section 2.9, we have

$$\begin{cases} \zeta = z - 1 = \dfrac{2i\pi}{\omega_1} u\Big(1 + \dfrac{i\pi}{\omega_1} u - \dfrac{2\pi^2}{\omega_1^2} u^2 + \cdots\Big) \\ \zeta^2 = (z-1)^2 = \dfrac{-4\pi^2}{\omega_1^2} u^2 \Big(1 + \dfrac{2i\pi}{\omega_1} u - \dfrac{7\pi^2}{3\omega_1^2} u^2 + \cdots\Big). \end{cases}$$

The function which corresponds to ρ_1 is an even elliptic function (since $\rho_1(1/z) = \rho_1(z)$), of lattice Λ, which admits a double pole at the origin with principal part given by

$$\zeta^{-2} + \zeta^{-1} + \frac{1}{12} = -\frac{\omega_1^2}{4\pi^2} u^{-2} + o(u).$$

Thus, we have the following result.

Theorem 2.12.1 *The loxodromic function ρ_1 of multiplicator q corresponds to $-\omega_1^2/4\pi^2\wp$, where $\wp$ denotes the Weierstrass function associated to Λ.*

It follows that

$$\frac{2i\pi}{\omega_1} z\rho_1'(z) = \frac{d}{du}\rho_1(z) = -\frac{\omega_1^2}{4\pi^2}\wp'(u),$$

so

$$z\rho_1'(z) = \left(\frac{\omega_1}{2i\pi}\right)^3 \wp'(u).$$

Finally if we plug the following expressions into equation (23):

$$\rho_1 = \left(\frac{\omega_1}{2i\pi}\right)^2 \wp, \qquad z\rho_1'(z) = \left(\frac{\omega_1}{2i\pi}\right)^3 \wp'(u),$$

we pass from the differential equation (23) to the equation of the theorem of Section 2.6. Thus we have

$$\begin{cases} 60G_4 = \left(\dfrac{2i\pi}{\omega_1}\right)^4 g_4 = \left(\dfrac{2i\pi}{\omega_1}\right)^4 \Big(\dfrac{1}{12} + 20 \displaystyle\sum_1^\infty \dfrac{m^3 q^m}{1-q^m}\Big) \\ 140G_6 = \left(\dfrac{2i\pi}{\omega_1}\right)^6 g_6 = \left(\dfrac{2i\pi}{\omega_1}\right)^6 \Big(-\dfrac{1}{216} + \dfrac{7}{3} \displaystyle\sum_1^\infty \dfrac{m^5 q^m}{1-q^m}\Big) \\ \Delta_1 = \left(\dfrac{2i\pi}{\omega_1}\right)^{12} \Delta = \left(\dfrac{2i\pi}{\omega_1}\right)^{12} q \displaystyle\prod_{m=1}^\infty (1-q^m)^{24}. \end{cases} \tag{27}$$

We will study these functions in more detail in Chapter 5.

COMMENTARY

Certain (young) readers may be surprised by the general form of the equation of a non-decomposable conic given in Section 2.1; it was taught in the last year of high school forty years ago [M-M]!

The double theory of elliptic functions which we gave in this chapter is contained in the book by Valiron [Va]. But any other book on functions of one complex variable can be used; cf. for instance the books by Siegel [Sie], by Whittaker and Watson [W-W], and by Jones and Singerman [J-S].

Let us also indicate certain books on the subject of elliptic **curves**, such as for instance those by Knapp [Kn] and by Silverman [Sil], which contain all of the essential material concerning the classical theory of elliptic functions.

For the history of this magnificent subject, consult the article by C. Houzel in [Hou 2] – and plunge into the complete works of Abel [Ab]!

Exercises and Problems for Chapter 2

2.1 Let e denote a primitive eighth root of unity ($e = (1+i)/\sqrt{2}$ for example), and make the variable change

$$r^2 = \frac{2(ev)^2}{1+(ev)^4} = \frac{2iv^2}{1-v^4}$$

in Fagnano's integral $\int dr/\sqrt{1-r^4}$.

(a) Prove that

$$\int_0^r \frac{dr}{\sqrt{1-r^4}} = (1+i)\int_0^v \frac{dv}{\sqrt{1-v^4}}.$$

(b) Using Fagnano's duplication, deduce from paragraph 2.2 that

$$\int_0^v \frac{dv}{\sqrt{1-v^4}} = (1-i)\int_0^u \frac{du}{\sqrt{1-u^4}}.$$

2.2 Take a "module" $k \in [0,1]$. Define the **complete elliptic integral of the first kind** by

$$K(k) := \int_0^{\pi/2} \frac{d\theta}{\sqrt{1-k^2\sin^2\theta}} = \int_0^1 \frac{dt}{\sqrt{(1-t^2)(1-k^2t^2)}}$$

and the **complete elliptic integral of the second kind** by

$$E(k) := \int_0^{\pi/2} \sqrt{1-k^2\sin^2\theta}\, d\theta = \int_0^1 \frac{\sqrt{1-k^2t^2}}{\sqrt{1-t^2}}\, dt.$$

To the module k, we associate the **complementary module** $k' = \sqrt{1-k^2}$ and the **complementary integrals**:

$$K'(k) := K(k'), \qquad E'(k) := E(k').$$

(a) Show that $K(0) = \pi/2,\ E(0) = \pi/2,\ E(1) = 1$.

(b) Show that the length of an ellipse with semi-axes a and b is given by $4aE'(b/a)$.

(c) Show that if $k < 1$, then

$$K(k) = \frac{\pi}{2} F\left(\frac{1}{2}, \frac{1}{2}; 1, k^2\right), \qquad E(k) = \frac{\pi}{2} F\left(-\frac{1}{2}, \frac{1}{2}; 1; k^2\right)$$

where $F(a, b; c; z)$ is **Gauss' hypergeometric function** defined by

$$F(a, b; c; z) = 1 + \frac{a.b}{1.c} z + \frac{a(a+1)b(b+1)}{1.2.c(c+1)} z^2 + \cdots$$

(d) Show that

$$\begin{cases} \dfrac{dE}{dk} = \dfrac{E - K}{k} \\ \dfrac{dK}{dk} = \dfrac{E - k'^2 K}{kk'}. \end{cases}$$

(e) For a and b such that $0 < b < a$, set

$$I(a, b) := \int_0^{\pi/2} \frac{d\theta}{\sqrt{a^2 \cos^2\theta + b^2 \sin^2\theta}} = \frac{1}{a} K'\left(\frac{b}{a}\right).$$

Taking $t := b \operatorname{tg} \theta$, show that

$$I(a, b) = \frac{1}{2} \int_{-\infty}^{\infty} \frac{dt}{\sqrt{(a^2 + t^2)(b^2 + t^2)}},$$

then setting $u := \frac{1}{2}(t - ab/t)$, justify that

$$I(a, b) = I\left(\frac{a+b}{2}, \sqrt{ab}\right).$$

(f) Keep the notation of (e) and set

$$\begin{aligned} &a_0 := a,\ a_1 := \frac{a+b}{2}, \ldots, a_{n+1} := \frac{a_n + b_n}{2}, \ldots \\ &b_0 := b,\ b_1 := \sqrt{ab}, \ldots, b_{n+1} := \sqrt{a_n b_n}, \ldots \end{aligned}$$

Show that $b_n \le a_n$, that the sequence (a_n) is monotone decreasing, that the sequence (b_n) is monotone increasing and that $(a_n - b_n) \longrightarrow 0$. Let $M(a, b)$ denote the common limit of these sequences (the **arithmetico-geometric mean** of a and b.).

(g) Deduce from (e) and (f) that $I(a, b) = (\pi/2)/M(a, b)$, and obtain the following result due to Gauss (1799):

$$\frac{1}{M(1, \sqrt{2})} = \frac{2}{\pi} \int_0^1 \frac{dt}{\sqrt{1 - t^4}}.$$

(h) Set $c := \sqrt{a^2 - b^2}$, $k := c/a$, $k' := b/a$. Deduce from (e) that

$$K(k) = \frac{1}{1+k} K\left(\frac{2\sqrt{k}}{1+k}\right) = \frac{2}{1+k'} K\left(\frac{1-k'}{1+k'}\right).$$

Differentiating the first equality with respect to k and using the second differential equation of (d), show that

$$E(k) = \frac{1+k}{2}E\left(\frac{2\sqrt{k}}{1+k}\right) + \frac{k'^2}{2}K(k) = (1+k')E\left(\frac{1-k'}{1+k'}\right) - k'K(k).$$

(i) Set

$$J(a,b) := \int_0^{\pi/2} \sqrt{a^2\cos^2\theta + b^2\sin^2\theta d\theta} = aE'\left(\frac{b}{a}\right),$$

and deduce from (h) the relation

$$2J(a_1,b_1) - J(a,b) = abI(a,b).$$

(j) Show that

$$J(a,b) = \left(a^2 - \sum_{n=0}^{\infty} 2^{n-1}c_n^2\right)I(a,b)$$

with $c_n^2 := a_n^2 - b_n^2$.

2.3 Using the results of Section 2.6, prove the identity

$$\begin{aligned}1 + G_4z^4 + 10G_6z^6 + \cdots + \frac{(2m-1)(2m-2)(2m-3)}{6}G_{2m}z^{2m} + \cdots \\ = (1 + 3G_4z^4 + 5G_6z^6 + \cdots + (2m-1)G_{2m}z^{2m} + \cdots)^2 - 5\,G_4z^4.\end{aligned}$$

Deduce the following expressions of G_8, G_{10} in terms of G_4 and G_6:

$$G_8 = \frac{3}{7}G_4^2, \qquad G_{10} = \frac{5}{11}G_4G_6.$$

2.4 Prove the result of Section 2.10:

$$g_6 = \frac{1}{216} - \frac{7}{3}\sum_1^{\infty}\frac{m^5s^m}{1-s^m}.$$

2.5 Using the expressions of g_4, g_6 and Δ given in Sections 2.10 and 2.11, deduce from the relation $\Delta = g_4^3 - 27g_6^2$ the relation

$$\left(1 + 240\sum_1^{\infty}\sigma_3(n)q^n\right)^3 - \left(1 - 504\sum_1^{\infty}\sigma_5(n)q^n\right)^2 = 1728s\prod_1^{\infty}(1-q^n)^{24}.$$

From this deduce that

$$\begin{cases}\sigma_3(n) \equiv \sigma_5(n) \bmod 4 \\ \sigma_3(n) \equiv \sigma_5(n) \bmod 3.\end{cases}$$

Check these congruences by a simple direct computation.

2.6 Using exercise (3), show that for every integer $n > 1$, we have

$$G_{2n} \in \mathbb{Q}[G_4, G_6].$$

Giving weight 4 to G_4 and 6 to G_6, what can we say about the weight of G_{2n}? Could this result be foreseen?

2.7 Let $S(z) = \sum_{-\infty}^{\infty} \alpha_n z^n$ be the Laurent expansion of $S(z)$ for $z \in \mathbb{C}^*$.

(a) Show that

$$\alpha_n = (-1)^n q^{n(n-1)/2} \alpha_0 \quad \text{if } n \geq 1.$$

We now propose to determine α_0.

(b) Show that if q and $x \in \mathbb{R}_+$, then

$$\alpha_0 = \lim_{x \longrightarrow +\infty} \frac{\prod_0^\infty (1 + q^n x)}{\sum_1^\infty q^{n(n-1)/2} x^n}.$$

(c) Setting $P(x) = \prod_0^\infty (1 + q^n x)$, show that

$$P\left(\frac{x}{q}\right) = \left(1 + \frac{x}{q}\right) P(x).$$

(d) Deduce from (c) that the Taylor expansion of $P(x)$ at the origin is

$$P(x) = \sum_0^\infty \beta_n x^n \quad \text{with } \beta_n = q^{n(n+1)/2} \prod_{j=1}^\infty \frac{1}{1 - q^j}.$$

(e) Deduce that the quotient of the coefficients of the same rank of $\sum_{-\infty}^{\infty} q^{n(n-1)/2} x^n$ and $\sum_0^\infty \beta_n x^n$ tends to $\prod_{j=1}^\infty 1/(1 - q^j)$.

(j) Deduce that $\alpha_0 = \prod_{j=1}^\infty 1/(1 - q^j)$.

(g) Extend this result to every complex number q such that $|q| < 1$.

2.8 Let $\wp$ be the Weierstrass function associated to a lattice Λ, and let x and $y \in \mathbb{C} \setminus \Lambda$ be such that $x + y \notin \Lambda$.

(i) Show that

$$\begin{vmatrix} \wp(x) & \wp'(x) & 1 \\ \wp(y) & \wp'(y) & 1 \\ \wp(x+y) & -\wp'(x+y) & 1 \end{vmatrix} = 0.$$

(ii) Suppose moreover that $x \pm y \notin \Lambda$. Show that

$$\wp(x+y) = \frac{1}{4}\left(\frac{\wp'(x) - \wp'(y)}{\wp(x) - \wp(y)}\right)^2 - \wp(x) - \wp(y).$$

(iii) With the hypotheses of (ii), show that

$$\wp(x+y) - \wp(x-y) = -\frac{\wp'(x)\wp'(y)}{(\wp(x) - \wp(y))^2}.$$

2.9 Assume that g_4 and g_6 are real and $g_4^3 - 27g_6^2 > 0$. Write

$$4X^3 - g_4X - g_6 = 4(X - e_1)(X - e_2)(X - e_3),$$

with $e_1 > e_2 > e_3$.

(a) Let $x > e_1$ and set

$$z = \int_x^{+\infty} (4t^3 - g_4t - g_6)^{-1/2}\, dt.$$

Show that if $\wp$ is the Weierstrass function of invariants g_4 and g_6, then

$$x = \wp(z + \alpha)$$

where α is a certain real constant.

(b) Letting x tend to $+\infty$, show that α is a pole of the function and deduce that

$$x = \wp(z).$$

(c) Set $\omega_1/2 := \int_{e_1}^{+\infty} (4t^3 - g_4t - g_6)^{-1/2}\, dt$. Show that $\omega_1/2$ is a half-period of $\wp$ such that $\wp(\omega_1/2) = e_1$.

(d) Set $\omega_3/2 := -i \int_{-\infty}^{e_3} (g_6 + g_4t - 4t^3)^{-1/2}\, dt$. Show that $\omega_3/2$ is a half-period of $\wp$ such that $\wp(\omega_3/2) = e_3$.

(e) Show that

$$\wp\left(z + \frac{\omega_1}{2}\right) - e_1 = \frac{(e_1 - e_2)(e_1 - e_3)}{\wp(z) - e_1}.$$

(f) Show that

$$\wp\left(\frac{\omega_1}{4}\right) - e_1 = \pm[(e_1 - e_2)(e_1 - e_3)]^{1/2}.$$

(g) Show that if $e_i = \wp(\omega_i/2)$, then

$$\wp'(z)\wp'\left(z + \frac{\omega_1}{2}\right)\wp'\left(z + \frac{\omega_2}{2}\right)\wp'\left(z + \frac{\omega_3}{2}\right) = 16[g_4^3 - 27g_6^2].$$

2.10 Let $\wp$ be the Weierstrass function of the lattice $\Lambda = \mathbb{Z}\omega_1 + \mathbb{Z}\omega_2$ with

$$\operatorname{Im}\left(\frac{\omega_2}{\omega_1}\right) > 0.$$

(a) Let ζ denote the function defined by

$$\begin{cases} \dfrac{d\zeta(z)}{dz} = -\wp(z) \\ \lim\limits_{z \to 0} [\zeta(z) - z^{-1}] = 0. \end{cases}$$

Show that

$$\zeta(z) = \frac{1}{z} + \sum_{\omega\in\Lambda}{}' \left[\frac{1}{z-\omega} + \frac{1}{\omega} + \frac{z}{\omega^2}\right]$$

where $\sum'_{\omega\in\Lambda} := \sum_{\omega\in\Lambda\setminus\{0\}}$.

(b) Show that ζ is a meromorphic odd function of z.

(c) Show that we have

$$\begin{cases} \zeta(z+\omega_1) = \zeta(z) + \eta_1 \\ \zeta(z+\omega_2) = \zeta(z) + \eta_2 \end{cases}$$

where η_1 and η_2 are constants.

(d) Integrating $\zeta(z)\,dz$ along the boundary of a period parallelogram, prove **Legendre's relation**:

$$\eta_1\omega_2 - \eta_2\omega_1 = 2i\pi.$$

(e) Show that if $x + y + z = 0$, then

$$[\zeta(x) + \zeta(y) + \zeta(z)]^2 + \zeta'(x) + \zeta'(y) + \zeta'(z) = 0$$

(the pseudo-addition theorem!).

2.11 Let ζ denote the function associated to the lattice Λ by the construction of Exercise 2.10, and define the **Weierstrass sigma function** by the conditions

$$\begin{cases} \dfrac{d}{dz}\log\sigma(z) = \zeta(z) \\ \lim\limits_{z\to 0}\dfrac{\sigma(z)}{z} = 1. \end{cases}$$

(a) Show that if $z \in \mathbb{C}\setminus\Lambda$, then

$$\sigma(z) = z\prod_{\omega}{}' \left[\left(1 - \frac{z}{\omega}\right)\exp\left(\frac{z}{\omega} + \frac{z^2}{2\omega^2}\right)\right]$$

where $\prod'_{\omega} := \prod_{\omega\in\Lambda\setminus\{0\}}$.

The analogy of this expansion with that of the sine function makes it natural to consider σ as a sort of pseudo-elliptic sine.

Show that this infinite product converges normally on every compact subset of $\mathbb{C}\setminus\Lambda$.

(b) Show that σ is an **odd entire function** admitting simple zeros at every point of Λ.

(c) Show that for $\nu \in \{1, 2\}$, we have

$$\sigma(z+\omega_\nu) = c_\nu e^{\eta_\nu z}\sigma(z), \quad c_\nu \in \mathbb{C}^*,$$

where $\eta_\nu := \zeta(z+\omega_\nu) - \zeta(z)$ (see Exercise 2.10). Taking $z = -\omega_\nu/2$, show that

$$c_\nu = -e^{\eta_\nu\omega_\nu/2}.$$

Deduce that

$$\sigma(z+\omega_\nu) = -e^{\eta_\nu(z+\omega_\nu/2)}\sigma(z).$$

(d) Set $q = e^{\pi i\omega_2/\omega_1}$ (so that $|q| < 1$). Show that the entire function g defined by

$$g(z) = e^{(\eta_1 z^2)/2\omega_1} \sin\left(\frac{\pi z}{\omega_1}\right) \prod_{n=1}^{\infty} \left[1 - 2q^{2n} \cos\frac{\pi z}{\omega_1} + q^{4n}\right]$$

admits the same zeros as σ, and that the function g/σ is elliptic for the lattice Λ.

(e) Using (d) and the Liouville theorem (theorem 2.4.1), show that

$$\sigma(z) = \frac{\omega_1}{\pi} e^{(\eta_1 z^2)/2\omega_1} \sin\left(\frac{\pi z}{\omega_1}\right) \prod_{n=1}^{\infty} \left[1 - 2q^{2n} \cos\frac{\pi z}{\omega_1} + q^{4n}\right].$$

(f) Let f be a non-constant elliptic function for the lattice Λ.
Let $a_1, \ldots, a_n$ (resp. $b_1, \ldots, b_m$) denote the (suitably repeated) zeros (resp. poles) of f in a period parallelogram. We recall that we have $n = m$ and that, up to adding a period to b_n, we have

$$a_1 + \cdots + a_n = b_1 + \cdots + b_n.$$

Show that the function h defined by

$$h(z) = \prod_{r=1}^{n} \frac{\sigma(z - a_r)}{\sigma(z - b_r)}$$

is an elliptic function of lattice Λ.
Deduce that $f = ch$, with $c \in \mathbb{C}^*$.

(g) Show that for y and $z \in \mathbb{C} \setminus \Lambda$, we have

$$\wp(z) - \wp(y) = -\frac{\sigma(z+y)\sigma(z-y)}{\sigma^2(z)\sigma^2(y)}.$$

(h) Show that for $z \in \mathbb{C} \setminus \Lambda$, we have

$$\wp'(z) = \frac{2\sigma(z+\omega_1/2)\sigma(z+\omega_2/2)\sigma(z-(\omega_1+\omega_2)/2)}{\sigma^3(z)\sigma(\omega_1/2)\sigma(\omega_2/2)\sigma((\omega_1+\omega_2)/2)}.$$

(i) Show that $\wp'(z) = -\sigma(2z)/\sigma^4(z)$.

(j) For $m \in \mathbb{N}$, show that

$$\begin{vmatrix} 1 & \wp(z_0) & \wp'(z_0) \cdots \wp^{(m-1)}(z_0) \\ 1 & \wp(z_1) & \wp'(z_1) \cdots \wp^{(m-1)}(z_1) \\ \vdots & & \\ 1 & \wp(z_m) & \wp'(z_m) \cdots \wp^{(m-1)}(z_m) \end{vmatrix}$$

$$= (-1)^{m(m-1)/2} 1!2! \cdots m! \frac{\sigma(z_0 + z_1 + \cdots + z_m) \prod \sigma(z_\lambda - z_\mu)}{\sigma^{m+1}(z_0) \cdots \sigma^{m+1}(z_m)}.$$

Problem I

The text at the end of this problem consists of pages 7–11 of a text by Abel. Please read these pages and answer the following questions.

(1) Briefly analyse the usage of functional notation in the text.

(2) Consider the lemniscate

$$(x^2+y^2)^2+y^2-x^2=0,$$

whose representation as a curve is given by

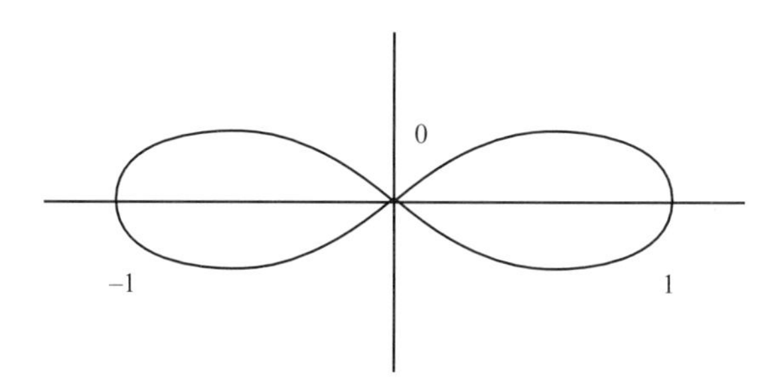

For $r \in [0,1]$, denote by $s(r)$ the length of the arc of the lemniscate running from $(0,0)$ to $\left(\sqrt{(r^2+r^4)/2}, \sqrt{(r^2-r^4)/2}\right)$.

How can we choose c, e and x in formula (2) so that $\alpha = s(r)$?

(3) Show that ω and $\widetilde{\omega}$ are finite.

(4) Using the formulae (10), show that if α and β are "indeterminates", then

(a) $\varphi(\alpha+\beta)+\varphi(\alpha-\beta)=2\varphi(\alpha)f(\beta)F(\beta)/R$

(b) $\varphi(\alpha+\beta)-\varphi(\alpha-\beta)=2\varphi(\beta)f(\alpha)F(\alpha)/R$

(c) $f(\alpha+\beta)+f(\alpha-\beta)=2f(\alpha)f(\beta)/R$

(d) $f(\alpha+\beta)-f(\alpha-\beta)=-2c^2\varphi(\alpha)\varphi(\beta)F(\alpha)F(\beta)/R$

(e) $F(\alpha+\beta)+F(\alpha-\beta)=2F(\alpha)F(\beta)/R$

(f) $F(\alpha+\beta)-F(\alpha-\beta)=2e^2\varphi(\alpha)\varphi(\beta)f(\alpha)f(\beta)/R$.

(5) Using the results of paragraph 1.1 of Abel's text, show how one can deduce, without any computation, formulae (b), (e) and (f) from formulae (a), (c) and (d).

(6) Taking $\beta = \pm\omega/2$, deduce the following relations from the formulae (10):

$$\varphi\left(a \pm \frac{\omega}{2}\right) = \pm\varphi\left(\frac{\omega}{2}\right)\frac{f(a)}{F(a)}$$

$$f\left(a \pm \frac{\omega}{2}\right) = \mp\frac{F(\omega/2)\varphi(a)}{\varphi(\omega/2)F(a)}$$

$$F\left(a \pm \frac{\omega}{2}\right) = \frac{F(\omega/2)}{F(a)}.$$

(7) Deduce that

$$\varphi\left(a+\frac{\omega}{2}\right)=\varphi\left(-a+\frac{\omega}{2}\right)$$
$$f\left(a+\frac{\omega}{2}\right)=-f\left(-a+\frac{\omega}{2}\right)$$
$$F\left(a+\frac{\omega}{2}\right)=F\left(-a+\frac{\omega}{2}\right).$$

(8) Show that

$$\varphi(a+\omega)=-\varphi(a)$$
$$f(a+\omega)=-f(a)$$
$$F(a+\omega)=F(a).$$

(9) What does Abel mean when he says that α and β are two "indeterminates"? Within what framework does he prove the relations (10)?
Do you have any idea of the origin of this proof?

(10) Show that the choice of $\beta=\pm(\widetilde{\omega}/2)i$ leads to the relations

$$\varphi(a+\widetilde{\omega}i)=-\varphi(a)$$
$$f(a+\widetilde{\omega}i)=f(a)$$
$$F(a+\widetilde{\omega}i)=-F(a).$$

(11) With the help of the results of (8) and (10) and the relations (10) of Abel's text, show that we can extend φ, f and F to the whole complex plane.

(12) Show that $\Lambda=2\omega\mathbb{Z}+2i\widetilde{\omega}\mathbb{Z}$ is a period lattice for φ, that $\Lambda'=2\omega\mathbb{Z}+i\widetilde{\omega}\mathbb{Z}$ is one for f and that $\Lambda''=\omega\mathbb{Z}+2i\widetilde{\omega}\mathbb{Z}$ is one for F.

Research on elliptic functions (Abel)

Journal für die reine und angewandte Mathematik, herausgegeben von Crelle, Bd.2,3. Berlin 1827, 1828.

For a long time, logarithmic functions, and the exponential and circular functions were the only transcendantal functions which attracted the attention of geometers. Only lately have we begun to consider some others. Among these, we distinguish the functions called elliptic, both for their beautiful analytic properties and for their applications in diverse branches of mathematics. The first idea of these functions was given by the immortal *Euler*, who proved that the separated equation

$$\frac{dx}{\sqrt{\alpha+\beta x+\gamma x^2+\delta x^3+\varepsilon x^4}}+\frac{dy}{\sqrt{\alpha+\beta y+\gamma y^2+\delta y^3+\varepsilon y^4}}=0$$

is algebraically integrable. After *Euler, Lagrange* added something, by giving his elegant theory of the transformation of the integral $\int(R\,dx)/\sqrt{(1-p^2x^2)(1-q^2x^2)}$, where R is a rational function of x. But the first, and if I am not mistaken, the only person who profoundly investigated the nature of these functions, is Mr. Legendre, who, first in a text on elliptic functions, and later in his excellent *Exercises in mathematics*, developed a number of elegant properties of these functions, and showed their applications. Since the publication of this work, nothing has been added to the theory of Mr. Legendre. I believe that the further research on these functions presented here will be welcomed with pleasure.

In general, by the denomination of elliptic functions, we understand every function contained in the integral

$$\int \frac{R\,dx}{\sqrt{\alpha + \beta x + \gamma x^2 + \delta x^3 + \varepsilon x^4}},$$

where R is a rational function and $\alpha, \beta, \gamma, \delta, \varepsilon$ are constant real quantities. Mr. Legendre proved that by suitable substitutions, we can always bring this integral to the form

$$\int \frac{P\,dy}{\sqrt{a + by^2 + cy^4}},$$

where P is a rational function of y^2. By means of suitable reductions, this integral can then be brought to the form

$$\int \frac{A + By^2}{C + Dy^2} \frac{dy}{\sqrt{a + by^2 + cy^4}},$$

and this to the form

$$\int \frac{A + B\sin^2\theta}{C + D\sin^2\theta} \frac{d\theta}{\sqrt{1 - c^2\sin^2\theta}},$$

where c is real and less than one.

It follows from this that every elliptic function can be brought to one of the three forms

$$\int \frac{d\theta}{\sqrt{1 - c^2\sin^2\theta}}, \qquad \int d\theta\sqrt{1 - c^2\sin^2\theta}, \qquad \int \frac{d\theta}{(1 + n\sin^2\theta)\sqrt{1 - c^2\sin^2\theta}},$$

which Mr. Legendre has called elliptic functions of the first, second and third kind. These are the three functions considered by Mr. Legendre, particularly the first, which has the simplest and most remarkable properties.

In this text, I propose to consider the inverse function, i.e. the function $\varphi\alpha$, determined by the equations

$$\alpha = \int \frac{d\theta}{\sqrt{1 - c^2\sin^2\theta}}, \qquad \sin\theta = \varphi\alpha = x.$$

The last equation gives

$$d\theta\sqrt{1 - \sin^2\theta} = d(\varphi\alpha) = dx,$$

so

$$\alpha = \int_0 \frac{dx}{\sqrt{(1 - x^2)(1 - c^2x^2)}}.$$

Mr. Legendre assumes c^2 positive, but I noted that if we assume c^2 negative, equal to $-e^2$, the formulae become simpler. Similarly, I write $1 - c^2x^2$ instead of $1 - x^2$ for greater symmetry, so that the function $\varphi\alpha = x$ will be given by the equation

$$\alpha = \int_0 \frac{dx}{\sqrt{(1 - c^2x^2)(1 + e^2x^2)}},$$

or

$$\varphi'\alpha = \sqrt{(1 - c^2\varphi^2\alpha)(1 + e^2\varphi^2\alpha)}.$$

For brevity's sake, I introduce two other functions of α, namely

$$f\alpha = \sqrt{1 - c^2\varphi^2\alpha}; \qquad F\alpha = \sqrt{1 + e^2\varphi^2\alpha}.$$

Several properties of these functions follow immediately from known properties of elliptic functions of the first kind, but others are more hidden. For example, one can show that the equations $\varphi\alpha = 0, f\alpha = 0, F\alpha = 0$ have an infinite number of roots all of which can be found. One of the most remarkable properties is that $\varphi(m\alpha), f(m\alpha)$ and $F(m\alpha)$ (for an integer m) can be expressed rationally in $\varphi\alpha, f\alpha, F\alpha$. Also, nothing is easier than finding $\varphi(m\alpha), f(m\alpha), F(m\alpha)$ when we know $\varphi\alpha, f\alpha, F\alpha$, but the inverse problem, namely determining $\varphi\alpha, f\alpha, F\alpha$ in terms of $\varphi(m\alpha), f(m\alpha), F(m\alpha)$, is more difficult, because it depends on an equation of high degree (namely of degree m^2).

The resolution of this equation is the main object of this text. First, we will see how to find all the roots, using the functions φ, f, F. Then we will consider the algebraic resolution of the equation in question, and we will succeed in obtaining the remarkable result that $\varphi(\alpha/m), f - (\alpha/m), F(\alpha/m)$ can be expressed in $\varphi\alpha$, $f\alpha$, $F\alpha$, by a formula which, with respect to α, contains no other irrationalities than radicals. This gives a very general class of equations which can be solved algebraically. It is notable that the expressions of the roots contain constant quantities, which in general cannot be expressed by algebraic quantities. These constant quantities depend on an equation of degree $m^2 - 1$. We will show, using algebraic functions, how we can reduce the resolution to that of an equation of degree $m + 1$. We will give several expressions of the functions $\varphi(2n + 1)\alpha, f(2n + 1)\alpha, F(2n + 1)\alpha$ in terms of $\varphi\alpha, f\alpha, F\alpha$. We will then deduce the values of $\varphi\alpha, f\alpha, F\alpha$ in terms of α. We will show that these functions can be decomposed into an infinite number of factors, and even an infinity of partial fractions.

Section I

Fundamental properties of the functions $\varphi\alpha, f\alpha, F\alpha$ (by N.H. Abel).

1

Assuming that

$$\varphi\alpha = x, \tag{1}$$

by virtue of what precedes, we find that

$$\alpha = \int_0^x \frac{dx}{\sqrt{(1 - c^2x^2)(1 + e^2x^2)}}. \tag{2}$$

From this we see that α, considered as function of x, is positive from $x = 0$ to $x = 1/c$. Thus, setting

$$\frac{\omega}{2} = \int_0^{1/c} \frac{dx}{\sqrt{(1 - c^2x^2)(1 + e^2x^2)}}, \tag{3}$$

it is obvious that $\varphi\alpha$ is positive and increasing from $\alpha = 0$ to $\alpha = \omega/2$, and we have

$$\varphi(0) = 0, \quad \varphi\left(\frac{\omega}{2}\right) = \frac{1}{c}. \tag{4}$$

As α changes sign, when we write $-x$ instead of x, the same holds for the function $\varphi\alpha$ with respect to α, and consequently we have the equation

$$\varphi(-\alpha) = -\varphi\alpha. \tag{5}$$

Substituting xi for x in (1) (where i, for simplicity, represents the imaginary quantity $\sqrt{-1}$) and βi denotes the value of α, we obtain

$$xi = \varphi(\beta i) \quad \text{and} \quad \beta = \int_0^x \frac{dx}{\sqrt{(1+c^2x^2)(1-e^2x^2)}}. \tag{6}$$

β is real and positive from $x = 0$ to $x = 1/e$, so setting

$$\frac{\widetilde{\omega}}{2} = \int_0^{1/e} \frac{dx}{\sqrt{(1-e^2x^2)(1+c^2x^2)}}, \tag{7}$$

x will be positive from $\beta = 0$ to $\beta = \widetilde{\omega}/2$, i.e. the function $(1/i)\varphi(\beta i)$ will be positive between the same limits. Setting $\beta = \alpha$ and $y = \varphi(ai)/i$, we find

$$\alpha = \int_0^y \frac{dy}{\sqrt{1-e^2y^2)(1+c^2y^2)}},$$

so we see that if we suppose c instead of e and e instead of c,

$$\frac{\varphi(\alpha i)}{i} \quad \text{becomes} \quad \varphi\alpha.$$

And as

$$f\alpha = \sqrt{1 - c^2\varphi^2\alpha},$$
$$F\alpha = \sqrt{1 + e^2\varphi^2\alpha},$$

we see that by changing c to e and e to c, $f(\alpha i)$ and $F(\alpha i)$ become respectively $F\alpha$ and $f\alpha$. Finally, equations (3) and (7) show that under the same transformation, ω and $\widetilde{\omega}$ become respectively $\widetilde{\omega}$ and ω.

By formula (7) we have $x = 1/e$ for $\beta = \widetilde{\omega}/2$, so by virtue of the equation $xi = \varphi(\beta i)$, we obtain

$$\varphi\left(\frac{\widetilde{\omega}i}{2}\right) = i.\frac{1}{e}. \tag{8}$$

2

By the above, we will have the values of $\varphi\alpha$ for every real value of α lying between $-(\omega/2)$ and $+(\omega/2)$, and for every imaginary value of the form βi of this quantity, if β is a quantity contained between the limits $-(\widetilde{\omega}/2)$ and $+(\widetilde{\omega}/2)$. We now need to find the value of this function for an arbitrary real or imaginary value of the variable. To do this, we will first establish the fundamental properties of the functions φ, f and F.

Having

$$f^2\alpha = 1 - c^2\varphi^2\alpha,$$
$$F^2\alpha = 1 + e^2\varphi^2\alpha,$$

we find, by differentiating,

$$f\alpha \cdot f'\alpha = -c^2\varphi\alpha \cdot \varphi'\alpha,$$
$$F\alpha \cdot F'\alpha = e^2\varphi\alpha \cdot \varphi'\alpha.$$

Now, by (2), we have

$$\varphi'\alpha = \sqrt{(1-c^2\varphi^2\alpha)(1+e^2\varphi^2\alpha)} = f\alpha \cdot F\alpha,$$

so substituting this value of $\varphi'\alpha$ into the two preceding equations, we find that the functions $\varphi\alpha$, $f\alpha$, $F\alpha$ are related by the equations

$$\begin{cases} \varphi'\alpha = f\alpha \cdot F\alpha, \\ f'\alpha = -c^2\varphi\alpha \cdot F\alpha, \\ F'\alpha = e^2\varphi\alpha \cdot f\alpha. \end{cases} \tag{9}$$

This said, I claim that letting α and β denote two indeterminates, we have

$$\begin{cases} \varphi(\alpha+\beta) = \dfrac{\varphi\alpha \cdot f\beta \cdot F\beta + \varphi\beta \cdot f\alpha \cdot F\alpha}{1+e^2c^2\varphi^2\alpha \cdot \varphi^2\beta}, \\ f(\alpha+\beta) = \dfrac{f\alpha \cdot f\beta - c^2\varphi\alpha \cdot \varphi\beta \cdot F\alpha \cdot F\beta}{1+e^2c^2\varphi^2\alpha \cdot \varphi^2\beta}, \\ F(\alpha+\beta) = \dfrac{F\alpha \cdot F\beta + e^2\varphi\alpha \cdot \varphi\beta \cdot f\alpha \cdot f\beta}{1+e^2c^2\varphi^2\alpha \cdot \varphi^2\beta}. \end{cases} \tag{10}$$

These formulae can be deduced immediately from known properties of elliptic functions (*Legendre, Exercises in Integral Calculus*), but we can also check them easily as follows.

Let r denote the right-hand side of first of equations (10). Then differentiating with respect to α, we have

$$\frac{dr}{d\alpha} = \frac{\varphi'\alpha \cdot f\beta \cdot F\beta + \varphi\beta \cdot F\alpha \cdot f'\alpha + \varphi\beta \cdot f\alpha \cdot F'\alpha}{1+e^2c^2\varphi^2 \cdot \alpha \cdot \varphi^2\beta} - \frac{(\varphi\alpha \cdot f\beta \cdot F\beta + \varphi\beta \cdot f\alpha \cdot F\alpha)2e^2c^2\varphi\alpha \cdot \varphi^2\beta \cdot \varphi'\alpha}{(1+e^2c^2\varphi^2\alpha \cdot \varphi^2\beta)^2}.$$

Substituting for $\varphi'\alpha, f'\alpha, F'\alpha$ their values given by equations (9), we find

$$\frac{dr}{d\alpha} = \frac{f\alpha \cdot F\alpha \cdot f\beta \cdot F\beta}{1+e^2c^2\varphi^2\alpha \cdot \varphi^2\beta} - \frac{2e^2c^2\varphi^2\alpha \cdot \varphi^2\beta \cdot f\alpha \cdot f\beta \cdot F\alpha \cdot F\beta}{(1+e^2c^2\varphi^2\alpha \cdot \varphi^2\beta)^2} + \frac{\varphi\alpha \cdot \varphi\beta \cdot (1+e^2c^2\varphi^2\alpha \cdot \varphi^2\beta)(-e^2F^2\alpha + e^2f^2\alpha) - 2e^2c^2\varphi\alpha \cdot \varphi\beta \cdot \varphi^2\beta \cdot f^2\alpha \cdot F^2\alpha}{(1+e^2c^2\varphi^2\alpha \cdot \varphi^2\beta)^2},$$

hence, substituting for $f^2\alpha$ and $F^2\alpha$ their values $1-e^2\varphi^2\alpha$, $1+e^2\varphi^2\alpha$, and reducing, we find that

$$dr/d\alpha = \frac{(1-e^2c^2\varphi^2\alpha \cdot \varphi^2\beta)[(e^2-c^2)\varphi\alpha \cdot \varphi\beta + f\alpha \cdot f\beta \cdot F\alpha \cdot F\beta] - 2e^2c^2\varphi\alpha \cdot \varphi\beta(\varphi^2\alpha + \varphi^2\beta)}{(1+e^2c^2\varphi^2\alpha \cdot \varphi^2\beta)^2}.$$

Now, α and β play symmetric roles in the expression of r, so we have the value of $dr/d\beta$, by permuting α and β in the value of $dr/d\alpha$. This shows that the expression of $dr/d\alpha$ does not change value, so we have

$$\frac{dr}{d\alpha} = \frac{dr}{d\beta}.$$

This partial differential equation shows that r is a function of $\alpha + \beta$, so we have

$$r = \psi(\alpha+\beta).$$

The form of the function ψ can be found by giving a value to β. Supposing for example that $\beta = 0$, and noting that $\varphi(0) = 0, f(0) = 1, F(0) = 1$, the two values of r become

$$r = \varphi\alpha \quad \text{and} \quad r = \psi\alpha,$$

so

$$\psi\alpha = \varphi\alpha,$$

hence

$$r = \psi(\alpha + \beta) = \varphi(\alpha + \beta).$$

The first of the formulae (10) thus holds. The other two formulae can be verified in the same manner.

3

A number of new relations can be deduced from (10). Let me give some of the most remarkable ones. To simplify, set

$$1 + e^2c^2\varphi^2\alpha \cdot \varphi^2\beta = R. \tag{11}$$

First, changing the sign of β, we obtain ...

Problem 2

Begin this problem by reading the following excerpt from another text by Abel, published in 1823. [Ab]

Value of the expression $\varphi(x + y\sqrt{-1}) + \varphi(x - y\sqrt{-1})$

Solutions of some problems using definite integrals

When φ is an algebraic, logarithmic, exponential or circular function, we can always, as we know, express the real value of $\varphi(x + y\sqrt{-1}) + \varphi(x - y\sqrt{-1})$ in real and finite form. If on the contrary φ preserves its generality, we have not, as far as I know, yet been able to express it in a real, finite form. We can do this using definite integrals, in the following way.

If we expand $\varphi(x + y\sqrt{-1})$ and $\varphi(x - y\sqrt{-1})$ by Taylor's theorem, we obtain

$$\varphi(x + y\sqrt{-1}) = \varphi x + \varphi' x \cdot y\sqrt{-1} - \frac{\varphi'' x}{1 \cdot 2} y^2 - \frac{\varphi''' x}{1 \cdot 2 \cdot 3} y^3 \sqrt{-1} + \frac{\varphi'''' x}{1 \cdot 2 \cdot 3 \cdot 4} y^4 + \cdots$$

$$\varphi(x - y\sqrt{-1}) = \varphi x - \varphi' x \cdot y\sqrt{-1} - \frac{\varphi'' x}{1 \cdot 2} y^2 + \frac{\varphi''' x}{1 \cdot 2 \cdot 3} y^3 \sqrt{-1} + \frac{\varphi'''' x}{1 \cdot 2 \cdot 3 \cdot 4} y^4 - \cdots,$$

so

$$\varphi(x + y\sqrt{-1}) + \varphi(x - y\sqrt{-1}) = 2\left(\varphi x - \frac{\varphi'' x}{1 \cdot 2} y^2 + \frac{\varphi'''' x}{1 \cdot 2 \cdot 3 \cdot 4} y^4 - \cdots\right).$$

To find the sum of this series, consider the series

$$\varphi(x + t) = \varphi x + t\varphi' x + \frac{t^2}{1 \cdot 2} \varphi'' x + \frac{t^3}{1 \cdot 2 \cdot 3} \varphi''' x + \cdots$$

Multiplying the two sides of this equation by $e^{-v^2t^2}\,dt$, and then taking the integral from $t=-\infty$ to $t=+\infty$, we find

$$\int_{-\infty}^{+\infty}\varphi(x+t)e^{-v^2t^2}\,dt=\varphi x\int_{-\infty}^{+\infty}e^{-v^2t^2}\,dt+\varphi' x\int_{-\infty}^{+\infty}e^{-v^2t^2}t\,dt$$

$$+\frac{1}{2}\varphi'' x\int_{-\infty}^{+\infty}e^{-v^2t^2}t^2dt+\cdots.$$

Now, $\int_{-\infty}^{+\infty}e^{-v^2t^2}t^{2n+1}\,dt=0$, so

$$\int_{-\infty}^{+\infty}\varphi(x+t)e^{-v^2t^2}\,dt=\varphi x\int_{-\infty}^{+\infty}e^{-v^2t^2}\,dt+\frac{\varphi'' x}{1\cdot 2}\int_{-\infty}^{+\infty}e^{-v^2t^2}t^2\,dt$$

$$+\frac{\varphi'''' x}{1\cdot 2\cdot 3\cdot 4}\int_{-\infty}^{+\infty}e^{-v^2t^2}t^4dt+\cdots$$

Consider the integral

$$\int_{-\infty}^{+\infty}e^{-v^2t^2}t^{2n}\,dt.$$

Let $t=a/v$, we have $e^{-v^2t^2}=e^{-a^2},\ t^{2n}=a^{2n}/v^{2n},\ dt=da/v$, so

$$\int_{-\infty}^{+\infty}e^{-v^2t^2}t^{2n}\,dt=\frac{1}{v^{2n+1}}\int_{-\infty}^{+\infty}e^{-a^2}a^{2n}\,da=\frac{\Gamma\big((2n+1)/2\big)}{v^{2n+1}},$$

i.e.

$$\int_{-\infty}^{+\infty}e^{-v^2t^2}t^{2n}\,dt=\frac{1\cdot 3\cdot 5\cdots(2n-1)\sqrt{\pi}}{2^n v^{2n+1}}=\frac{\sqrt{\pi}}{v^{2n+1}}A_n.$$

Substituting this value above, we obtain

$$\int_{-\infty}^{+\infty}\varphi(x+t)e^{-v^2t^2}\,dt=\frac{\sqrt{\pi}}{v}\left(\varphi x+\frac{A_1}{2}\frac{\varphi'' x}{v^2}+\frac{A_2}{2\cdot 3\cdot 4}\frac{\varphi'''' x}{v^4}+\cdots\right).$$

Multiplying by $e^{-v^2y^2}v\,dv$ and taking the integral from $v=-\infty$ to $v=+\infty$, we obtain

$$\frac{1}{\sqrt{\pi}}\int_{-\infty}^{+\infty}e^{-v^2y^2}v\,dv\int_{-\infty}^{+\infty}\varphi(x+t)e^{-v^2t^2}\,dt$$

$$=\varphi x\int_{-\infty}^{+\infty}e^{-v^2y^2}\,dv+\frac{A_1\varphi'' x}{2}\int_{-\infty}^{+\infty}e^{-v^2y^2}\frac{dv}{v^2}+\cdots$$

Let $vy=\beta$. Then

$$\int_{-\infty}^{+\infty}e^{-v^2y^2}v^{-2n}dv=y^{2n-1}\int_{-\infty}^{+\infty}e^{-\beta^2}\beta^{-2n}d\beta.$$

Now,

$$\int_{-\infty}^{+\infty}e^{-\beta^2}\beta^{-2n}d\beta=\Gamma\left(\frac{1-2n}{2}\right)=\frac{(-1)^n2^n\sqrt{\pi}}{1\cdot 3\cdot 5\ldots(2n-1)}=\frac{(-1)^n\sqrt{\pi}}{A_n},$$

so

$$\int_{-\infty}^{+\infty}e^{-v^2y^2}v^{-2n}\,dv=\frac{(-1)^n\sqrt{\pi}\,y^{2n-1}}{A_n},$$

and consequently

$$A_n \int_{-\infty}^{+\infty} e^{-v^2y^2} v^{-2n}\, dv = (-1)^n\, y^{2n-1} \sqrt{\pi}.$$

Substituting this value and dividing by $\sqrt{\pi}/2y$, we obtain

$$\frac{2y}{\pi} \int_{-\infty}^{+\infty} e^{-v^2y^2} v\, dv \int_{-\infty}^{+\infty} \varphi(x+t) e^{-v^2t^2}\, dt = 2\left(\varphi x - \frac{\varphi'' x}{2} y^2 + \frac{\varphi'''' x}{2\cdot 3\cdot 4} y^4 - \cdots\right).$$

The right-hand side of this equation is equal to

$$\varphi(x + y\sqrt{-1}) + \varphi(x - y\sqrt{-1}),$$

so

$$\varphi(x + y\sqrt{-1}) + \varphi(x - y\sqrt{-1}) = \frac{2y}{\pi} \int_{-\infty}^{+\infty} e^{-v^2y^2} v\, dv \int_{-\infty}^{+\infty} \varphi(x+t) e^{-v^2t^2}\, dt.$$

Setting $x = 0$, we have

$$\varphi(y\sqrt{-1}) + \varphi(-y\sqrt{-1}) = \frac{2y}{\pi} \int_{-\infty}^{+\infty} e^{-v^2y^2} v\, dv \int_{-\infty}^{+\infty} \varphi t \cdot e^{-v^2t^2}\, dt.$$

Now, answer the following questions concerning Abel's text.

(1) In this text by Abel, published in 1823, the author says on line 4 that he considers a function φ which "preserves its generality".

Can you interpret the subsequent lines, to explain Abel's conception of a general function?

(2) Show that if $v > 0$, then

$$\int_{-\infty}^{\infty} e^{-v^2t^2} t^{2n}\, dt = \frac{1\cdot 3\cdot 5\cdots(2n-1)}{2^n \cdot v^{2n+1}} \sqrt{\pi}.$$

(3) Abel takes $e(t) = e^{(t)}$. Do we have the right to take $\varphi(t) = e^t$? Using the last formula of the text, show that

$$\cos\, y = \frac{y}{\pi} \int_{-\infty}^{\infty} e^{-v^2t^2} \left(\int_{-\infty}^{\infty} e^{t-v^2t^2}\, dt\right) v\, dv$$

deduce that $y > 0$ when $|y| < \pi/2$.

(4) Show that we have

$$\int_{-\infty}^{\infty} e^{t-v^2t^2}\, dt = \frac{\sqrt{\pi}}{v} e^{1/4v^2}$$

when $v > 0$.

(5) Show that the right-hand side of the equation in (3) does not make sense. Find an error in Abel's reasoning (this error was pointed out by Eric Brier).

$$\cos\, y = \frac{1}{\sqrt{\pi}} \int_{-\infty}^{\infty} e^{-t^2 + y^2/4t^2}\, dt.$$

(6) Recall that the Bernoulli numbers B_n are defined by the relation

$$\frac{z}{e^z - 1} = \sum_{n=0}^{\infty} B_n \frac{z^n}{n!}.$$

Show that

$$\frac{z}{2}\operatorname{cotg}\frac{z}{2}-1=\sum_{n=1}^{\infty}(-1)^n B_{2n}\frac{z^{2n}}{2n!}$$

(recall that $\operatorname{cotg} u=\cos u/\sin u$).

(7) Show that

$$\operatorname{cotg}\frac{z}{2}-\frac{2}{z}=\sum_{k\geq 1}\frac{4z}{z^2-(2k\pi)^2}.$$

(8) Using the preceding question, show that

$$\frac{B_{2n}}{(2n)!}=\frac{(-1)^{n-1}}{2^{2n-1}\pi^{2n}}\left(1+\frac{1}{2^{2n}}+\frac{1}{3^{2n}}+\cdots\right).$$

(9) Using the expansion

$$\frac{1}{e^t-1}=e^{-t}+e^{-2t}+e^{-3t}+\cdots,\quad \text{if } t>0,$$

show that if $n\geq 1$, then

$$\int_0^{\infty}\frac{t^{2n-1}}{e^t-1}dt=\sum_{k=1}^{\infty}\int_0^{\infty}e^{-kt}t^{2n-1}dt.$$

(10) Prove that

$$\int_0^{\infty}e^{-kt}t^{2n-1}\,dt=\frac{(2n-1)!}{k^{2n}}.$$

(11) Deduce that

$$B_n=(-1)^{n-1}\frac{2n}{2^{2n-1}}\int_0^{\infty}\frac{t^{2n-1}}{e^{\pi t}-1}dt.$$

(12) Let us admit the fact that the successive derivatives of the analytic function φ satisfy the condition

$$|\varphi^{(n)}(x)|\leq C^{te}$$

for every $n\geq 0$ and for every $x\in[a,b]$. Show that we have

$$\sum_{r=a}^{b-1}\varphi(r)=\int_a^b\varphi(x)\,dx+\sum_{k=1}^{\infty}\frac{B_k}{k!}(\varphi^{(k-1)}(b)-\varphi^{(k-1)}(a))$$

(where it is understood that a and b lie in $\mathbb{Z}$). Show that

$$\left[\int_0^{\infty}\frac{\varphi(x+(t/2)\sqrt{-1})-\varphi(x-(t/2)\sqrt{-1})}{2\sqrt{-1}}\frac{dt}{e^{\pi t}-1}\right]_a^b$$
$$=\sum_{r=a}^{b-1}\varphi(r)-\int_a^b\varphi(x)\,dx+\frac{1}{2}[\varphi(x)]_a^b.$$

(13) Taking $\varphi(x)=e^x$, deduce from (12) the relation

$$\int_0^{\infty}\frac{\sin(t/2)}{e^{\pi t}-1}dt=\frac{1}{e-1}-\frac{1}{2}.$$

(This result is due to Abel).

3

NUMBERS AND GROUPS

The aim of this chapter is to provide the reader with at least rudimentary notions of two important mathematical objects: the field of algebraic numbers, and its automorphism group.

The goal to the chapter is a structure theorem which is **algebraic** in nature, namely Steinitz' theorem stated in Section 3.4. But as we go, we will use **analytic** methods to construct an infinity of interesting fields (the p-adic fields discovered by Kurt Hensel[31] in 1902) which will enable us to construct as many distinct but isomorphic images of "the" field of algebraic numbers. Like every mathematical object, this one is defined up to isomorphism, and it should not be surprising that it keeps turning up in different disguises.

Finally, we will study the group of automorphisms of a field, and we will conclude the chapter with an introduction to the representations of such a group, and to Galois theory[32].

3.1 ABSOLUTE VALUES ON $\mathbb{Q}$

Our goal here is to describe all the absolute values defined over $\mathbb{Q}$. However, in order to be able to use the concept of absolute value in other cases, we first need an abstract definition.

Definition 3.1.1 *Let A be an arbitrary ring with unit. An* **absolute value** *on A is a map $A \to \mathbb{R}$, written $x \mapsto |x|$, satisfying the conditions*

(i) $|x| \geq 0$ *for every $x \in A$ and $|x| = 0$ if and only if $x = 0$.*

(ii) $|xy| = |x|\,|y|$.

(iii) *There exists a constant C (depending on the absolute value) such that for every $(x, y) \in A^2$, we have*

$$|x + y| \leq C \sup\{|x|, |y|\}.$$

[31] K. Hensel 1861–1941.

[32] E. Galois 1811–1832.

Remark 3.1.1

(1) When $C = 1$, we say that the absolute value is **ultrametric**, and the inequality (iii) is then called the **ultrametric inequality**.

(2) Certain absolute values satisfy the **triangle inequality**

(iv) $|x + y| \leq |x| + |y|$.

We will see in the exercises that this happens if and only if $C \leq 2$. We call these **absolute values "triangular"**.

(3) Let $|\ \ |$ denote an absolute value, and let $\alpha > 0$. Then the map $x \mapsto |x|^\alpha$ is also an absolute value. This absolute value is said to be **equivalent** to the first one (see Problem 1, page 164).

Thus, we see that every absolute value is equivalent to a triangular absolute value. In particular, the "norms" introduced in Chapter 1 are equivalent to triangular absolute values.

Example 3.1.1 **(1)** Every field k is equipped with the absolute value $|\ \ |_0$ defined by $|x|_0 = 1$ if $x \neq 0$ and $|0|_0 = 0$. This absolute value is called the **trivial absolute value** on k.

(2) If A is equal to $\mathbb{Z}$ or $\mathbb{F}_q[t]$, the map $x \mapsto |x|_\infty$ defined by

$$\begin{cases} |x|_\infty = 0 & \text{if } x = 0 \\ |x|_\infty = \operatorname{Card}(A/(x)) & \text{if } x \neq 0 \end{cases} \tag{1}$$

is an absolute value on A. We extend it to the field of fractions k of A by the formula

$$\left|\frac{x}{y}\right|_\infty = \frac{|x|_\infty}{|y|_\infty} \quad \text{if } y \neq 0.$$

This absolute value is called the "**absolute value at infinity**" of $\mathbb{Q}$ or of $\mathbb{F}_q(t)$.

To justify this terminology, consider the case where $A = \mathbb{F}_q[t]$, and suppose that $x = a_m t^m + \cdots + a_0$ with $a_m \neq 0$. As each element of $A/(x)$ can be represented by a unique polynomial of degree $< m$, we see that

$$|x|_\infty = q^m = q^{\deg_t(x)}.$$

Now, if $y = b_n t^n + \cdots + b_0$ with $b_n \neq 0$, then

$$\left|\frac{x}{y}\right|_\infty = q^{m-n} = q^{\deg_t(x/y)}, \tag{2}$$

and we easily check that $|\ \ |_\infty$ is an absolute value on $\mathbb{F}_q(t)$. We can give a second expression of $|x/y|$ by noting that $\mathbb{F}_q(t) = \mathbb{F}_q(\theta)$ with $\theta = 1/t$ ("**uniformising parameter at infinity**"). We then have

$$\frac{x}{y} = \theta^{n-m} \cdot \frac{a_m + \cdots + a_0\theta^m}{b_n + \cdots + b_0\theta^n} = \theta^{n-m} \cdot z,$$

where z is a rational function in θ which admits neither zeros nor poles at $\theta = 0$ (i.e. "at infinity"). Thus, we see that $n - m$ is the valuation of x/y at θ, and we set

$$n - m = v_\infty\left(\frac{x}{y}\right).$$

Therefore

$$\left|\frac{x}{y}\right|_\infty = q^{-v_\infty(x/y)}. \tag{3}$$

(3) P-adic absolute value on $\mathbb{F}_q(t)$.

What we did in example (2) with θ can be repeated with an arbitrary irreducible polynomial $P(t) \in \mathbb{F}_q[t]$. Fix a monic irreducible polynomial $P(t) \in \mathbb{F}_q[t]$, and for every $z \in \mathbb{F}_q(t)^*$, let $v_p(z)$ denote the exponent of P in the decomposition of z into prime factors. We then set

$$|z|_P := q^{-nv_P(z)}, \tag{4}$$

where n denotes the degree of P.

When $z = 0$, we write $v_p(0) = \infty$, hence $|0|_P = 0$. Clearly relations (i) and (ii) are satisfied by $|\ |_P$. As for (iii), it follows from the inequality

$$v_P(x + y) \geq \inf\{v_P(x), v_P(y)\}.$$

Thus we have the ultrametric inequality

(v) $$|x + y|_P \leq \sup\{|x|_P, |y|_P\}.$$

Remark 3.1.2 One may be surprised to see that the degree n of P occurs in formula (4). In fact, this is due to a normalisation designed to ensure the validity of the **product formula**: for every $x \in \mathbb{F}_q(t)^*$,

$$|x|_\infty \cdot \prod_{P \in \mathcal{P}} \cdot |x|_P = 1, \tag{5}$$

where $\mathcal{P}$ denotes the set of monic irreducible polynomials of $\mathbb{F}_q[t]$. Applying (5) to $x = P$, we understand the need to introduce n in (4).

(4) The p-adic absolute value on $\mathbb{Q}$.

The construction of example (3) can easily be transposed to $\mathbb{Q}$.

Fix a (positive) prime number p, and for every $z \in \mathbb{Q}^*$, let $v_p(z)$ denote the exponent of p in the decomposition of z into prime factors. We set

$$|z|_p = p^{-v_p(z)}, \qquad |0|_p = 0, \tag{6}$$

and we check, as above, that $|\ |_p$ satisfies the relations (i), (ii), (iv) and (v).

Remark 3.1.3 We can also check the product formula (5) for every $x \in \mathbb{Q}^*$, where now $\mathcal{P}$ denotes the set of (positive) prime numbers p in $\mathbb{Z}$.

The following result is immediate.

Lemma 3.1.1 *Let α, β, γ be three real numbers > 0. If for every $n \in \mathbb{N}$ we have $\gamma^n \leq \alpha n + \beta$, then $\gamma \leq 1$.*

Proposition 3.1.1 *Let A be a commutative ring with unit 1_A, and let $| \ |$ be an absolute value on A. Then the following conditions are equivalent.*

(1) *The absolute value $| \ |$ is ultrametric.*
(2) *For every $n \in \mathbb{Z}$, we have $|n \cdot 1_A| \leq 1$.*

Proof. We may assume without loss of generality that the absolute value is triangular.

(1) If (i) is satisfied and if $n \in \mathbb{N}$, we have

$$|n.1_A| \leq \sup\{|1_A|, \ldots, |1_A|\} = 1.$$

Furthermore, it is clear that $|-1| = 1$, so this property extends to $\mathbb{Z}$.
(2) Let x and $y \in A$. If (ii) is satisfied, then for every $n \in \mathbb{N}$ we have

$$\begin{aligned}|x+y|^n = |(x+y)^n| &= \left|\sum_{i=0}^{n} \binom{n}{i} x^i y^{n-i}\right| \\ &\leq |x|^n + |x|^{n-1}|y| + \cdots + |y|^n \leq (n+1)M^n,\end{aligned}$$

where $M = \sup\{|x|, |y|\}$.
Thus we have

$$\left(\frac{|x+y|}{M}\right)^n \leq n+1,$$

so $\gamma := |x+y|/M \leq 1$ by lemma 3.1.1. □

Corollary 3.1.1 *Every absolute value on A which is bounded on $\mathbb{Z} \cdot 1_A$ is ultrametric.*

Proof. Let M be the bound of $|n 1_A|$. Then for every $\nu \in \mathbb{N}$, we have

$$|n^\nu .1_A| = |n 1_A|^\nu \leq M,$$

so $|n 1_A| \leq 1$. □

Example 3.1.2

(1) Every absolute value on $\mathbb{Q}$ which is bounded on $\mathbb{Z}$ is ultrametric.
(2) If k is a field of positive characteristic, every absolute value on k is ultrametric.

Ostrowski's Theorem 3.1.1 *Every non-trivial absolute value on $\mathbb{Q}$ is either $| \ |_\infty^\alpha$, or $| \ |_p^\alpha$ with $0 < \alpha$.*

Proof. Let $| \ |$ be a triangular absolute value on $\mathbb{Q}$.

(1) For every $n \in \mathbb{N}$, we have

$$|n| = |1 + \cdots + 1| \leq n.$$

Now let a and b be two integers in $\mathbb{N}^*$. We can expand b^ν in powers of a, and we have

$$b^\nu = c_0 + c_1 a + \cdots + c_n a^n$$

with $0 \leq c_i < a$ and $c_n \neq 0$.

Since $a^n \leq b^\nu$, we have $n \leq \nu(\log b/\log a)$.

We deduce from the expression of b^ν in base a that

$$\begin{aligned} |b|^\nu &\leq |c_0| + |c_1|\,|a| + \cdots + |c_n|\,|a|^n \\ &< a(1 + |a| + \cdots + |a|^n) \leq a(n+1)M^n, \end{aligned}$$

where $M = \sup\{1, |a|\}$. Therefore, for every $\nu \in \mathbb{N}$, we have

$$|b|^\nu < a\left(\nu\frac{\log b}{\log a} + 1\right)\left(M^{\log b/\log a}\right)^\nu,$$

i.e. $\gamma^\nu < \alpha\nu + \beta$ with

$$\alpha = a\frac{\log b}{\log a}, \qquad \beta = a, \qquad \gamma = \frac{|b|}{M^{\log b/\log a}}.$$

The above lemma implies that $\gamma \leq 1$, hence

$$|b| \leq \sup\left\{1, |a|^{\log b/\log a}\right\}.$$

(2) If $| \ |$ is not ultrametric, there exists $b \in \mathbb{N}$ such that $|b| > 1$ and the inequality of (1) shows that for every $a > 1$, we have

$$|b|^{1/\log b} \leq |a|^{1/\log a}.$$

For reasons of symmetry, we deduce that

$$|b|^{1/\log b} = |a|^{1/\log a}.$$

Thus if $|b| = b^\alpha$, we see that $|a| = a^\alpha$.

Furthermore, $|n| \leq n$ for $n \in \mathbb{N}$ implies that $0 < \alpha \leq 1$.

(3) If the absolute value $| \ |$ is ultrametric, then $|a| \leq 1$ for every $a \in \mathbb{Z}$.

We see easily that $I = \{a \in \mathbb{Z};\ |a| < 1\}$ is a prime ideal of $\mathbb{Z}$. If the absolute value is not trivial, this ideal is different from (0). Thus there exists a prime number $p \in \mathbb{N}$ such that $I = (p)$. Now, if $x \in \mathbb{Q}^*$, we can write

$$x = p^{v_p(x)} \cdot \frac{u}{v}, \qquad u, v \in \mathbb{Z},$$

where u and v do not contain the number p in their prime decompositions.

We deduce that u and v do not lie in I, so $|u| = |v| = 1$.
Finally, we obtain

$$|x| = |p|^{v_p(x)} = |p|_p^{\alpha v_p(x)} = |x|_p^{\alpha}.$$

Since $|p| < 1$, we see that $\alpha > 0$. □

The proof of the analogue of Ostrowski's theorem for $\mathbb{F}_q(t)$ can be found in the exercises. Here is its statement.

Theorem 3.1.2 *Every non-trivial absolute value on $\mathbb{F}_q(t)$ is either $| \ |_\infty^\alpha$ or $| \ |_P^\alpha$, with $0 < \alpha$ and P monic irreducible in $\mathbb{F}_q[t]$.*

3.2 COMPLETION OF A FIELD EQUIPPED WITH AN ABSOLUTE VALUE

We **know** that the field of real numbers $\mathbb{R}$ contains the field of the rationals $\mathbb{Q}$, and that $\mathbb{R}$ is equipped with an absolute value $| \ |$ which induces the absolute value at infinity $| \ |_\infty$ on $\mathbb{Q}$.

When a field K is equipped with a triangular absolute value $| \ |$, we can define a distance d on K by the formula

$$d(x, y) = |x - y|, \tag{1}$$

and this distance, in turn, lets us define a topology on K (see [Di,1]).

If K is the field of real numbers, we **know** that $\mathbb{Q}$ is dense for this topology (its closure is K) and we **know** that K is complete. We now propose to find a field K possessing all the above properties for every absolute value of $\mathbb{Q}$. With this goal in mind, we introduce a generalised form of the construction of $\mathbb{R}$ given by Cantor[33].

Definition 3.2.1 *Let k be a field equipped with a triangular absolute value $| \ |$.*

A sequence (a_n) of elements of k is said to be a **Cauchy sequence** *if and only if for every $\varepsilon > 0$, there exists an integer $N(\varepsilon)$ such that $p \geq N(\varepsilon)$ and $q \geq N(\varepsilon)$ imply that $|a_p - a_q| \leq \varepsilon$.*

Example 3.2.1

(1) Every sequence of elements of k which converges to an element of k is a Cauchy sequence.

(2) We know that the converse is false for the absolute value at infinity on $\mathbb{Q}$; there exists a Cauchy sequence of rationals which converges to $\sqrt{2} \in \mathbb{R} \setminus \mathbb{Q}$.

(3) The same remark holds for the p-adic absolute values on $\mathbb{Q}$ and the P-adic ones on $\mathbb{F}_q(t)$, see the exercises.

[33] G. Cantor 1845–1918.

Thus, we see that $\mathbb{Q}$ (resp. $\mathbb{F}_q(t)$) is not "complete" for any of its non-trivial triangular absolute values. Let us give the precise definition of this notion (which is extremely useful in analysis) in the case which interests us (cf. [Di 1]).

Definition 3.2.2 *Let k be a field equipped with a triangular absolute value.*
We say that k is **complete** *if every Cauchy sequence of elements of k converges to an element of k.*

Example 3.2.2 Every field is complete for the trivial absolute value.

Given a triangular absolute value $|\ |$ on k, our goal is to construct a field K containing k which is complete for an extension $|\ |_k$ of the absolute value of k.

The following lemma is proved in Exercise 3.10.

Lemma 3.2.1 *Let A be the ring of maps from $\mathbb{N}$ to k. Then*

(1) *the Cauchy sequences form a subring C of A,*
(2) *the sequences converging to zero form an ideal N of C,*
(3) *the quotient ring C/N is a field K.*

Our next objective is to equip K with an absolute value $|\ |_K$ extending that of k.

Corollary 3.2.1 *Let $(a_n) \in C$ be a Cauchy sequence of elements of k.*
Then $(|a_n|)$ is a Cauchy sequence of real numbers.

Proof. Indeed, the triangle inequality implies that

$$\begin{cases} |a_p| - |a_q| \leq |a_p - a_q| \\ |a_q| - |a_p| \leq |a_p - a_q|. \end{cases}$$

Thus if p and q are greater than $N(\varepsilon)$, we have

$$||a_p| - |a_q|| \leq \varepsilon.$$

□

Since $\mathbb{R}$ is complete, the sequence $(|a_n|)$ has a limit; if $(a_n) \in C$, we obtain the following result.

Lemma 3.2.2 *For $(a_n) \in C$, set*

$$|(a_n)|_c := \lim(|a_n|) \in \mathbb{R}.$$

Then

(1) *the map $(a_n) \mapsto |(a_n)|_c$ satisfies conditions (ii) and (iv) of Section 3.1;*
(2) *it induces an absolute value $|\ |_K$ on K.*

Proof. **(1)** Since $|a_n + b_n| \leq |a_n| + |b_n|$, we have

$$\lim(|a_n + b_n|) \leq \lim(|a_n|) + \lim(|b_n|),$$

so $|a + b|_c \leq |a|_c + |b|_c$, where a (resp. b, c) means (a_m) (resp. (b_m), (c_m)).

Similarly, we have

$$\lim(|a_n b_n|) = \lim(|a_n|) \lim(|b_n|),$$

so

$$|ab|_c = |a|_c |b|_c.$$

(2) If, in particular, $a \in N$, then $|a|_c = 0$. Thus if $\bar{x} = x + N = y + N \in K$, then $y = x + a$, with $a \in N$, so

$$|y|_c \leq |x|_c + |a|_c = |x|_c.$$

By symmetry, $|x|_c \leq |y|_c$, so $|x|_c = |y|_c$. It follows moreover that

$$\begin{cases} |\bar{x} + \bar{y}|_K = |x + y|_c \leq |x|_c + |y|_c = |\bar{x}|_K + |\bar{y}|_K \\ |\bar{x}\bar{y}|_K = |xy|_c = |x|_c |y|_c = |\bar{x}|_K |\bar{y}|_K. \end{cases}$$

Finally, $|\bar{x}|_K = 0$ implies $|x|_c = 0$, hence $x \in N$ and $\bar{x} = \bar{0} \in K$. □

We can now state the main result of Section 3.2.

Theorem 3.2.1

(1) *If we identify k with the set of constant Cauchy sequences, the image of this set by the canonical map $C \twoheadrightarrow C/N = K$ is a field $\tilde{k}$ isomorphic to k.*

(2) *The field $\tilde{k}$ is dense in K for the topology associated to $|\ |_K$.*

(3) *The field K is complete for this topology.*

Proof. **(1)** Let $x \in k$, and let $\sigma(x) \in C$ denote the constant Cauchy sequence equal to x; clearly $\sigma(k)$ is a field $\tilde{k}$ isomorphic to k.

Furthermore, $\sigma(k) \cap N = \{0\}$, so the image of $\sigma(k)$ under the map $C \twoheadrightarrow C/N$ is a field $\tilde{k}$ which is isomorphic to k.

(2) Let $\bar{x} = (x_n) + N \in K$, with $x_n \in k$ for every $n \in \mathbb{N}$. We want to show that for every $\varepsilon > 0$, there exists $y \in k$ such that

$$|\bar{x} - \overline{\sigma(y)}|_K \leq \varepsilon.$$

Now, there exists $N(\varepsilon)$ such that $p \geq N(\varepsilon)$ and $q \geq N(\varepsilon)$ imply that $|x_p - x_q| \leq \varepsilon$, so we can take $y = x_{N(\varepsilon)}$.

(3) Let $(\bar{x}_n)$ be a Cauchy sequence of K. We know that for every $n \in \mathbb{N}$, there exists $y_n \in k$ such that $|\bar{x}_n - \overline{\sigma(y_n)}|_K \leq 1/n$.

Let us show that (y_n) is a Cauchy sequence of k. Indeed let $\varepsilon > 0$ and $N(\varepsilon)$ be such that

$$p \text{ and } q \geq N(\varepsilon) \quad \Longrightarrow \quad |\bar{x}_p - \bar{x}_q| \leq \frac{\varepsilon}{3}.$$

Then for p and $q \geq N(\varepsilon)$, we have

$$|y_p - y_q| \leq |\overline{\sigma(y_p)} - \bar{x}_p|_K + |\bar{x}_p - \bar{x}_q|_K + |\bar{x}_q - \overline{\sigma(y_q)}|_K \leq \varepsilon.$$

Now let $\bar{y} = \overline{(y_n)} \in K$; for every $n \in \mathbb{N}$, we have $|\bar{x}_n - \bar{y}|_K \leq 1/n$, so $\lim_{n\to\infty} \bar{x}_n = \bar{y}$. □

Definition 3.2.3 *Let k be a field equipped with a triangular absolute value $|\ |$. A field K containing k and equipped with a triangular absolute value $|\ |_K$ is called a* **completion** *of k if*

(i) *for every $x \in k$, we have $|x|_K = |x|$,*
(ii) *K is complete for the absolute value $|\ |_K$,*
(iii) *k is dense in K for the topology associated to $|\ |_K$.*

Important Remark 3.2.1 When $k = \mathbb{Q}$ is equipped with the p-adic absolute value $|\ |_p$, the preceding construction gives us a completion K equipped with an absolute value $|\ |_K$.

Since the set of absolute values of the elements of $\mathbb{Q}^*$ is the **discrete** group $\langle p \rangle$, it follows from the preceding construction that the set of absolute values of K^* is the same group. In this case, we say that the absolute value is **discrete** (even though o is an accumulation point of $\langle p \rangle$!).

In Chapter 6 and in the exercises, we will find other ways to construct a completion of k. It is natural to ask if all these completions are isomorphic. The answer is given by the following result.

Theorem 3.2.2 *Let k_1 and k_2 be two fields equipped with triangular absolute values $|\ |_1$ and $|\ |_2$. Assume that there exists an isomorphism $\varphi : k_1 \to k_2$ such that $|\varphi(x)|_2 = |x|_1$. Then if K_1 and K_2 are completions of k_1 and k_2, there exists an isomorphism $\phi : K_1 \to K_2$ such that*

(i) $\phi|_{k_1} = \varphi$,
(ii) $|\phi(x)|_{K_2} = |x|_{K_1}$ *for every $x \in K_1$.*

Proof. Let $x \in K_1$.

To construct $\phi(x)$, consider a sequence (x_n) of elements of k_1 such that $|x - x_n|_1 \leq 1/n$ for every n (which is possible since k_1 is dense in K_1). The sequence (x_n) is then a Cauchy sequence, so this is also true for $(\varphi(x_n))$. Consequently there exists $y \in K_2$ such that $\lim_{n\to\infty} \varphi(x_n) = y$.

Set $\phi(x) := y$.

We easily check that ϕ is an isomorphism of rings with unit, and also that assertions (i) and (ii) hold. □

3.3 THE FIELD OF P-ADIC NUMBERS

The notions introduced in the preceding paragraph now enable us to give a definition of "the" field of p-adic numbers.

Definition 3.3.1 *Let $p > 0$ be a prime, and let $| \ |_p$ be the p-adic absolute value of $\mathbb{Q}$. The completion of $\mathbb{Q}$ for the absolute value $| \ |_p$ is called the* **field of p-adic numbers**, *and is denoted by $\mathbb{Q}_p$.*

Thus, this field is defined (here!) as a ring quotient, the quotient of C/N. But we would like to define each p-adic number by a **unique** representative modulo N, just as we represent a real number by an infinite sequence of decimals; the idea of the latter is due to Stevin, although its "uniqueness" is a little tricky!

First, we need to mention an important criterion of convergence of series.

Theorem 3.3.1 *Let K be a field which is complete for an* **ultrametric** *absolute value $| \ |$.*

In order for the series $\sum_{n \gg -\infty} a_n$, of general term $a_n \in K$, to converge in K, it is necessary and sufficient that its general term a_n tends to zero.

Remark 3.3.1 The notation $\sum_{n \gg -\infty} a_n$ means that there exists an integer N (depending on the series) such that $a_n = 0$ if $n < -N$. Similarly, $\prod_{n \gg -\infty} u_n$ means that there exists an integer N such that $u_n = 1$ if $n < -N$.

Proof.

(1) The necessity of the condition is a general property of normed vector spaces.

(2) To show the converse, we consider the sequence of partial sums

$$m \longmapsto s_m = \sum_{-\infty \leq n \leq m} a_n;$$

we must show that (s_m) is a Cauchy sequence. So, let $\varepsilon > 0$. Since a_n tends to zero, there exists M such that $n \geq M$ implies that $|a_n| \leq \varepsilon$. Now if $q \geq p \geq M$, we have

$$|s_q - s_p| = |a_{p+1} + \cdots + a_q| \leq \sup \left\{ |a_{p+1}|, \ldots, |a_q| \right\} \leq \varepsilon. \qquad \square$$

Definition 3.3.2 *Let K be a field which is complete for an absolute value $| \ |$.*

An infinite product $\prod_{n \gg -\infty} u_n$, of general term $u_n \in K^*$, is said to be **convergent** *in K^* if there exists $\ell \in K^*$ such that the products $\pi_m = \prod_{-\infty < n \leq m} u_n$ tend to ℓ.*

The preceding criterion admits an important analogue for infinite products.

Theorem 3.3.2 *Let K be a field which is complete for an* **ultrametric** *absolute value $| \ |$.*

For the product $\prod_{n \gg -\infty} u_n$ of general term $u_n \in K^*$ to converge, it is necessary and sufficient that u_n tend to 1.

Proof. **(1)** The necessity of the condition follows from the fact that

$$\frac{\pi_m}{\pi_{m-1}} = u_m,$$

and if $\pi_m \to \ell \in K^*$, the left-hand side tends to 1.

(2) Since $\pi_m = u_0 + \sum_{n=1}^{m}(\pi_n - \pi_{n-1})$, we are led to show that the series of general term

$$a_n = \pi_n - \pi_{n-1} = \pi_{n-1}(u_n - 1)$$

converges.

Set $u_n - 1 = v_n$.

Since v_n tends to zero, there exists N such that $n \geq N$ implies that $|v_n| < 1$, hence $|u_n| = 1$.

Thus, for $n \geq N$, we have

$$|\pi_{n-1}| = |\pi_n| = |\pi_{n+1}| = \cdots$$

Denoting this number by A, we see that for $n \geq N$, we have

$$|a_n| = A|v_n|.$$

Since $v_n \to 0$, theorem 3.3.1 implies that the series of general term a_n converges. □

Definition 3.3.3 *A* **local ring** $\mathcal{O}$ *is a commutative ring with unit which has a single non-zero maximal ideal* $\mathcal{M}$. *The field* $\mathcal{O}/\mathcal{M}$ *is called the* **residue field** *of* $\mathcal{O}$.

Example 3.3.1 If $h > 1$, the ring $\mathbb{Z}/p^h\mathbb{Z}$ is a local ring; in particular, if $h = 1$ it is a field and its single maximal ideal is zero.

Theorem 3.3.3 *Let* $\mathbb{Z}_p := \{x \in \mathbb{Q}_p;\ |x| \leq 1\}$. *Then*

(1) $\mathbb{Z}_p$ *is a local ring whose maximal ideal is* $p\mathbb{Z}_p$.
(2) *The group of units of* $\mathbb{Z}_p$ *is* $\mathbb{U}_p = \{x \in \mathbb{Z}_p;\ |x| = 1\}$.
(3) *The residue field of* $\mathbb{Z}_p$ *is* $\mathbb{F}_p = \mathbb{Z}/p\mathbb{Z}$.

Proof. Properties (ii) and (iii) (with $C = 1$) of the absolute value $|\ |$ show that $\mathbb{Z}_p$ is a ring with unit.

Let us show that $\mathbb{U}_p = \{z \in \mathbb{Z}_p;\ |x| = 1\}$.

Indeed, if $|u| = 1$ we know that $u \neq 0$, so there exists $v \in \mathbb{Q}_p$ such that $uv = 1$. But if $|u| = 1$, then $|v| = 1$ so $v \in \mathbb{Z}_p$.

Conversely, if $u \in \mathbb{U}_p$, there exists $v \in \mathbb{Z}_p$ such that $uv = 1$. It follows that $|uv| = 1$. As $|u| \leq 1$ and $|v| \leq 1$, we must have $|u| = 1$.

Now let I be an ideal of $\mathbb{Z}_p$. If I is not contained in $p\mathbb{Z}_p$, then I contains a unit u since $p\mathbb{Z}_p = \{x \in \mathbb{Z}_p;\ |x| < 1\}$, so there exists $v \in \mathbb{Z}_p$ such that $uv = 1$, but $uv \in I$ so $I = \mathbb{Z}_p$. This shows that $p_p^{\mathbb{Z}}$ is **maximal**.

Clearly $\mathbb{F}_p$ is a subfield of $\mathbb{Z}_p/p\mathbb{Z}_p$, since the image of $\mathbb{Z}$ under the canonical projection

$$\mathbb{Z}_p \xrightarrow{\pi} \mathbb{Z}_p/p\mathbb{Z}_p$$

is isomorphic to $(\mathbb{Z}/\ker\pi \cap \mathbb{Z}) = \mathbb{Z}/p\mathbb{Z}$.

Thus, it suffices to show that $\#(\mathbb{Z}_p/p\mathbb{Z}_p) \leq p$. For this we will see that the image of $\mathbb{U}_p$ contains at most $p-1$ elements.

Indeed, if $u \in \mathbb{U}_p$, then u is the limit of a sequence of rationals r_n, so we have

$$|u - r_n| < 1$$

for n greater than some fixed N.

Thus we have $|r_n| = 1$ for $n \geq N$, so r_n is congruent modulo p to a rational integer a not divisible by p; this follows from Bachet's theorem (usually called Bézout's theorem). But we also have

$$\{|u - r_n| < 1 \text{ and } |r_n - a| < 1\} \implies |u - a| < 1,$$

so $u \equiv a \mod p\mathbb{Z}_p$.

As $\#\{a + p\mathbb{Z}_p;\ a \in \mathbb{Z}\} = p$, we are done. □

Definition 3.3.4 *A system of* **digits** $\mathcal{S}$ *of* $\mathbb{Q}_p$ *consists of zero and a system of representatives of the elements of* $\mathbb{U}_p$ *modulo* $p\mathbb{Z}_p$.

Example 3.3.2 **(1)** The most frequent example is $\mathcal{S} = \{0, 1, \ldots, p-1\}$.

(2) When p is odd, $\mathcal{S} = \{-(p-1)/2, \ldots, -1, 0, 1, \ldots, (p-1)/2\}$ is more symmetric and sometimes preferable.

(3) The best system of digits consists of the **Teichmüller digits**

$$\mathcal{S} = \{0, \omega_p(\bar{1}), \ldots, \omega_p(\overline{p-1})\},$$

where $\omega_p(x)$ is the unique $p-1^{th}$ root of unity of $\mathbb{Q}_p$ which is congruent to x modulo p, so that

$$\mathbb{F}_p^* \xrightarrow{\omega_p} S^* = S\backslash\{0\}$$

is an isomorphism of multiplicative groups.

The computation of $\omega_p(x)$ is easy; we note that if $x = \overline{u}$ with $u \in \mathbb{U}_p$, then

$$\omega_p(x) = \lim_{n\longrightarrow\infty} (u^{p^n}).$$

Indeed, we know by Fermat's little theorem that $x^{p^n} = x$ for every $n \in \mathbb{N}$, so

$$u^p = u + pa_1$$

with $a_1 \in \mathbb{Z}_p$. By induction on n, it follows that

$$u^{p^n} = u + p^n a_n$$

with $a_n \in \mathbb{Z}_p$. Thus we see that (u^{p^n}) is a Cauchy sequence, and passing to the limit, we have

$$\omega_p(x)^p = \omega_p(x).$$

(4) When $p = 2$, the system $\mathcal{S}$ of example (1) is a Teichmüller system. When $p = 3$, the same holds for that of example (2). For $p > 3$, the three systems of digits are distinct.

Theorem 3.3.4 *Let $\mathcal{S}$ be a system of digits of $\mathbb{Q}_p$.*
Every element $x \in \mathbb{Q}_p$ can be written uniquely in the form

$$x = \sum_{n \gg -\infty} a_n p^n \tag{1}$$

with $a_n \in \mathcal{S}$, and every series of the form $\sum_{n \gg -\infty} a_n p^n$ converges to a p-adic number.

Proof. The second assertion follows from theorem 3.3.1 since

$$\lim_{n \longrightarrow \infty} (a_n p^n) = 0.$$

Now let $x \in \mathbb{Q}_p$. Up to dividing x by $|x|_p$, we can assume that $x \in \mathbb{Z}_p$.

We know that there exists a number $a_0 \in \mathcal{S}$ such that $x \equiv a_0$ modulo $p\mathbb{Z}_p$, and this determines the first term of the expansion of x.

Then, we set $x_1 = x - a_0$ and we obtain the second term of the expansion of x since $|x_1|_p \leq 1/p$. Proceeding this way successively, we see that $|x_n|_p \leq 1/p^n$ and

$$x = a_0 + a_1 p + \cdots + a_{n-1} p^{n-1} + x_n.$$

We obtain the result by passing to the limit as $n \to \infty$. □

Remark 3.3.2

(1) We deduce from theorem 3.3.4 that $\mathbb{Q}_p$, like $\mathbb{R}$, is not countable.

(2) In order to compute the sum of two series of the form (1), it suffices to know how to write the sum of two arbitrary digits a and $b \in \mathcal{S}$; to compute the product of these series, it suffices to know how to write ab.

Example 3.3.3

(1) If $\mathcal{S} = \{0, 1, \ldots, p-1\}$, we know that $a+b \in \mathcal{S}$ if $a+b < p$ and $a+b = (a+b-p)+p$ if $a + b \geq p$.

For the product, the expression of ab is a polynomial of degree ≤ 1 in p with coefficients in $\mathcal{S}$. In particular, we have

$$(p-1)(p-1) = 1 + (p-2)p.$$

(2) If $\mathcal{S} = \{-(p-1)/2, \ldots, -1, 0, 1, \ldots, (p-1)/2\}$, we see that $a + b \in \mathcal{S}$ if a and b do not have the same sign. If they do have the same sign ε, then

$$a + b = (a + b - \varepsilon p) + \varepsilon p.$$

This system is useful to write the opposite of a number, for example (1), whereas in the first system we get the horrible expression

$$-1 = (p-1) + (p-1)p + (p-1)p^2 + \cdots$$

(3) For the Teichmüller digits, we always have $ab \in \mathcal{S}$.

3.4 ALGEBRAIC CLOSURE OF A FIELD

In this part, the word "field" is always taken to mean "commutative field".

Recall that if we are given an extension of fields $K \subset L$, an element $x \in L$ is said to be **algebraic** over K if x is a root of a non-zero polynomial of $K[X]$, and the extension $K \subset L$ is said to be **algebraic** if every element x in L is algebraic over K.

Recall also that a field K is said to be **algebraically closed** if every non-constant polynomial in $K[X]$ has a root in K (it follows that **all** its roots then lie in K).

Definition 3.4.1 *Given a field K, we say that an extension L of K (i.e. a field $L \supset K$) is an* **algebraic closure** *of K if*

(i) *the extension L/K is algebraic,*
(ii) *the field L is algebraically closed.*

Example 3.4.1 The field of complex numbers $\mathbb{C}$ is an algebraic closure of $\mathbb{R}$; the **degree** of this extension (which is the dimension of $\mathbb{C}$ considered as an $\mathbb{R}$-vector space) is equal to 2.

It is natural to ask if every field K has an algebraic closure L. The answer to this problem is positive, if we admit the **axiom of choice**; it is the object of the following statement (see [Bou 1]).

Steinitz'[34] Theorem 3.4.1

(i) *Every commutative field has an algebraic closure.*
(ii) *If there exists an isomorphism $\varphi : K_1 \twoheadrightarrow K_2$, and if $\bar{K}_1$ and $\bar{K}_2$ are algebraic closures of K_1 and K_2, then there exists an isomorphism $\phi : \bar{K}_1 \twoheadrightarrow \bar{K}_2$ such that $\phi\,|_{K_1} = \varphi$.*

Example 3.4.2

(1) The field of rational numbers $\mathbb{Q}$ admits an algebraic closure $\bar{\mathbb{Q}}$.

Since the polynomial $X^n - 2 \in \mathbb{Q}[X]$ is irreducible for every $n \geq 1$ by Eisenstein's criterion, we see that $[\bar{\mathbb{Q}} : \mathbb{Q}] = \infty$.

[34] E. Steinitz 1871–1928.

(2) The field $\mathbb{F}_p$ admits an algebraic closure $\bar{\mathbb{F}}_p$. Since for every $n \in \mathbb{N}$, the roots of the polynomial $X^{p^n} - X$ in $\bar{\mathbb{F}}_p$ are distinct, $\bar{\mathbb{F}}_p$ is infinite, so $[\bar{\mathbb{F}}_p : \mathbb{F}_p] = \infty$.

(3) The field $\mathbb{Q}_p$ admits an algebraic closure $\overline{\mathbb{Q}_p}$. Since the polynomial $X^n - p \in \mathbb{Q}_p[X]$ is irreducible for every $n \geq 1$ by Eisenstein's criterion, we see that $[\bar{\mathbb{Q}}_p : \mathbb{Q}_p] = \infty$.

(4) If K_1 and K_2 are two subfields of $\bar{\mathbb{Q}}$ and if $\varphi : K_1 \twoheadrightarrow K_2$ is an isomorphism, then φ extends to an automorphism ϕ of $\bar{\mathbb{Q}}$. Indeed, $\bar{\mathbb{Q}}$ is a common algebraic closure of K_1 and K_2.
It follows that $\mathrm{Aut}(\bar{\mathbb{Q}})$ is infinite since the group of automorphisms of the field generated by the roots of $X^n - 1$ over $\mathbb{Q}$ is isomorphic to $(\mathbb{Z}/n\mathbb{Z})^*$ for every $n \in \mathbb{N}$.

We often read that "the algebraic closure of $\mathbb{Q}$ consists of the set of complex numbers which are algebraic over $\mathbb{Q}$". This statement can be explained as follows.

Definition 3.4.2 *Let $K \subset L$ be a field extension. We say that K is* **algebraically closed inside** *L (or that L is regular over K) if every element of L which is algebraic over K lies in K.*

Proposition 3.4.1 *Let $K \subset L$ be a field extension. The set M of elements of L which are algebraic over K is an intermediate field between K and L. Moreover, M is algebraically closed inside L.*

Definition 3.4.3 *We say that M is the* **algebraic closure** *of K inside L.*

Proof of the proposition. Since $M \subset K(M)$ and $K(M)$ is an algebraic extension of K, we see that $M = K(M)$.

Now, if $x \in L$ is algebraic over M, the extension $M \subset M(x)$ is algebraic, so x is a root of some polynomial $X^n + a_{n-1}X^{n-1} + \cdots + a_0 \in M[X]$. As each a_i is algebraic over K, the extension $K(a_1, \ldots, a_{n-1})$ is a finite extension of K. By the tower rule, we have

$$[K(a_1, \ldots, a_{n-1}, x) : K] = [K(a_1, \ldots, a_{n-1}, x) : K(a_1, \ldots, a_{n-1})] \times [K(a_1, \ldots, a_{n-1}) : K] < \infty,$$

so the degree of the extension $K \subset K(x)$, which divides the degree of the extension $K \subset K(a_1, \ldots, a_{n-1}, x)$, is finite, and $x \in M$. □

Corollary 3.4.1 *Let $K \subset L$ be a field extension. If L is algebraically closed, then the algebraic closure M of K inside L is actually an algebraic closure of K.*

Example 3.4.3

(1) The algebraic closure of $\mathbb{Q}$ inside $\mathbb{C}$ is an algebraic closure of $\mathbb{Q}$, so it is isomorphic to $\bar{\mathbb{Q}}$. We show in Problem 4 that $\mathrm{Aut}(\mathbb{C})$ is infinite. If the complex number z belongs to $\bar{\mathbb{Q}}$, then clearly its orbit under the action of $\mathrm{Aut}(\mathbb{C})$ is **finite** (its cardinal is the **degree** of z over $\mathbb{Q}$). Conversely, it can be shown that if the orbit of z under the action of $\mathrm{Aut}(\mathbb{C})$ is **finite**, then z belongs to $\bar{\mathbb{Q}}$ (again see Problem 4, page 167).

(2) For every prime number p, the algebraic closure of $\mathbb{Q}$ inside $\bar{\mathbb{Q}}_p$ is an algebraic closure of $\mathbb{Q}$, so it is isomorphic to $\bar{\mathbb{Q}}$. Thus we obtain an infinity of distinct constructions of the algebraic closure of $\mathbb{Q}$.

Remark 3.4.1 **(1)** A well-known reasoning (Exercise 3.17) shows that $\bar{\mathbb{Q}}$ is countable. It follows that

$$[\mathbb{C} : \bar{\mathbb{Q}}] = \infty,$$

and for every prime p,

$$[\bar{\mathbb{Q}}_p : \bar{\mathbb{Q}}] = \infty.$$

(2) We saw earlier that $[\bar{\mathbb{Q}}_p : \mathbb{Q}_p] = \infty$. One can also show that we can extend the p-adic absolute value of $\mathbb{Q}_p$ to $\bar{\mathbb{Q}}_p$. However, we note that $\bar{\mathbb{Q}}_p$ is not complete for this extension of $| \ |_p$. We can thus also consider

$$\mathbb{C}_p := \hat{\bar{\mathbb{Q}}}_p \underset{\neq}{\supset} \bar{\mathbb{Q}}_p,$$

and ask ourselves if $\hat{\bar{\mathbb{Q}}}_p$ is algebraically closed. The answer is "yes"! For all these results, [Ca 1] p. 149–151 or [Sch] are good references.

(3) The same phenomenon occurs for the algebraic closure of the field K_P, where P is an irreducible polynomial of $\mathbb{F}_q[t]$ or $P = \infty$ and $K = \mathbb{F}_q(t)$. We can show that $\bar{K}_P$ is not complete (see [Sch], pp. 43–44). However, its completion $\mathbb{C}_P := \hat{\bar{K}}_P$ is algebraically closed.

Theorem 3.4.2 *For every prime number p, the equation F_p (resp. G_p) locally admits non-trivial solutions, i.e. there exists a solution (x, y, z) of this equation in $\mathbb{Z}_\ell$ such that $xyz \neq 0$ (resp. $xyz \neq 0$ and $x \neq y$) for every prime ℓ.*

Proof. We start with the case $\ell = p$.

Consider the equations

$$x^p + y^p + z^p = 0 \qquad (F_p)$$

$$x^p + y^p + 2z^p = 0. \qquad (G_p)$$

Choose $x = 1$ and $z = p^h$ with $h \geq 1$. Then we can take $y = -(1 + p^{ph})^{1/p} \in \bar{\mathbb{Q}}_p$ (resp. $y = -(1 + 2p^h)^{1/p}$) and try to make sense of this expression by using analysis.

We easily check that the binomial series

$$(1 + x)^{1/p} = 1 + \frac{1}{p}x + \frac{1}{2!}\frac{1}{p}\left(\frac{1}{p} - 1\right)x^2 + \cdots$$

converges for every $x \in \hat{\bar{\mathbb{Q}}}_p$ such that $|x|_p < p^{p/(1-p)}$ (see Exercise 3.20).

Taking $x = p^{ph}$ (resp. $2p^{ph}$), we see that the sum of this series belongs to $\mathbb{Z}_p$.

Now we pass to the case $\ell \neq p$, which is actually the easier case. Take $x = 1$ and $z = \ell$. Then we can take $y = -(1+\ell)^{1/p} \in \bar{\mathbb{Q}}_\ell$ (resp. $y = -(1+2\ell)^{1/p}$) and we see that the binomial series converges in $\mathbb{Z}_\ell$. □

3.5 GENERALITIES ON THE LINEAR REPRESENTATIONS OF GROUPS

Recall that the **prime subfield** P_K of a field K is the smallest subfield of K. If the field K is of characteristic zero, then $P_K = \mathbb{Q}$, and if K is of characteristic p, then $P_K = \mathbb{F}_p$.

When σ is an automorphism of K, we will show that $\sigma(P_K) = P_K$ and $\sigma|_{P_K} = \mathrm{id}_{P_K}$. Indeed, $P_K \cap \sigma(P_K)$ is a field contained in P_K, so $P_K \subset \sigma(P_K)$. Considering σ^{-1}, we now see that $\sigma(P_K) \subset P_K$, which gives the first assertion. The second comes from the fact that $\sigma(1) = 1$ and 1 generates the ring $\mathbb{Z} \cdot 1_K$. As $\sigma|_{\mathbb{Z}\cdot 1_K} = \mathrm{id}_{\mathbb{Z}\cdot 1_K}$ we see that $\sigma|_F = \mathrm{id}_F$ where F denotes the field of fractions of $\mathbb{Z} \cdot 1_K$, i.e. P_K. Let us conclude these remarks with the following definition.

Definition 3.5.1 *Let $F \subset K$ be a field extension. If an automorphism σ of K satisfies the conditions*

(i) $\sigma(F) = F$,
(ii) $\sigma|_F = \mathrm{id}_F$,

then we say that σ is an **F-automorphism** *of K.*

Proposition 3.5.1 *Let K be a (commutative) field of prime subfield P_K. Then every automorphism of K is a P_K-automorphism of K.*

In particular, $\mathrm{Aut}(K) \subset GL_{P_K}(K)$, *where $GL_{P_K}(K)$ denotes the group of the P_K-linear bijections of K.*

Example 3.5.1

(1) Every automorphism of $\bar{\mathbb{Q}}$ belongs to $GL_{\mathbb{Q}}(\bar{\mathbb{Q}})$.
(2) Every automorphism of $\bar{\mathbb{F}}_p$ belongs to $GL_{\mathbb{F}_p}(\bar{\mathbb{F}}_p)$.

Understanding $\mathrm{Aut}(\bar{\mathbb{Q}})$ is one of the great problems of number theory, and one of the ways of studying it is to consider, not $GL_{\mathbb{Q}}(\bar{\mathbb{Q}})$ which is too big, but rather the **representations of degree** n of $\mathrm{Aut}(\bar{\mathbb{Q}})$, i.e. the homomorphisms

$$\rho : \mathrm{Aut}\,(\bar{\mathbb{Q}}) \xrightarrow{\rho} GL_F(V)$$

where V denotes a vector space of dimension n over some commutative field F.

Definition 3.5.2 *Let G be a group and F a commutative field.* A **representation of** G **defined over** F *is a homomorphism*

$$G \xrightarrow{\rho} GL_F(V).$$

Two representations ρ and ρ' are said to be **isomorphic** if there exists an F-linear bijection $\varphi : V \twoheadrightarrow V'$ such that for every $s \in G$, we have

$$\rho'(s) \circ \varphi = \varphi \circ \rho(s).$$

The dimension of V is called the **degree** of the representation, and the **character** χ_ρ of the representation is the map $s \mapsto$ trace $(\rho(s))$ from G to F.

Example 3.5.2 **(1)** If V is a line, we say that the representation is **one-dimensional**. Its character χ_ρ is then a homomorphism

$$G \longrightarrow F^*.$$

(2) If $K = \mathbb{Q}(i)$ with $i^2 + 1 = 0$, then Aut $K = \langle c \rangle$, where c denotes the complex conjugation defined by $c(i) = -i$.

Since $\mathbb{Q}(i)$ is a $\mathbb{Q}$-vector space of dimension 2, we can take $V = \mathbb{Q}(i)$ and $F = \mathbb{Q}$. If we consider the basis $(1, i)$ for $\mathbb{Q}(i)$, we have

$$\rho(\mathrm{id}) = \begin{pmatrix} 1 & 0 \\ 0 & 1 \end{pmatrix}, \qquad \rho(c) = \begin{pmatrix} 1 & 0 \\ 0 & -1 \end{pmatrix},$$

so $\chi_\rho(\mathrm{id}) = 2$ and $\chi_\rho(c) = 0$. We see in particular that χ_ρ is not a homomorphism of F or F^*!

Note that the image of ρ is the subgroup of $GL_{\mathbb{Q}}(\mathbb{Q}(i))$ generated by $\rho(c)$, and that $\rho(c)$ has two distinct eigenvalues. Thus, we see that

$$\mathbb{Q}(i) = D_1 \oplus D_2,$$

where $D_1 = \mathrm{Vec}_{\mathbb{Q}}(1)$ and $D_2 = \mathrm{Vec}_{\mathbb{Q}}(i)$ are two lines of V which are globally invariant under the linear automorphisms of the image of ρ.

(3) If $K = \mathbb{F}_4$, then Aut $(\mathbb{F}_4) = \{\mathrm{id}_{\mathbb{F}_4}, \sigma\}$ where σ denotes the Frobenius automorphism $\sigma(x) = x^2$. Since $\mathbb{F}_4$ is an $\mathbb{F}_2$-vector space of dimension 2, we can take $V = \mathbb{F}_4$ and $F = \mathbb{F}_2$.

If we take the basis $(1, \varepsilon)$ of F_4, where ε is a root of the (irreducible) polynomial $X^2 + X + 1 \in \mathbb{F}_2[X]$, then we see that

$$\rho(\mathrm{id}) = \begin{pmatrix} 1 & 0 \\ 0 & 1 \end{pmatrix}, \qquad \rho(\sigma) = \begin{pmatrix} 1 & 1 \\ 0 & 1 \end{pmatrix},$$

and it follows that $\chi_\rho(\mathrm{id}) = 0$ and $\chi_\rho(\sigma) = 0$.

Moreover, we see that $D = \mathbb{F}_2$ is the only line of $\mathbb{F}_4$ which is invariant under the image of ρ.

(4) Assume that $\mathbb{Q}(i) \subset \bar{\mathbb{Q}}$ (resp. $\mathbb{F}_4 \subset \bar{\mathbb{F}}_2$). Then $\mathbb{Q}(i)$ (resp. $\mathbb{F}_4$) is globally invariant under every element of Aut $(\bar{\mathbb{Q}})$ (resp. Aut $\bar{\mathbb{F}}_2$).

It follows that ρ lifts to a representation of degree 2 of Aut $(\bar{\mathbb{Q}})$ (resp. Aut $\bar{\mathbb{F}}_2$) by the formula $\tau \mapsto \rho(\tau|_{\mathbb{Q}(i)})$ (resp. $\tau \mapsto \rho(\tau\,|_{\mathbb{F}_4})$).

Proposition 3.5.2 *Let χ be the character of a representation ρ of degree n of G, defined over the field of complex numbers $\mathbb{C}$. We have*

(i) $\chi(e) = n$, *where e is the identity element of G,*
(ii) *If σ is of finite order in G, then* $\chi(\sigma^{-1}) = \overline{\chi(\sigma)}$,
(iii) $\chi(\tau\sigma\tau^{-1}) = \chi(\sigma)$ *for every* $\tau \in G$.

Proof. Property (i) is obvious.

Property (ii) follows from the fact that $\rho(\sigma)$ is of finite order in $GL_{\mathbb{C}}(V)$, so it is diagonalisable. Its eigenvalues λ_i are roots of unity and we have

$$\chi(\sigma^{-1}) = \mathrm{Tr}\,\rho(\sigma^{-1}) = \lambda_1^{-1} + \cdots + \lambda_n^{-1} = \bar{\lambda}_1 + \cdots + \bar{\lambda}_n = \overline{\mathrm{Tr}\,\rho(\sigma)} = \overline{\chi(\sigma)}.$$

(iii) We know that if f and g are in $GL_C(V)$, then

$$\mathrm{Tr}(fgf^{-1}) = \mathrm{Tr}(g),$$

so $\mathrm{Tr}(\rho(\tau)\rho(\sigma)\rho(\tau)^{-1}) = \mathrm{Tr}(\rho(\sigma))$, which gives the result. □

Definition 3.5.3 *A representation $G \to GL_F(V)$ is said to be* **irreducible** *if $V \neq \{0\}$ and if no subvector-space of V (except for the trivial subspaces $\{0\}$ and V) is stable under all the linear automorphisms of $\rho(G)$.*

Example 3.5.3

(1) Every representation of degree one is irreducible.
(2) The representations in examples (2) and (3) above are *reducible*.

The following result is a fundamental structure theorem in representation theory (see [Se 2] p. 18 or Exercise 3.23).

Maschke's Theorem 3.5.1 *Let $\rho : G \to GL_F(V)$ be a linear representation of a finite group G defined over a field F whose characteristic does not divide the order of G.*

If V has a subspace W which is stable under $\rho(G)$, then there is a subspace W′ in V which is a supplement of W (i.e. $V = W \oplus W'$), and is stable under $\rho(G)$.

Example 3.5.4

(1) In example (2), the field $F = \mathbb{Q}$ is of characteristic zero, and we saw that $V = D_1 \oplus D_2$.
(2) In example (3), the field $F = \mathbb{F}_2$ is of characteristic 2 and 2 divides the order of G. We saw that D is the only invariant line inside V, so there is no supplementary subspace.

Theorem 3.5.2 *If the characteristic of F does not divide the order of G, then every representation of G is a direct sum of irreducible representations. In other words, there exist $W_1, \ldots, W_r$ in V, invariant under $\rho(G)$ and irreducible, such that*

$$V = W_1 \oplus \cdots \oplus W_r.$$

Proof. We prove this result by induction on $\dim_F(V) < \infty$.

(1) If $\dim_F V = 0$, then V is a direct sum of the empty family of stable irreducible subspaces.

(2) Suppose $\dim_F V > 0$.

If V is irreducible, we are done.

Otherwise, V contains a stable subspace V' such that $\{0\} \underset{\neq}{\subset} V' \underset{\neq}{\subset} V$.

By Maschke's theorem, V contains a subspace V'' which is a supplement of V', and invariant under $\rho(G)$. Since

$$\dim_F V' < \dim_F V, \qquad \dim_F V'' < \dim_F V,$$

we can apply the induction hypothesis to V' and V'' and we are done. □

Definition 3.5.4 *We say that the character of an irreducible representation is an* **irreducible character**.

Corollary 3.5.1 *The character of a representation of finite degree of a finite group G defined over a field F whose characteristic does not divide the order of G is a linear combination with positive integral coefficients of irreducible characters.*

Example 3.5.5 **(1)** In example (1), we had $V = D_1 \oplus D_2$. The representation

$$G = \langle c \rangle \xrightarrow{\rho_1} GL_{\mathbb{Q}}(D_1)$$

is the unit representation $c \mapsto \mathrm{id}_{D_1}$, and the representation

$$G \xrightarrow{\rho_2} GL_{\mathbb{Q}}(D_2)$$

is defined by $c \mapsto -\mathrm{id}_{D_2}$.

Let χ_1 and χ_2 be the respective characters of these representations. Then

$$\chi = \chi_1 + \chi_2.$$

(2) A particularly important example is the **regular representation** of G.

We construct the space $V = \oplus_{s \in G} F \overrightarrow{s}$, where $\overrightarrow{s}$ is a non-zero vector symbolically attached to $s \in G$, and we define the action of G on V by the formula

$$\rho(t)(\overrightarrow{s}) = \overrightarrow{ts}.$$

We thus obtain a representation

$$G \xrightarrow{\rho} GL_F(V)$$

of degree $g = \#G$.

Assuming that the characteristic of F does not divide g, we have

$$\chi_\rho = n_1\chi_1 + \cdots + n_h\chi_h, \tag{1}$$

where $\chi_1, \ldots, \chi_h$ are irreducible characters whose respective degrees we want to determine.

Proposition 3.5.3

(i) *The character χ_{reg} of the regular representation is given by*

$$\chi_{reg}(e) = g = \#G \quad and \quad \chi_{reg}(s) = 0 \qquad if\ s \neq e.$$

(ii) *If $s \in G \setminus \{e\}$, then*

$$\sum_{i=1}^{i=h} n_i \chi_i(s) = 0.$$

Proof. (ii) follows from (i) and (i) follows from the fact that the matrix of $\rho(s)$ in the basis $(\overrightarrow{t})_{t\in G}$ has no non-zero terms on its diagonal when $s \neq e$. □

When the base field F is the field of complex numbers $\mathbb{C}$, we can embed the characters in a vector space $\mathcal{H}$ equipped with a Hermitian product.

Definition 3.5.5 *The vector space of* **central functions** *$\mathcal{H}$ is defined by*

$$\mathcal{H} = \{f : G \longrightarrow \mathbb{C}; f(tst^{-1}) = f(s) \text{ for every } s \text{ and } t \in G\}.$$

For f and $g \in \mathcal{H}$, set

$$(f, g) := \frac{1}{\#G} \sum_{s\in G} f(s)\overline{g(s)}.$$

The essential results concerning $\mathcal{H}$ are collected in the following statement (see [Se 2] or [J-L]).

Theorem 3.5.3

(i) *The vector space $\mathcal{H}$ equipped with the above scalar product is a Hermitian space for which the irreducible characters $\chi_1, \ldots, \chi_h$ appearing in the decomposition (1) of the regular character form an orthonormal basis.*

(ii) *Two representations having the same character are isomorphic.*

Corollary 3.5.2 *Let χ be the character of a representation ρ of G. Then (χ, χ) is a positive or zero integer, and χ is irreducible if and only if $(\chi, \chi) = 1$.*

Proof. Let us write $\chi = m_1\chi_1 + \cdots + m_h\chi_h$. We have

$$(\chi, \chi) = m_1^2 + \cdots + m_h^2, \tag{2}$$

which is a positive or zero integer.

If χ is irreducible, then just one m_i must be different from zero, and it must be equal to 1. Indeed, if ρ' is the representation

$$m_1\rho_1 \oplus \cdots \oplus m_h\rho_h$$

where ρ_i admits the character χ_i and the direct sum means that the space of the representation is

$$V_1^{m_1} \oplus \cdots \oplus V_h^{m_h},$$

we see that ρ and ρ' have the same character. By (ii) of theorem 3.5.3, ρ' must be isomorphic to ρ, so it must be irreducible, and the space of the representation must be one of the V_i. □

Corollary 3.5.3 *The number of irreducible characters of G (defined over $\mathbb{C}$) is equal to the number of conjugacy classes of G.*

Proof. Indeed, corollary 1 shows that every irreducible character is a χ_i, with $1 \leq i \leq h$ and $h = \dim_{\mathbb{C}} \mathcal{H}$. Moreover it is clear that the characteristic functions of the conjugacy classes of G form a basis of $\mathcal{H}$. □

Corollary 3.5.4 *The degrees of the χ_i are equal to their multiplicity n_i in the decomposition (1) of the regular character.*

Proof. We deduce from formula (1) that we indeed have

$$n_i = (\chi_{\text{reg}}, \chi_i) = \frac{1}{g} \sum_{s \in G} \chi_{\text{reg}}(s)\overline{\chi_i(s)} = \frac{g}{g}\,\overline{\chi_i(e)}.$$

Thus $\chi_i(e) = n_i$. □

Corollary 3.5.5 *The following properties are equivalent:*

(i) *G is commutative.*

(ii) *All its irreducible representations (defined over $\mathbb{C}$) are of degree 1.*

Proof. **(i)** Let g be the order of G, h the number of its conjugacy classes and $n_1, \ldots, n_h$ the degrees of the different irreducible representations of G.

By formula (1), we know that we have

$$(\chi_{\text{reg}}, \chi_{\text{reg}}) = n_1^2 + \cdots + n_h^2.$$

But by definition, we also have

$$(\chi_{\text{reg}}, \chi_{\text{reg}}) = \frac{1}{g} \sum_{s \in G} \chi_{\text{reg}}(s)\overline{\chi_{\text{reg}}(s)}.$$

By Proposition 3.5.3, we see that the right-hand side of this equation is equal to g, so we obtain

$$g = n_1^2 + \cdots + n_h^2. \tag{3}$$

(ii) Since $n_i \geq 1$ for every i, we see that we have the equivalence

$$(g = h) \iff n_1 = n_2 = \cdots = n_h = 1.$$ □

Remark 3.5.1 **(1)** Thus, we see that the irreducible characters of G provide precious information about this group, since there are as many of them as there are conjugacy classes of G.

In particular, corollary 3.5.5 shows how they can be used to characterise commutative groups (without distinguishing between their isomorphism classes, when the orders are equal).

More generally, the number of irreducible characters of degree 1 of G is equal to the index $(G : G')$, where G' denotes the commutator subgroup of G (see Exercise 3.26).

(2) The theory of linear representations of finite groups can also be presented as the theory of modules of finite type defined over the **group algebra** of G.

The underlying F-vector space V of this algebra $F[G]$ is the space of the regular representation, and its multiplication table is given simply by

$$\vec{s} \cdot \vec{t} = \vec{st}\,.$$

Thus, we see that $\vec{e}$ is a unit of $F[G]$ and that $F[G]$ is associative.

The regular representation is just the representation of G by "left multiplication" in $F[G]$;

$$\rho \begin{cases} G \longrightarrow GL_F(F[G]) \\ t \longmapsto (h_{\vec{t}} : \vec{x} \longmapsto \vec{t} \cdot \vec{x}). \end{cases}$$

An arbitrary representation ρ is just an $F[G]$-**module structure over its representation space**. This explains why the regular representation is particularly important (its space is just $F[G]$).

(3) When the characteristic of F divides $\#G$, the theory is much more difficult (see [Se 2]).

3.6 GALOIS EXTENSIONS

The goal of this second-to-last section is to determine and study the finite extensions of $\mathbb{Q}$ contained in $\bar{\mathbb{Q}}$, which are invariant under every automorphism of $\bar{\mathbb{Q}}$. It is obvious that there exist a great many such extensions (it suffices to adjoin all the roots of any irreducible polynomial $P(X) \in \mathbb{Q}[X]$ to $\mathbb{Q}$), but what are the representations of Aut $(\bar{\mathbb{Q}})$ which we then obtain?

In order to obtain statements adaptable to a varied set of situations, we need to extend our conceptual framework. Fix an extension $K \subset \Omega$.

Definition 3.6.1 *Let $K \subset \Omega$ be a field extension. The* **Galois group** Gal(Ω/K) *of the extension Ω/K is the group of K-automorphisms of Ω. Thus, we have*

$$\mathrm{Gal}(\Omega/K) = \{\sigma \in \mathrm{Aut}(\Omega);\ \sigma(x) = x \textit{ for every } x \in K\}.$$

Example 3.6.1 If $K = P_\Omega$, we see that $\mathrm{Gal}(\Omega/K) = \mathrm{Aut}\,\Omega$.

3.6.1 The Galois Correspondence

Let $\mathcal{K}$ (resp. $\mathcal{G}$) denote the set of fields L such that $K \subset L \subset \Omega$ (resp. of groups H such that $\{\mathrm{id}_\Omega\} \subset H \subset \mathrm{Gal}(\Omega/K)$). We propose to show that $\mathcal{K}$ and $\mathcal{G}$ correspond to each other via a certain duality which resembles the theory of orthogonality in linear algebra.

Definition 3.6.2 Let $H \in \mathcal{G}$.
The **field of invariants** *of H in Ω is defined by*

$$\Omega^H = \{x \in \Omega;\ \sigma(x) = x \text{ for every } \sigma \in H\}.$$

Now consider the two following maps:

$$\begin{cases} \mathcal{K} \xrightarrow{\varphi} \mathcal{G} \\ L \longmapsto L^0 = \mathrm{Gal}(\Omega/L) \end{cases} \qquad \begin{cases} \mathcal{G} \xrightarrow{\psi} \mathcal{K} \\ H \longrightarrow {}^0H = \Omega^H. \end{cases}$$

Example 3.6.2

$$\varphi(\Omega) = \{\mathrm{id}_\Omega\}, \quad \varphi(K) = \mathrm{Gal}(\Omega/K), \quad \psi(\{\mathrm{id}_\Omega\}) = \Omega, \quad \psi(\mathrm{Gal}(\Omega/K)) = ?$$

Proposition 3.6.1

(i) *φ and ψ are decreasing maps for the inclusion relations in $\mathcal{K}$ and $\mathcal{G}$.*
(ii) *For every $L \in \mathcal{K}$, we have ${}^0(L^0) = \psi \circ \varphi(L) \supset L$, and for every $H \in \mathcal{G}$, we have $({}^0H)^0 = \varphi \circ \psi(H) \supset H$.*
(iii) *Moreover, we have $\varphi \circ \psi \circ \varphi = \varphi$ and $\psi \circ \varphi \circ \psi = \psi$.*

Proof. As the proof is the same for φ and ψ, we give it only for φ.

(i) If $L_1 \subset L_2$, then

$$L_2^0 = \{\sigma \in \mathrm{Gal}(\Omega/K);\ \sigma(x) = x \text{ for every } x \in L_2\} \subset L_1^0.$$

(ii) Let us show that $L \subset {}^0(L^0)$.
It suffices to see that for every $x \in L$, we have

$$\sigma(x) = x \quad \text{for every } \sigma \in L^0.$$

But this follows from the definition of L^0.

(iii) Let us show that $\varphi \circ \psi \circ \varphi = \varphi$, or that for every $L \in \mathcal{K}$, we have

$$\varphi\left({}^0(L^0)\right) = L^0.$$

By (ii), we have ${}^0(L^0) \supset L$, so by (i), $\varphi({}^0(L^0)) \subset \varphi(L) = L^0$.
Now set $L^0 = H$; by (ii) we have

$$\left({}^0H\right)^0 \supset H,$$

which is the inverse inclusion. □

Corollary 3.6.1 *Let $\mathcal{K}'$ and $\mathcal{G}'$ be the images of ψ and φ.*

Then φ and ψ induce mutually inverse, decreasing bijections of $\mathcal{K}'$ onto $\mathcal{G}'$.

Corollary 3.6.2 *A necessary and sufficient condition for $H \in \mathcal{G}$ to lie in $\mathcal{G}'$ is that $\varphi(\psi(H)) = H$.*

A necessary and sufficient condition for $L \in \mathcal{K}$ to lie in $\mathcal{K}'$ is that $\psi(\varphi(L)) = L$.

Proof. These conditions are clearly sufficient, and they are also necessary by corollary 3.6.1. □

Definition 3.6.3 *If $H \in \mathcal{G}'$, we say that H is a* **stationary** *subgroup of* Gal(Ω/K).

If $L \in \mathcal{K}'$, we say that L is a **stationary** *subfield of Ω/K.*

We say that the extension Ω/K is **Galois** *if it is algebraic and K is a stationary subfield of Ω/K.*

Remark 3.6.1 Thus, in order for the extension Ω/K to be Galois, it is necessary and sufficient for it to be algebraic and to have $\psi \circ \varphi(K) = K$. As $\varphi(K) = \text{Gal}(\Omega/K)$, this means that

$$\Omega^{\text{Gal}(\Omega/K)} = \{x \in \Omega;\ \sigma(x) = x \text{ for every } \sigma \in \text{Gal}(\Omega/K)\} = K.$$

This gives the following important criterion.

Criterion 3.6.1 *The extension Ω/K is Galois if and only if it is algebraic and if for every $x \in \Omega \setminus K$, there exists $\sigma \in$ Gal(Ω/K) such that $\sigma(x) \neq x$.*

Example 3.6.3 We always have $\{e\} \in \mathcal{G}'$, but we do not always have $K \in \mathcal{K}'$. For example, $\mathbb{Q}$ is not stationary in $\mathbb{R}/\mathbb{Q}$ since Aut $(\mathbb{R}) = \{\text{id}_{\mathbb{R}}\}$.

Indeed, every $\sigma \in$ Aut $(\mathbb{R})$ is strictly increasing (if $y > x$ we have $y - x = z^2$ with $z \in \mathbb{R}^*$, so $\sigma(y) - \sigma(x) = \sigma(z)^2 > 0$). Now, for every $x \in \mathbb{R}$ and every $n \in \mathbb{N}$, there exists $r \in \mathbb{Q}$ such that

$$r < x < r + 10^{-n}.$$

Since $\mathbb{Q}$ is the prime field of $\mathbb{R}$, it follows that

$$r < \sigma(x) < r + 10^{-n},$$

hence $|\sigma(x) - x| < 10^{-n}$ for every n, so $\sigma(x) = x$.

Proposition 3.6.2 *Let K be a field of characteristic zero, and let Ω an algebraic closure of K.*

Then the extension Ω/K is Galois.

Proof. We will apply criterion 3.6.1.

Let $x \in \Omega \setminus K$; then x is algebraic over K.

Let $P(X) \in K[X]$ be the minimal polynomial of x over K; then $\deg P > 1$ (otherwise we would have $x \in K$) and all its roots in Ω are simple because the characteristic is zero.

Let $y \neq x$ be a second root of $P(x)$ in Ω. We know that the map

$$\sigma : Q(x) \longmapsto Q(y),$$

where $Q(X) \in K[X]$, is a K-isomorphism of the field $K(x)$ onto the field $K(y)$.

By Steinitz' theorem, σ extends to an isomorphism τ of Ω, and we have

$$\tau \in \mathrm{Gal}(\Omega/K) \quad \text{and} \quad \tau(x) \neq \tau(y). \qquad \square$$

Examples and counterexamples 3.6.4

(1) Although $\bar{\mathbb{Q}}/\mathbb{Q}$ is Galois, $(\mathbb{R} \cap \bar{\mathbb{Q}})/\mathbb{Q}$ is not, for the reason explained in the above example.

(2) Although $\mathbb{Q}$ is stationary in $\mathbb{C}/\mathbb{Q}$, this last extension is not algebraic and cannot be Galois (in the sense introduced above).

3.6.2 Questions of Dimension

As in linear algebra, this duality becomes much clearer when we make a finiteness hypothesis on the degrees of the extensions or the indices of the groups.

Theorem 3.6.1 *Let $K \subset L_1 \subset L_2 \subset \Omega$ be a tower of extensions with $[L_2 : L_1] = n < \infty$. Then we have $(\varphi(L_1) : \varphi(L_2)) \leq n$.*

Proof. We use induction on n.

(1) When $n = 1$, the property is obvious.

(2) Suppose that there exists an intermediate field L' such that $L_1 \underset{\neq}{\subset} L' \underset{\neq}{\subset} L_2$. Then the induction hypothesis gives

$$\begin{cases} (\varphi(L_1) : \varphi(L')) \leq [L' : L_1] \\ (\varphi(L') : \varphi(L_2)) \leq [L_2 : L'], \end{cases}$$

and, multiplying these inequalities term by term, we obtain the result.

(3) If there exists no intermediate field between L_1 and L_2, we necessarily have $L_2 = L_1(a)$ for every $a \in L_2 \setminus L_1$.

The isotropy group of a in the action of $\varphi(L_1)$ on L_2 is equal to $\varphi(L_2)$, since the action of $\sigma \in \varphi(L_1)$ is determined by the knowledge of $\sigma(a)$. Thus, we see that the cardinal of the orbit of a, under the action of $\varphi(L_1)$ is equal to $(\varphi(L_1) : \varphi(L_2))$. Let $P(X)$ be the irreducible polynomial of a on L_1; we know that $\deg P(X) = [L_2 : L_1]$.

But the orbit of a is contained in the set of roots of $P(X)$, so

$$(\varphi(L_1) : \varphi(L_2)) \leq [L_2 : L_1]. \qquad \square$$

Theorem 3.6.1 admits a symmetric statement which we give here.

Theorem 3.6.2 *Let* $\{e\} \subset H_1 \subset H_2 \subset \mathrm{Aut}(\Omega)$ *with* $(H_2 : H_1) = n < \infty$. *Then we have*

$$[\psi(H_1) : \psi(H_2)] \leq n.$$

Proof. **(1)** Note first that the left cosets of H_2 modulo H_1 act on $\psi(H_1)$.

Indeed, let $C = \sigma H_1$ be a left coset, and let $x \in \psi(H_1)$; then for every $\tau \in C$, we have $\tau = \sigma h$ with $h \in H_1$, hence $\tau(x) = \sigma(h(x)) = \sigma(x)$. We write $C(x)$ for this common value $\sigma(x)$.

(2) We reason by the absurd.

If $[\psi(H_1) : \psi(H_2)] > n$, there exist $a_1, \ldots, a_{n+1} \in \psi(H_1)$, linearly independent over $\psi(H_2)$.

Now consider the following equations in $(\lambda_1, \ldots, \lambda_{n+1}) \in \psi(H_1)^{n+1}$:

$$\begin{cases} \lambda_1 C_1(a_1) + \cdots + \lambda_{n+1} C_1(a_{n+1}) = 0 \\ \vdots \qquad\qquad\qquad\qquad \vdots \\ \lambda_1 C_n(a_1) + \cdots + \lambda_{n+1} C_n(a_{n+1}) = 0, \end{cases} \tag{1}$$

where $C_1, \ldots, C_n$ are the n left cosets of H_2 modulo H_1.

The system (1) contains $n + 1$ unknowns and n equations: thus it admits a non-trivial solution in $\psi(H_1)^{n+1}$. Up to changing the notation, we can assume that it is

$$(1, \lambda_2, \ldots, \lambda_r, 0, \ldots, 0), \tag{2}$$

with the **smallest possible** value for r.

The rest of the proof is based on the following trick: if we show that for every $\sigma \in H_2$, $(1, \sigma(\lambda_2), \ldots, \sigma(\lambda_r), 0, \ldots, 0)$ is a solution of (1), then we will have

$$\sigma(\lambda_2) = \lambda_2, \ldots, \sigma(\lambda_r) = \lambda_r;$$

otherwise the difference of these solutions would also be a non-trivial solution of (1), and it would admit a greater number of zeros than (2), which is impossible. Thus all the λ_i would be invariant under H_2, and $a_1, \ldots, a_{n+1}$ would not be linearly independent over $\psi(H_2)$, which is absurd.

(3) Thus, it suffices to see that for every $\sigma \in H_2$, $(\sigma(\lambda_1), \ldots, \sigma(\lambda_{n+1}))$ is a solution of (1).

Let us apply σ to the equations of (1); we obtain

$$\begin{cases} \sigma(\lambda_1)\sigma C_1(a_1) + \cdots + \sigma(\lambda_{n+1})\sigma C_1(a_{n+1}) = 0 \\ \vdots \qquad\qquad\qquad\qquad\qquad \vdots \\ \sigma(\lambda_1)\sigma C_n(a_1) + \cdots + \sigma(\lambda_{n+1})\sigma C_n(a_{n+1}) = 0. \end{cases} \tag{3}$$

As the **set** $\{\sigma C_1, \ldots, \sigma C_n\}$ is exactly $\{C_1, \ldots, C_n\}$, we obtain the **same** system (up to the order of the equations), and the proof is complete. □

Corollary 3.6.3

(i) *Let* $K \subset L_1 \subset L_2 \subset \Omega$.
If L_1 *is stationary and if* $[L_2 : L_1] = n < \infty$, *then* L_2 *is stationary and we have* $(\varphi(L_1) : \varphi(L_2)) = n$.

(ii) *Let* $\{e\} \subset H_1 \subset H_2 \subset \mathrm{Gal}(\Omega/K)$.
If H_1 *is stationary and if* $(H_2 : H_1) = n < \infty$, *then* H_2 *is stationary and we have* $[\psi(H_1) : \psi(H_2)] = n$.

Proof. We only prove (i) here.
From the above theorems, we deduce that

$$[L_2 : L_1] \geq (\varphi(L_1) : \varphi(L_2)) \geq [\psi \circ \varphi(L_2) : \psi \circ \varphi(L_1)].$$

By hypothesis, $\psi \circ \varphi(L_1) = L_1$, so

$$[L_2 : L_1] \geq [\psi \circ \varphi(L_2) : L_1].$$

As $\psi \circ \varphi(L_2) \supset L_2$, they are equal. □

Corollary 3.6.4

(i) *If the extension* $K \subset \Omega$ *is* **Galois**, *then for every* **finite** *extension* L *of* K *contained in* Ω, $L \subset \Omega$ *is* **Galois** *and we have* $[L : K] = (G : \varphi(L))$.

(ii) *All the* **finite** *subgroups of* $\mathrm{Gal}(\Omega/K)$ *are stationary, and if* $H \in \mathcal{G}$ *is finite,* Ω/Ω^H *is* **Galois** *of group* H.

Proof. For (i), we take $L_1 = K$, and for (ii), $H_1 = \{e\}$. □

Fundamental Theorem of Galois Theory 3.6.3 *Let* $K \subset \Omega$ *be a* finite **Galois** *extension.*

(i) *Then we have*

$$[\Omega : K] = \#\mathrm{Gal}(\Omega/K).$$

(ii) *Let* $K \subset L_1 \subset L_2 \subset \Omega$ *be a tower of extensions; we have*

$$(\varphi(L_1) : \varphi(L_2)) = [L_2 : L_1].$$

(iii) *Let* $\{e\} \subset H_1 \subset H_2 \subset \mathrm{Gal}(\Omega/K)$ *be a tower of groups; we have*

$$[\psi(H_1) : \psi(H_2)] = (H_2 : H_1).$$

Criterion 3.6.2 *Let* Ω/K *be a* finite *field extension.*
The extension Ω/K *is* **Galois** *if and only if we have*

$$[\Omega : K] = \#\mathrm{Gal}(\Omega/K).$$

Proof.

(1) The condition is necessary by the fundamental theorem.

(2) Let us show that it is sufficient. Set $L := \psi(G)$ with $G := \mathrm{Gal}(\Omega/K)$; then Ω/L is Galois by construction. But we have $[\Omega : L] = (G : \{e\})$, so $[\Omega : L] = [\Omega : K]$, which implies that $L = K$. □

3.6.3 Stability

In spite of its admirable æsthetic qualities the above result does not really explain how to go about finding all the subextensions of $K \subset \Omega$ which are stable under every automorphism of $\mathrm{Gal}(\Omega/K)$.

Definition 3.6.4 *Let $K \subset L \subset \Omega$ be a tower of extensions.*
We say that L is **stable** *relative to Ω/K if for every $\sigma \in \mathrm{Gal}(\Omega/K)$, we have $\sigma(L) = L$.*

Proposition 3.6.3 *Let $K \subset L \subset \Omega$ be a tower of extensions, and let $G = \mathrm{Gal}(\Omega/K)$.*

(i) *If L is* **stable** *relative to Ω/K, then $\varphi(L)$ is a* **normal** *subgroup of G.*

(ii) *If H is a* **normal** *subgroup of G, then $\psi(H)$ is* **stable** *relative to Ω/K.*

Proof.

(i) Let $\sigma \in \varphi(L)$; we must show that for every $\tau \in G$, we have $\tau\sigma\tau^{-1} \in \varphi(L)$. This means that for every $x \in L$, $\tau\sigma\tau^{-1}(x) = x$, or rather $\sigma(\tau^{-1}(x)) = \tau^{-1}(x)$. As $\tau^{-1}(x) \in L$ by hypothesis, this is obvious.

(ii) Let $x \in \psi(H)$; we must show that for every $\tau \in G$, we have $\tau(x) \in \psi(H)$. This means that for every $\sigma \in H$, we must have $\sigma(\tau(x)) = \tau(x)$, i.e. $\tau^{-1}\sigma\tau(x) = x$. But since $\tau^{-1}\sigma\tau \in H$ by hypothesis, this is again obvious. □

Proposition 3.6.4 *Let $K \subset L \subset \Omega$ be a tower of field extensions. If Ω/K is Galois and if L is stable, then L/K is also Galois.*

Proof. We need to show that $\psi \circ \varphi(K) = K$ in the extension L/K, so that if $x \in L\backslash K$, there exists $\bar{\sigma} \in \mathrm{Gal}(L/K)$ such that $\bar{\sigma}(x) \neq x$.

But since Ω/K is Galois, there exists $\sigma \in \mathrm{Gal}(\Omega/K)$ such that $\sigma(x) \neq x$; thus it suffices to take $\bar{\sigma} = \sigma|_L$, which is indeed a K-automorphism of L, since L is stable. □

3.6.4 Conclusions

We will now give a more precise description of the Galois extensions.

Proposition 3.6.5 *If Ω/K is Galois and if $P(X) \in K[X]$ is an irreducible polynomial having a root $x \in \Omega$, then $P(X)$ decomposes into a product of distinct linear factors in $\Omega[X]$.*

Proof. Let $\{x_1, \ldots, x_r\}$ be the orbit of x under the action of $G := \mathrm{Gal}(\Omega/K)$: all the x_i are roots of $P(X)$, so we have $r \leq \deg(P)$.

Set $Q(X) = (X - x_1) \ldots (X - x_r) \in \Omega[X]$.

As the coefficients of $Q(X)$ are invariant under the action of G and Ω/K is Galois, they lie in K.

Now, $P(X)$ and $Q(X)$ are two polynomials in $K[X]$, both of which have x as a root. As $P(X)$ is irreducible, $P(X)$ must divide $Q(X)$, and $r \geq \deg P(x)$. □

Corollary 3.6.5 *If Ω/K is Galois and if $P(X) \in K[X]$ is an irreducible polynomial which has a root in Ω, then all the roots of $P(X)$ are simple.*

Corollary 3.6.6 *Let $K \subset L \subset \Omega$ be a tower of extensions; assume L/K is Galois. Then L is stable relative to Ω/K.*

Proof. Let $x \in L$ and $\sigma \in \mathrm{Gal}(\Omega/K)$; we must show that $\sigma(x) \in L$. This can be seen by applying proposition 3.6.5 to the irreducible polynomial of x over K. □

Theorem 3.6.4 *Let Ω/K be a* **finite Galois** *extension of fields with Galois group G, and let L be such that $K \subset L \subset \Omega$. Then*

(i) *L/K is Galois if and only if* $\mathrm{Gal}(\Omega/L)$ *is a normal subgroup of G.*
(ii) *In this case,* $\mathrm{Gal}(L/K) \cong G/\mathrm{Gal}(\Omega/L)$.

Proof. **(i)** Since Ω/K finite, L/K is as well; thus it is algebraic.

If L/K is Galois, corollary 3.6.3 shows that L/K is stable relative to Ω/K, and proposition 3.6.3 shows that $H = \varphi(L)$ is a normal subgroup of G.

Conversely, we know that $L = \psi(H)$ since L is stationary (since Ω/K is finite Galois). If moreover H is a normal subgroup of G, then L is stable by proposition 3.6.3 (ii). And by proposition 3.6.4, L/K is then Galois.

(ii) Consider the homomorphism h given by

$$\begin{cases} G \xrightarrow{h} \mathrm{Gal}(L/K) \\ \sigma \longmapsto \sigma|_L \end{cases}.$$

It is clear that $\mathrm{Ker}\, h = H$, so $\mathrm{Im} h \cong G/H$.

It remains only to show that h is surjective, which follows from the fundamental theorem of Galois theory on the one hand, since

$$[L : K] = (\varphi(K) : \varphi(L)) = (G : H) = \# \operatorname{Im} h,$$

and from criterion 3.6.2 on the other hand, since

$$[L : K] = \#\mathrm{Gal}(L/K).$$

□

Theorem 3.6.5 *Let Ω be an algebraic closure of K, and let $K \subset L \subset \Omega$ with $[L : K] < \infty$. Then L/K is Galois if and only if L is generated by the roots of a polynomial $Q(X) \in K[X]$ all of whose roots are simple (i.e. a "separable" polynomial).*

Proof. **(1)** Let us show that the condition is necessary.

Since L/K is finite, we have $L = K(\alpha_1, \dots, \alpha_r)$ with α_i algebraic over K. Let $P_1, \dots, P_r$ be the irreducible polynomial of $\alpha_1, \dots, \alpha_r$; we may assume that the s first $P_1, \dots, P_s$ are distinct. We then see that L/K is generated by the roots of $P_1, \dots, P_s$, since these decompose completely in $L[X]$ by proposition 3.6.5 and their roots are simple (corollary 3.6.3). Thus we take $Q = P_1 \cdots P_s$.

(2) We will prove that the condition is sufficient by induction on $[L : K] = n$, but without assuming that K is fixed in Ω.

Note first that if $L = K(\alpha_1, \dots, \alpha_r)$, where the α_i are the roots of $Q(X) \in K[X]$, then L is stable in Ω/K.

If $[L : K] = 1$, there is nothing to prove.

Assume that the assertion holds for all the K' and L' such that $[L' : K'] < n$, and consider an irreducible polynomial $P(X) = (X - \alpha_1) \cdots (X - \alpha_s)$ which divides $Q(X)$ in $K[X]$.

If $s = n$, then $L = K(\alpha_1) = \cdots K(\alpha_n)$ and as there exists $\sigma_i \in \mathrm{Gal}(\Omega/K)$ such that $\sigma_i(\alpha_1) = \alpha_i$ for $1 \leq i \leq n$, we see that $\#\mathrm{Gal}(L/K) \geq [L : K]$. Criterion 3.6.2 then shows that L/K is Galois.

If $s < n$, we set $K' = K(\alpha_1)$.

Then $[L : K'] = r < n$, so L/K' is Galois by the induction hypothesis: there exist r distinct automorphisms $\tau_1, \dots, \tau_r$ of $\mathrm{Gal}(L/K')$ such that $\tau_j(\alpha_1) = \alpha_1$ and $\tau_j|_K = \mathrm{id}_K$.

But we also know (by Steinitz' theorem) that there exist s automorphisms ρ_i of Ω such that $\rho_i(\alpha_1) = \alpha_i$ for $1 \leq i \leq s$.

Since L is stable, they induce distinct automorphisms σ_i of L such that $\sigma_i(\alpha_1) = \alpha_i$.

Thus, we see that $\mathrm{Gal}(L/K)$ contains at least $n = rs$ elements, namely the $\tau_j \circ \sigma_i$, which are all distinct.

Indeed, if $i \neq i'$, it is clear that $\tau_j \circ \sigma_i \neq \tau_j \circ \sigma_i$, and if $i = i'$ but $j \neq j'$, it is also clear that $\tau_j \circ \sigma_i \neq \tau_{j'} \circ \sigma_i$.

Criterion 3.6.2 then implies that L/K is Galois. □

Counterexample 3.6.5 Let t be transcendental over $\mathbb{F}_p$. We take $K = \mathbb{F}_p(t)$, $\Omega = \overline{\mathbb{F}_p(t)}$ and $Q(X) = X^p - t \in K[X]$.

The polynomial $Q(X)$ is irreducible, but has only one root in Ω, since if $\alpha^p = t$, then $Q(X) = (X - \alpha)^p$. Proposition 3.6.5 then implies that $K(\alpha)/K$ cannot be Galois.

Example 3.6.6 Let p be a prime number. We saw in Chapter 1 that the cyclotomic polynomial

$$\Phi_p(X) = X^{p-1} + \cdots + 1 \in \mathbb{Q}[X]$$

is irreducible.

Let ζ be a root of this polynomial in $\bar{\mathbb{Q}}$; then we know that ζ generates the group $\mu_p(\bar{\mathbb{Q}})$ of the p-th roots of unity. Theorem 3.6.5 then shows that the extension $\mathbb{Q}(\zeta)/\mathbb{Q}$ is Galois. Furthermore, $\mathrm{Gal}(\bar{\mathbb{Q}}/\mathbb{Q})$ acts on $\langle\zeta\rangle$, and $\langle\zeta\rangle = \{\zeta^n; n \in \mathbb{Z}/p\mathbb{Z}\}$ can be considered as a line D in the field $\mathbb{F}_p$.

We thus obtain a natural representation

$$\mathrm{Gal}(\bar{\mathbb{Q}}/\mathbb{Q}) \xrightarrow{\rho_p} GL_{\mathbb{F}_p}(D),$$

whose character is called the cyclotomic character χ_p.

The representation of the preceding example has kernel equal to the group $\varphi(\mathbb{Q}(\zeta)) = \{\sigma \in \mathrm{Aut}\,\bar{\mathbb{Q}}; \sigma(\zeta) = \zeta\}$, which is a subgroup of **finite** index (equal to $p-1$) in $G_{\mathbb{Q}} := \mathrm{Aut}\,(\bar{\mathbb{Q}})$. This suggests the following vocabulary.

Definition 3.6.5 Let K be an arbitrary field and $\bar{K}$ an algebraic closure of K.
A **representation** $\mathrm{Gal}(\bar{K}/K) \xrightarrow{\rho} GL_F(V)$ *is said to be* **continuous** *if its kernel is a subgroup of finite index of* $\mathrm{Gal}(\bar{K}/K)$.

By the above, we see that if we let H denote the kernel of ρ, the group H is a normal subgroup of $\mathrm{Gal}(\bar{K}/K)$ and $\psi(H)/K$ is a finite Galois extension: we say that ρ factors through $\psi(H)$.

3.7 RESOLUTION OF ALGEBRAIC EQUATIONS

The complete investigation of the resolution of algebraic equations falls well outside the scope of this book. We restrict ourselves to simply stating some general principles which we will then apply to equations of degree three.

3.7.1 Some General Principles

Let K be a commutative field of characteristic p, equal either to a prime number or to zero, and let $P(X) \in K[X]$ be an irreducible polynomial of degree n. Take an algebraic closure Ω of K, and consider the smallest subfield L of Ω which contains K and all the roots of $P(X)$ in Ω. If these roots are distinct, we saw (Section 3.6, theorem 3.6.5) that L/K is Galois; we denote its Galois group by G.

Example 3.7.1 **(1)** Let $P(X) = X^3+a_2X^2+a_4X+a_6 \in K[X]$ be an irreducible polynomial of degree 3. Then the roots x_1, x_2, x_3 of $P(X)$ are distinct if and only if $P(X)$ and $P'(X) = 3X^2 + 2a_2X + a_4$ have no common root, i.e. if and only if $P(X)$ does not divide $P'(X)$.

This last condition would be possible only if $P'(X) \equiv 0$, i.e. $p = \mathrm{characteristic}(K) = 3$ *and* $a_2 = a_4 = 0$. The last case occurs, for example, when $P(X) = X^3 + pX + q \in \mathbb{F}_3(t)[X]$ and $p = 0$; Eisenstein's criterion shows that $P(X)$ is irreducible if $q \in \mathbb{F}_3[t]$ and $v_t(q) = 1$.

(2) In any case, $[L : K] \in \{3, 6\}$ since we have a tower of extensions $K \subset K(x_1) \subset L$ and $[L : K(x_1)] \in \{1, 2\}$.

If L/K is not separable $[L : K] = 3$, but we will see that we can have $[L : K] = 3$ even if L/K is separable.

Theorem 3.7.1 *Assume that L/K is Galois, and let $S = \{x_1, \ldots, x_n\}$ denote the set of roots of $P(X)$ in Ω. Let $\mathfrak{S}_n$ denote the symmetric group of S. Then the map*

$$G \xrightarrow{\theta} \mathfrak{S}_n$$
$$\sigma \longmapsto \sigma|_S$$

is an injective morphism of G **to** *$\mathfrak{S}_n$. Moreover, the image of G acts* **transitively** *on S.*

Proof. The only non-trivial assertion is the transitivity of the action of $\varphi(G)$.

Let x_i and x_j be two elements of S. We know that

$$K(x_i) \cong K[X]/(P) \cong K(x_j).$$

Thus, there exists a K-isomorphism from $K(x_i)$ onto $K(x_j)$. By Steinitz' theorem, this isomorphism extends to a K-automorphism of Ω, and its restriction to L is an element σ in G such that $\sigma(x_i) = x_j$. □

Example 3.7.2 If $n = 3$ and if L/K is Galois, then G is isomorphic to a subgroup of order 3 or 6 of $\mathfrak{S}_3$. Thus we have either $G \cong \mathfrak{S}_3$ or $G \cong \mathfrak{A}_3$ (they both act transitively on S!), where $\mathfrak{A}_3$ denotes the alternating group of order 3.

Corollary 3.7.1 *If L/K is Galois, then n divides $\#G$.*

Definition 3.7.1 *The* **discriminant** *of the polynomial $P(X)$ is the product*

$$D = -1^{(n(n-1))/2} \prod_{i \neq j} (x_i - x_j).$$

Remark 3.7.1 Since D is a symmetric function of the roots of $P(X)$, the theorem of symmetric functions tells us that

$$D \in P_K[\text{coefficients of } P],$$

where P_K denotes the prime subfield of K.

If the characteristic of K is zero, we even have

$$D \in \mathbb{Z}[\text{coefficients of } P].$$

Example 3.7.3 If $P(X) = X^3 + a_2X^2 + a_4X + a_6$, then

$$D = P'(x_1)P'(x_2)P'(x_3) = -4a_2^3a_6 + a_2^2a_4^2 + 18a_2a_4a_6 - 4a_4^3 - 27a_6^2. \tag{1}$$

In particular, the discriminant of $X^3 + pX + q$ is given by

$$D = -(27q^2 + 4p^3). \tag{2}$$

Theorem 3.7.2 *Assume that $p \neq 2$ and that L/K is Galois. Then $\theta(G) \subset \mathfrak{A}_n$ if and only if $D \in {}^2K^*$ (i.e. D is a square in K^*).*

Proof. Set

$$E = \prod_{i<j}(x_i - x_j) \in L,$$

so that $E^2 = D$.

Let $\sigma \in G$ and let $\varepsilon(\sigma)$ be the signature of $\theta(\sigma) \in \mathfrak{S}_n$. Then

$$\sigma(E) = \varepsilon(\sigma)E,$$

so $\theta(G) \subset \mathfrak{A}_n$ is equivalent to $E \in K$. □

Corollary 3.7.2 *Assume that $p \neq 2$ and $P(X) = X^3 + a_2X^2 + a_4X + a_6$ is separable. Then the following propositions are equivalent:*

- **(i)** $G = \mathfrak{A}_3$
- **(ii)** $[L : K] = 3$
- **(iii)** *$K(x_i)/K$ is Galois for $i \in \{1, 2, 3\}$.*

Example 3.7.4 The extension $\mathbb{Q}(\sqrt[3]{2})/\mathbb{Q}$ is not Galois, since the discriminant of $X^3 - 2$ is -108, which is not the square of a rational number (it was obvious anyway!).

Definition 3.7.2 *Let G be a group. We say that G is* **solvable** *if there exists a chain of subgroups*

$$G = G_0 \supset G_1 \supset G_2 \supset \cdots \supset G_\ell = \{e\}$$

such that G_{i+1} is **normal** *in G_i and G_i/G_{i+1} is* **Abelian**, *and this holds for all $i \in \{0, \ldots, \ell - 1\}$.*

The next result is the core of this section.

Galois' Theorem 3.7.3 *Let K be a field of characteristic zero. The equation $P(X) = 0$ is solvable by radicals if and only if the Galois group G of L/K is solvable.*

Remark 3.7.2

- **(1)** This theorem can be extended to fields of positive characteristic as long as their characteristic does not divide the orders of the radicals or the indices $[G_i : G_{i+1}]$.
- **(2)** For a proof, see [Kap] pp. 32–36 or [VW1] pp. 172–175.

Example 3.7.5 Every equation of degree three is solvable by radicals. Indeed, for $P(X)$ irreducible, it is enough to note that $\mathfrak{A}_3$ and $\mathfrak{S}_3$ are solvable groups.

Corollary (Ruffini [35]–Abel[36] Theorem) 3.7.4 *Let K be a field and $u_1, u_2, \ldots, u_n$ indeterminates. Then the general equation of degree n,*

$$P(X) = X^n + u_1 X^{n-1} + \cdots + u_n = 0,$$

is not solvable by radicals when $n \geq 5$.

Proof. Indeed, the Galois group of this equation, which is isomorphic to $\mathfrak{S}_n$, is not solvable when $n \geq 5$ (see [Tau] p. 41). □

Remark 3.7.3

(1) What we give here is Galois' proof of the Ruffini–Abel theorem. For the original proof, see [Ab].

(2) Note that if $n \leq 4$, then the general equation of degree n is solvable by radicals, since $\mathfrak{S}_n$ is solvable.

3.7.2 Resolution of the Equation of Degree Three

Let us give a brief presentation of the resolution of this equation by a method attributed to **Tartaglia**[37] and **Cardan**[38], although apparently this resolution is really due to **Scipione del Ferro**[39].

This method can be applied to the general polynomial

$$X^3 + a_2 X^2 + a_4 X + a_6 \tag{3}$$

when the characteristic of the field K **is not equal to 2 or 3**.

The translation $X \mapsto X - a_2/3$ brings this equation to the form

$$X^3 + pX + q. \tag{4}$$

Since the characteristic of K is not equal to 3, $\bar{K}$ contains two primitive cubic roots of unity, which we denote by j and j^2.

Scipione del Ferro and his disciples remarked that we have the identity

$$(X - u - v)(X - uj - vj^2)(X - uj^2 - vj) = X^3 - 3uvX - (u^3 + v^3), \tag{5}$$

[35] P. Ruffini (1765–1822).
[36] N.H. Abel (1802–1829).
[37] N. Tartaglia (1500–1557).
[38] G. Cardan (1501–1576).
[39] Scipione del Ferro (1465–1526).

and they were led to solve the equations

$$3uv = -p, \qquad u^3 + v^3 = -q.$$

We see that u^3 and v^3 must be roots of the equation

$$Y^2 + qY - \left(\frac{p}{3}\right)^3 = 0. \tag{6}$$

Thus, if the characteristic of K is not equal to 2, we have

$$u = \sqrt[3]{\frac{-q \pm \sqrt[2]{q^2 + 4(p/3)^3}}{2}}, \qquad v = -\frac{p}{3u}, \tag{7}$$

and we obtain the three roots of (5).

Remark 3.7.4 A priori, it seems that this method should give too many solutions. But the indetermination of the root of u^3 only has the effect of cyclically permuting the three roots of (5), and the indetermination of the $\pm$ in (7) only permutes u and v and exchanges the last two roots of (5).

Explanation of this resolution when the characteristic of K does not divide 6.
We saw in the first part of this section that if (3) is irreducible over K, its Galois group is either $\mathfrak{A}_3$ or $\mathfrak{S}_3$, according to whether D is or is not a square in K^*.

Let us begin by adjoining $\sqrt{D}$ to K; set $R = K(\sqrt{D}) \subset \bar{K}$. The polynomial (3) remains irreducible over R since $[R : K]$ is equal to 1 or 2, and its Galois group is now $\mathfrak{A}_3$.

Let x_1, x_2, x_3 denote the roots of (4) in $\bar{K}$, and let us introduce the **Lagrange resolvents**[40]

$$\begin{cases} r_0 = x_1 + x_2 + x_3 = 0 \\ r_1 = x_1 + jx_2 + j^2 x_3 \in \bar{K} \\ r_2 = x_1 + j^2 x_2 + jx_3 \in \bar{K}. \end{cases}$$

We know that $R(x_1) = R(x_2) = R(x_3)$ because D is a square in R^*, and we see that (3) remains irreducible over $R(j)$ since $[R(j) : R]$ is equal to 1 or 2. Set $M = R(j)$.

Let $\sigma = (x_1, x_2, x_3)$ be a generator of $\mathrm{Gal}(M(x_1)/M) \cong \mathfrak{A}_3$. Then we have

$$\begin{cases} \sigma(r_1) = x_2 + jx_3 + j^2 x_1 = j^2 r_1 \\ \sigma(r_2) = x_2 + j^2 x_3 + jx_1 = jr_2. \end{cases}$$

which shows that r_1^3 and r_2^3 are invariant under σ, so they belong to M.

[40] L. de Lagrange (1736–1813).

Let us define $\sqrt{D}$ by

$$\sqrt{D} := (x_1 - x_2)(x_1 - x_3)(x_2 - x_3) = s - t$$

with

$$s = x_1^2 x_2 + x_2^2 x_3 + x_3^2 x_1, \qquad t = x_1 x_2^2 + x_2 x_3^3 + x_3 x_1^2.$$

We then find that

$$\begin{aligned} r_1^3 &= x_1^3 + x_2^3 + x_3^3 + 3js + 3j^2 t + 6x_1 x_2 x_3 \\ &= -3q + 3js + 3j^2(s - \sqrt{D}) - 6q \\ &= -9q - 3s - 3j^2\sqrt{D}. \end{aligned}$$

Now, the equations

$$\begin{cases} s + t = 3q \\ s - t = \sqrt{D} \end{cases}$$

give $2s = 3q + \sqrt{D}$, hence

$$2r_1^3 = -27q \pm 3\sqrt{-3D}. \tag{8}$$

But equation (6) implies that

$$\begin{cases} 2u^3 = -q \pm \sqrt{-D/27} \\ 2v^3 = -q \mp \sqrt{-D/27}. \end{cases}$$

Thus, we see that $r_1^3 \in \{(3u)^3, (3v)^3\}$.

Furthermore, we have

$$r_1 r_2 = x_1^2 + x_2^2 + x_3^2 - (x_2 x_3 + x_3 x_1 + x_1 x_2) = -3p,$$

so $r_1^3 = (3u)^3$ (resp. $(3v)^3$) implies $r_2^3 = (3v)^3$ (resp. $(3u)^3$). As

$$\begin{cases} 3x_1 = r_1 + r_2 \\ 3x_2 = j^2 r_1 + j r_2 \\ 3x_3 = j r_1 + j^2 r_2, \end{cases}$$

we obtain the formulae of Tartaglia and Cardan.

Remark 3.7.5 This explanation goes back to Lagrange [Vui].

COMMENTARY

It seems probable that Newton was delighted to be able to manipulate formal series as though they were decimal expansions of real numbers (which were first introduced by Stevin). It is an established fact that Hensel and Landsberg introduced the $\mathfrak{p}$-adic expansions of algebraic numbers (in 1902), as analogues of the Newton–Puiseux[41] expansions of algebraic functions. Thus the theory made a full, and remarkably fruitful, circle!

Another fundamental idea, which was inspired in the mind of Hasse, in the 1920s, by a postcard from Hensel, and which he named the "Hasse principle", consists in asking if a Diophantine problem (assumed to have no solution) admits p-adic solutions for every p. If there exists a prime number ℓ for which the problem has no solution, we say that the problem is "trivial". We saw (theorem 3.4.2 of Section 3.4) that Fermat's problem is a "non-trivial" problem (as we may imagine!), since in fact Fermat's equation admits non-trivial solutions in every p-adic field, and in $\mathbb{R}$, but not in the field of rational numbers. We say that a given problem P satisfies the **Hasse principle** if the existence of solutions to P in all the $\mathfrak{p}$-adic completions of a global field k implies the existence of a solution in k (a global field is a finite extension of $\mathbb{Q}$ or a finite separable extension of $\mathbb{F}_p(t)$). Thus Fermat's equation does not satisfy the Hasse principle.

The reader can find useful complements to Sections 3.1 and 3.2 in the books by Cassels [Ca 1] and by Descombes [Des].

For Section 3.3, we should add the books by Schikhoff [Sch] and by Amice [Am].

For Steinitz' theorem, the reader can consult Bourbaki [Bou 1], and Cassels [Ca 1].

Linear representations of groups form a vast and marvellous subject which was born in the mind of Galois ("the last theorem of Galois") before being taken up and studied by the German school at the turn of the century (by Frobenius in particular). The books by Serre [Se 2] and James and Liebeck [J-L] are excellent basic references for the subject.

For Galois theory, we recommend the books by Artin [Ar 1] or Kaplansky [Kap], and for its history the books by Edwards [Ed] and by Verriest [Ve]. Note also that for certain authors (see [Tau] p. 354), a "Galois extension" Ω/K is only an extension in which K is stationary. In this text, we have adopted a more classical point of view by also requiring that Ω/K be algebraic.

Exercises and Problems for Chapter 3

3.1 Show that if a commutative ring with unit is equipped with an absolute value, then it is an integral domain. Show that the absolute value of this ring extends to its fraction field.

3.2 Let A be a commutative ring equipped with an absolute value $|\ |$ satisfying condition (iii) of the definition of absolute value (Section 3.1) for a certain constant C.

- **(a)** Show that if $\alpha > 0$, then the function $|\ |^{\alpha}$ is an absolute value satisfying (iii) for the constant C^{α}.
- **(b)** Show that if the absolute value is triangular, then we can take $C = 2$.
- **(c)** We now want to show that if $C \leq 2$, the absolute value is triangular.

[41] V. Puiseux 1820–1883.

Show that for every $h \in \mathbb{N}$, we have

$$|2^h| \leq C^h \leq 2^h.$$

Writing every positive integer $n \leq 2^h$ as a sum of n terms equal to 1 and $2^h - n$ terms equal to 0, show that

$$|n| \leq 2^h.$$

Writing $n = 2^{h-1} + r$ with $0 \leq r < 2^{h-1}$, show that

$$|n| \leq 2^h \leq 2n.$$

Let x and y be any elements of A and $n = 2^h - 1$. Show that

$$|(x+y)^n| = \left|\sum_{i=0}^{n} \binom{n}{i} x^i y^{n-i}\right| \leq C^h \sup\left\{\left|\binom{n}{i} x^i y^{n-i}\right|\right\}.$$

Deduce the following inequality from the relation $|m| \leq 2m$:

$$|(x+y)^n| \leq C^h \sup\left\{2\binom{n}{i} |x^i y^{n-i}|\right\}.$$

Giving a suitable upper bound for the right-hand term, show that

$$|(x+y)^n| \leq 2^{h+1}(|x| + |y|)^n$$

and deduce that

$$|x+y| \leq 2^{(h+1)/(2^h-1)}(|x| + |y|).$$

Conclude by letting h tend to infinity (a prototype of this kind of reasoning can be found in the works of A. Cauchy).

3.3 Prove the product formula (5):

$$|x|_\infty \prod_{P \in \mathcal{P}} |x|_P = 1$$

when $x \in \mathbb{F}_q(t)^*$.
Start by checking the case where x is an irreducible polynomial $Q \in \mathbb{F}_q[t]$, then extending by "linearity", writing every x in the form

$$x = \lambda \prod_{Q \in P} Q^{v_Q(x)}$$

with $\lambda \in \mathbb{F}_q^*$ and $v_Q(x) \in \mathbb{Z}$ almost all zero.

3.4 Let F be an arbitrary commutative field and $F(t)$ the field of rational functions with coefficients in F. Let $\mathcal{P}$ denote the set of monic irreducible polynomials in $F[t]$.

(α) Following the procedure given in the main text, show that every absolute value $|\ |$ on $F(t)$ which is trivial on F and bounded on $F[t]$ is associated to an irreducible polynomial $P(t) \in F[t]$, via the formula

$$|x| = \rho_P^{-v_P(x)},$$

where ρ_P is a fixed real number > 1 and $v_P(x)$ is the exponent of P in the decomposition of x as a product of monic irreducible polynomials.

(β) Show that if $|\ |$ is not bounded on $F[t]$, we have

$$|x| = \rho_\infty^{-v_\infty(x)},$$

where $v_\infty(x)$ is the degree of x.

(γ) Consider the family of absolute values $\{|\ |_P\}_{P \in \{\infty\} \cup \mathcal{P}}$, associated to real numbers ρ_P and ρ_∞ greater than 1. Show that in order for this family to satisfy the product formula (5), it is necessary and sufficient that for every $P \in \mathcal{P}$, we have

$$\rho_P = \rho_\infty^{\deg P}.$$

3.5 Let F be a commutative field; consider $A = F[X, Y]$.

(α) Let $\geq$ indicate the lexicographic order on $\mathbb{Z} \times \mathbb{Z}$ defined by

$$(m, n) \geq (m', n') \iff \begin{cases} m > m' & \text{or} \\ m = m' & \text{and} \quad n \geq n'. \end{cases}$$

For $\sum c_{\alpha_1,\alpha_2} X^{\alpha_1} Y^{\alpha_2} \neq 0$, set

$$v\left(\sum_{\substack{\alpha \in \mathbb{Z}\times\mathbb{Z} \\ \text{almost all zero}}} c_{\alpha_1,\alpha_2} X^{\alpha_1} Y^{\alpha_2}\right) = \inf\{(\alpha_1, \alpha_2);\, c_{\alpha_1,\alpha_2} \neq 0\}$$

and $v(0) = \infty$. Show that if ρ is a real number > 1, the map

$$P = \sum c_{\alpha_1,\alpha_2} X^{\alpha_1} Y^{\alpha_2} \mapsto \rho^{-v(P)}$$

is an ultrametric absolute value on A.

(β) Same question, but now replacing the lexicographic order by

$$(m, n) \geq (m', n') \iff m + n\sqrt{2} \geq m' + n'\sqrt{2}.$$

3.6 Set $\mathbb{Q}(i) = \{a + bi,\, a \text{ and } b \in \mathbb{Q}\} \subset \mathbb{C}$.
Show that $a + bi \mapsto |a + bi| = a^2 + b^2$ is an absolute value on $\mathbb{Q}(i)$.
Is it triangular ?
Same question for $|a + b\sqrt{-5}| = a^2 + 5b^2$ defined over $\mathbb{Q}(\sqrt{-5}) \subset \mathbb{C}$.

3.7 Show that the sequence defined by

$$\begin{cases} x_1 = 2 \\ x_{n+1} = x_n - \dfrac{f(x_n)}{2x_n} \end{cases}$$

with $f(x) = x^2 - 2$ is a Cauchy sequence of rational numbers for the absolute value $| \ |_\infty$. Is it convergent in $\mathbb{Q}$?

3.8 Show that the sequence defined by

$$\begin{cases} x_1 = 2 \\ x_{n+1} = x_n - \dfrac{f(x_n)}{2x_n} \end{cases}$$

with $f(x) = x^2 + 1$ is a Cauchy sequence of rational numbers for the absolute value $| \ |_5$. Is it convergent in $\mathbb{Q}$?

3.9 Show that the polynomial $P(t) = t^2 + 1$ is irreducible in $\mathbb{F}_3[t]$.
Consider the sequence defined by

$$\begin{cases} x_1 = t \\ x_{n+1} = x_n - \dfrac{f(x_n)}{2x_n} \end{cases}$$

with $f(x) = x^2 + 1$.
Show that (x_n) is a Cauchy sequence for the absolute value $| \ |_P$. Is it convergent in $\mathbb{F}_3(t)$?

3.10 Let k be a field equipped with a triangular absolute value $| \ |$.
Let B denote the k-algebra of bounded sequences of elements of k, i.e. of maps

$$x : \mathbb{N} \longrightarrow k$$

for which there exists a constant $C_x \in \mathbb{R}$ such that

$$|x_n| \leq C_x$$

for every $n \in \mathbb{N}$.

(α) Show that the set C of Cauchy sequences in k for the absolute value $| \ |$ is a sub-k-algebra of B.
(β) Show that the set N of Cauchy sequences which tend to zero is an ideal of C.
(γ) Show that if x is a Cauchy sequence which does not belong to N, then there exists a Cauchy sequence y such that $xy = 1$. Deduce that N is a maximal ideal of C and that C is a local ring.

3.11 We now proceed to state and prove a version of **Hensel's lemma** .
Let $f(X) \in \mathbb{Z}_p[X]$, and let $f'(X) \in \mathbb{Z}_p[X]$ be its derivative. If $a_1 \in \mathbb{Z}_p$ is such that

$$|f(a_1)|_p < |f'(a_1)|_p^2,$$

then there exists $a \in \mathbb{Z}_p$ such that

$$|a - a_1|_p \leq \frac{|f(a_1)|_p}{|f'(a_1)|_p} \quad \textit{and} \quad f(a) = 0.$$

(α) Show that we have

$$f(X+Y) = f(X) + f_1(X)Y + \cdots + f_n(X)Y^n$$

(where n denotes the degree of f), that the $f_i(X)$ lie in $\mathbb{Z}_p[X]$ and that $f_1(X) = f'(X)$.

(β) Set $b_1 = -f(a_1)/f'(a_1)$.
Show that $b_1 \in \mathbb{Z}_p$ and that we have

$$\begin{cases} |f(a_1+b_1)|_p \leq |b_1|_p^2 < |f(a_1)|_p \\ |f'(a_1+b_1) - f'(a_1)|_p \leq |b_1|_p < |f'(a_1)|_p. \end{cases}$$

Deduce that $|f'(a_1+b_1)|_p = |f'(a_1)|_p$ and that the number $a_2 = a_1 + b_1$ still satisfies the conditions of the statement.

(γ) Define a_n by induction on n.
Show that (a_n) converges in $\mathbb{Z}_p$ to a root a of $f(X)$, and that this root satisfies the inequality of the statement.

(δ) Show that a is the only root of $f(X)$ which lies in the closed ball

$$\left\{x \in \mathbb{Q}_p;\ |x - a_1|_p \leq \frac{|f(a_1)|_p}{|f(a_1')|_p}\right\}.$$

3.12 Using Exercise 3.11, prove the following assertions:

(α) Let p be an odd prime.
Assume that there exist b and a_1 in $\mathbb{U}_p = \mathbb{Z}_p^*$ such that $|a_1^2 - b| < 1$. Then b is a square in $\mathbb{Z}_p$.

(β) If $p \neq 2$, then $\mathbb{Q}_p^*/{}^2\mathbb{Q}_p^* \cong (\mathbb{Z}/2\mathbb{Z})^2$ where ${}^2\mathbb{Q}_p^* := \{x^2;\ x \in \mathbb{Q}_p^*\}$.

(γ) If $b \in \mathbb{Z}_2$ is congruent to 1 modulo 8, then b is a square in $\mathbb{Z}_2$.

(δ) $\mathbb{Q}_2^*/{}^2\mathbb{Q}_2^* \cong (\mathbb{Z}/2\mathbb{Z})^3$.

(ε) Let $b \in \mathbb{U}_3$. Then

$$(b \text{ is a cube in } \mathbb{Z}_3) \iff (b \equiv \pm 1 \bmod 9).$$

3.13 Show that the completion of $\mathbb{F}_q(t)$ for its t-adic absolute value is the field $\mathbb{F}_q((t))$ of formal series in t with coefficients in $\mathbb{F}_q$.

3.14 Give a version of Hensel's lemma (Exercise 3.11) in for the t-adic completion $\mathbb{F}_q((t))$ of $\mathbb{F}_q(t)$.

3.15 Let $b = b_0 + b_1 t + b_2 t^2 + \cdots \in \mathbb{F}_q[[t]] = A$. Show that if n is not divisible by the characteristic of $\mathbb{F}_q$ and if $b_0 \neq 0$, then we have the equivalence

$$(b \text{ is an } n^{th} \text{power in } A) \iff (b_0 \text{ is an } n^{th} \text{ power in } \mathbb{F}_q)$$

3.16 Let $|\ \ |_\infty$ be the absolute value at infinity of $\mathbb{F}_q[t]$. What is the completion of $\mathbb{F}_q(t)$ for the extension of this absolute value to $\mathbb{F}_q(t)$?

3.17 We propose to show that $\bar{\mathbb{Q}}$ is countable.

(α) Show that the set E_n of monic polynomials of $\mathbb{Q}[X]$ of degree n is countable.

(β) Deduce that the set A_n of algebraic numbers $x \in \bar{\mathbb{Q}}$ of degree n is countable.

(γ) Show that $\bar{\mathbb{Q}} = \bigcup_{n \geq 1} A_n$ and deduce that $\bar{\mathbb{Q}}$ is countable.

3.18 Irrationality of π

Define the number π as the smallest real positive root of the function $x \mapsto \sin x$. Let us show that we cannot have $\pi = a/b$ with a and $b \in \mathbb{N}^*$.

Set

$$\begin{cases} f_n(x) = \dfrac{x^n(a-bx)^n}{n!} \\ F_n(x) = f_n(x) - f_n^{(2)}(x) + f_n^{(4)}(x) - \cdots + (-1)^n f_n^{(2n)}(x). \end{cases}$$

(α) Show that $f_n^{(k)}(0)$ and $f_n^{(k)}(\pi)$ lie in $\mathbb{Z}$, so $F_n(0)$ and $F_n(\pi)$ lie in $\mathbb{Z}$.

(β) Show that $\int_0^{\pi} f_n(x) \sin x \, dx = F_n(\pi) - F_n(0) \in \mathbb{Z}$.

(γ) Show that for n sufficiently large and $x \in [0, \pi]$, we have

$$0 < f_n(x) \sin x < \frac{1}{\pi}.$$

(δ) Obtain a contradiction.

3.19 The p-adic exponential

We propose to study the convergence of the series $\sum_{n=0}^{\infty} \frac{x^n}{n!}$ for $x \in \mathbb{C}_p$, an algebraically closed field of characteristic zero which is also complete for an absolute value $| \ |_p$ which extends the absolute value p-adic of $\mathbb{Q}$.

(α) Let $n \in \mathbb{N}$; write it in base p as

$$n = a_0 + a_1 p + \cdots + a_s p^s$$

with $a_i \in \{0, 1, \ldots, p-1\}$ and $a_s \neq 0$.

Set $s_n := \sum_{j=0}^{s} a_j$ (sum of the digits of n). Show that

$$|n!|_p = \sum_{j=1}^{\infty} \left[\frac{n}{p^j}\right] = \frac{n - s_n}{p-1}.$$

(β) Set $E = \{x \in \mathbb{C}_p; |x|_p < p^{1/1-p}\}$.

Show that the series $\exp x$ converges for every $x \in E$.

(δ) Show that if x and y are in E, then $x + y$ is in E and we have

$$\exp(x+y) = \exp(x)\exp(y).$$

3.20 Binomial series

Assume that $a \in \mathbb{Z}_p$, and set

$$(1+x)^a := \sum_{n=0}^{\infty} \binom{a}{n} x^n,$$

where $\binom{a}{0} = 1$ and $\binom{a}{n} = a(a-1)\ldots(a-n+1)/n!$.

(α) Show that for every $a \in \mathbb{Z}_p$, we have

$$\left|\binom{a}{n}\right|_{p \leq 1}.$$

(β) Conclude that the series $(1 + x)^a$ is convergent for every $x \in B$, where B denotes the open ball

$$B = \{x \in \mathbb{C}_p;\ |x|_p < 1\}.$$

(γ) Show that the map

$$\begin{cases} x \longmapsto (1+x)^a \\ B \longrightarrow \mathbb{C}_p \end{cases}$$

is continuous for the topology associated to $|\ |_p$.

(δ) Check that if $a \in \mathbb{Z}$ and $x \in B$, we recover the usual value of $(1 + x)^a$ (we say that we have p-adically interpolated the function defined over $\mathbb{Z}$).

(ε) Take a fixed $x \in B$.

Show that there exists a unique continuous function $\mathbb{Z}_p \to \mathbb{C}_p$ whose restriction to $\mathbb{Z}$ is the usual value of $(1 + x)^a$ (use the fact that $\mathbb{Z}$ is dense in $\mathbb{Z}_p$).

3.21 Let $n \in \mathbb{N}$ and let $\mathfrak{S}_n$ be the group of permutations of $1, 2, \dots, n$.

(α) If $\sigma = (a_1, \dots, a_p)$ is a cycle of $\mathfrak{S}_n$ and if τ is an arbitrary element of $\mathfrak{S}_n$, show that

$$\tau\sigma\tau^{-1} = (\tau(a_1), \dots, \tau(a_p)).$$

(β) Deduce a way of determining the conjugacy classes of $\mathfrak{S}_n$.

(γ) Show that $\mathfrak{S}_3$ is the group of isometries of an equilateral triangle.

(δ) Use (γ) to determine the three irreducible characters of $\mathfrak{S}_3$.

3.22 Schur's Lemma

(α) Let $G \xrightarrow{\rho_i} GL_K(V_i)$, for $i \in \{1, 2\}$, be two *irreducible* representations of finite degree of the same group G.
Assume that $\varphi : V_1 \to V_2$ is a K-linear map such that for every $s \in G$, we have

$$\varphi \circ \rho_1(s) = \rho_2(s) \circ \varphi.$$

Show that if $\varphi \neq 0$, φ is a K-linear isomorphism between ρ_1 and ρ_2.

(β) Assume now that K is algebraically closed, and that $V_1 = V_2 = V$. Show that φ is a homothety (multiplication by a scalar) of V.

3.23 Proof of Maschke's theorem
Let G be a finite group and K a commutative field whose characteristic does not divide the order n of G.

(α) Reduce the proof of Maschke's theorem to showing that if a $K[G]$-module V contains a $K[G]$-submodule U, then there is a $K[G]$-module W in V which is supplementary to U in the sense that

$$V = U \oplus W.$$

(β) Let U be a submodule, and let W_0 be a K-subvector space supplementary to U in V. Let π denote the projection $V \twoheadrightarrow U$, which has kernel W_0. Show that if for $x \in V$ we set

$$\theta(x) = \frac{1}{n} \sum_{s \in G} s^{-1}[\pi(sg)],$$

then the map $\theta : V \to V$ is a $K[G]$-endomorphism.

(γ) Show that for every $t \in G$ and $y \in U$, we have

$$\pi(ty) = ty.$$

Deduce that $\theta(y) = y$ and $\theta^2 = \theta$.

(δ) Set $W = \operatorname{Ker} \theta$, and show that W is a $K[G]$-submodule of V such that

$$V = U \oplus W.$$

3.24 Let $\rho : G \to GL_n(\mathbb{C})$ be a representation of a finite group G.

(i) Show that for every $g \in G$, we have

$$|\chi(g)| \leq \chi(1) = n.$$

(ii) Show that if $|\chi(g)| = \chi(1)$, then

$$\rho(g) = \lambda I_n = \lambda \rho(e)$$

with $\lambda \in \mathbb{C}^*$.

(iii) Deduce that

$$\operatorname{Ker} \rho = \{g \in G; \chi(g) = \chi(e)\}.$$

3.25 Recall that every finite Abelian group G is a finite product of cyclic groups $C_{n_i} \cong \mathbb{Z}/n_i\mathbb{Z}$:

$$G \cong C_{n_1} \times \cdots \times C_{n_r}$$

(i) Let c_i be a generator of C_{n_i}. Set

$$g_i = (1, 1, \ldots, c_i, \ldots 1)$$

where 1 represents the identity element of C_{n_j} for every j.
Show that the set $\{g_1, \ldots, g_r\}$ generates G_r, i.e. that

$$G = \langle g_1, \ldots, g_r \rangle.$$

(ii) Consider $\hat{G} = Hom(G, \mathbb{C}^*)$. Show that every element $\rho \in \hat{G}$ is determined by $(\rho(g_1), \ldots, \rho(g_r))$, and that for every i, $\rho(g_i) \in \mu_{n_i}(\mathbb{C})$, where $\mu_{n_i}(\mathbb{C})$ denotes the set of roots of $X^{n_i} - 1$ in $\mathbb{C}$.

(iii) Show that $\hat{G} \cong \mu_{n_1}(\mathbb{C}) \times \cdots \times \mu_{n_r}(\mathbb{C}) \cong G$.

(iv) Show that $\hat{G}$ is in bijection with the equivalence classes of representations of degree 1 of G over the field of complex numbers.

3.26 Let G be an arbitrary group.

(i) Let $G \xrightarrow{g} A$ be a homomorphism from a group G to an Abelian group A. Show that the kernel of g contains the *set of commutators* of G, i.e. the set of elements of the form $[x, y] = xyx^{-1}y^{-1}$, x and $y \in G$.

(ii) Let the *commutator subgroup* G' of G be the subgroup of G generated by the set of commutators. Show that G' is a normal subgroup of G, and that G/G' is Abelian. Let φ denote the canonical homomorphism $G \twoheadrightarrow G/G'$. Show that φ has the following *universal property*: for every homomorphism $G \xrightarrow{g} A$, there exists a unique homorphism $G/G' \xrightarrow{h} A$ such that

$$\begin{array}{ccc} G & \xrightarrow{\varphi} & G/G' \\ & g \searrow & \downarrow h \\ & & A \end{array}$$

(iii) Show that if ρ is a representation of degree 1 of G, the kernel of ρ contains G', and that there exists a unique representation ρ' of degree 1 of G/G' such that $\rho = \rho' \circ \varphi$. Deduce the existence of a bijective correspondence between the representations of degree 1 of G and those of G/G'.

(iv) Assume now that G is finite, and consider the *complex* representations of G. Show (using Exercise 3.25) that the number of equivalence classes of representations of degree 1 of G (i.e. of "linear characters" of G) is equal to $(G : G')$.

3.27 Hasse principle for conics

Let a, b and c be three non-zero integers in $\mathbb{Z}$, and consider the projective conic of equation

$$ax^2 + by^2 + cz^2. \qquad (L_{a,b,c})$$

We will establish a necessary and sufficient condition for this equation to admit a rational point. We saw in problem 7 of Chapter 1 that we may assume that a, b and c are square-free and pairwise relatively prime.

I. Show that if $L_{a,b,c}$ admits a rational point, then it admits a real point and points in every p-adic field (this is trivial!).

II. We want to show the converse, using problem 7 of Chapter 1.

(1) Suppose that p divides c and that (u, v, w) is a p-adic point of $L_{a,b,c}$.
Show that we can suppose that $\sup(|u|_p, |v|_p, |w|_p) = 1$, and show that then we have $|u|_p = |v|_p = 1$.

(2) Deduce that $-ab$ is a square modulo p.

(3) Considering all the prime divisors p of c, show that $-ab$ is a square modulo c.

(4) Conclude.

Problem I

Let $|\ |$ be an arbitrary absolute value defined over a field k, and let $a \in k$.

We say that a subset V of k is a neighbourhood of a for this absolute value, if there exists $\varepsilon > 0$ such that

$$V \supset \{x \in k;\ |x-a| < \varepsilon\}.$$

(1) Show that this definition equips k with a topology (attached to $|\ |$).

(2) What is this topology when $|\ |$ is the trivial absolute value?

(3) Let $\alpha > 0$. Show that the topologies associated to $|\ |$ and $|\ |^\alpha$ are the same.

(4) Show that if the absolute values $|\ |_1$ and $|\ |_2$ defined over k induce the same topology, then there exists $\alpha > 0$ such that $|\ |_2 = |\ |_1^\alpha$. Show that we have the equivalence

$$|x|_1 < 1 \iff |x|_2 < 1.$$

Now, taking $x = a^m b^n$, show that if $ab \neq 0$, then

$$\frac{\log |a|_2}{\log |a|_1} = \frac{\log |b|_2}{\log |b|_1}.$$

(5) We say that the topology $\mathcal{T}_1$ is **finer** than the topology $\mathcal{T}_2$ if for every $a \in k$, we have

$$(\lim_{\mathcal{T}_1}(x_n) = a) \quad \Rightarrow \quad (\lim_{\mathcal{T}_2}(x_n) = a).$$

Show that if the absolute value $|\ |_1$ is not trivial and if it induces a topology $\mathcal{T}_1$ which is finer than the topology $\mathcal{T}_2$ induced by $|\ |_2$, then there exists $\alpha > 0$ such that $|\ |_2 = |\ |_1^\alpha$.

Deduce that the topologies $\mathcal{T}_1$ and $\mathcal{T}_2$ are the same.

(6) Two absolute values $|\ |_1$ and $|\ |_2$ defined over k are said to be "equivalent" if there exists $\alpha > 0$ such that $|\ |_2 = |\ |_1^\alpha$.

Show by induction on n that if the absolute values $|\ |_1, \ldots, |\ |_n$ are non-trivial and pairwise inequivalent, then there exists $a \in k$ such that

$$|a|_1 > 1 \quad \text{and} \quad |a|_i < 1 \ \text{for } 2 \le i \le n.$$

(7) We propose to deduce from question (6) the following result, called the "theorem of weak independence of absolute values".

Take n elements $a_1, \ldots, a_n$ of k and n non-trivial inequivalent absolute values $|\ |_1, \ldots, |\ |_n$. Then for every $\varepsilon > 0$, there exists $x \in k$ such that $|x - a_i|_i < \varepsilon$ for $i = 1, \ldots, n$.

To prove this, use elements $c_i \in k$ such that

$$|c_i|_i > 1, \quad |c_i|_j < 1 \qquad \text{if } j \neq i.$$

For every $\nu \in \mathbb{N}$, set

$$x_\nu = \sum_{i=1}^{n} \frac{c_i^\nu}{1 + c_i^\nu} a_i,$$

and then take $x = x_\nu$ with ν sufficiently large.

(8) Assume that $k = \mathbb{Q}$ and take n distinct primes $p_1, \ldots, p_n$.

Take n elements $a_1, \ldots, a_n \in \mathbb{Z}$ and let $\varepsilon > 0$. Show that there exists $x \in \mathbb{Q}$ such that

$$|x - a_i|_{p_i} < \varepsilon \quad \text{for } i = 1, \ldots, n.$$

Deduce the following property (known as the Chinese remainder theorem): *there exists $y \in \mathbb{Z}$ such that*

$$|y - a_i|_{p_i} < \varepsilon \quad \text{for } i = 1, \ldots, n.$$

The Chinese remainder theorem is a special case of the "theorem of strong independence of absolute values" in global fields.

Problem 2

(Ostrowski's Theorem)

Let k be a commutative field equipped with an Archimedean absolute value $\|\ \|$, i.e. an absolute value which is not bounded on the subring generated by 1_k.

We propose to show that there exists an isomorphism $j : k \to \mathbb{C}$ of commutative rings with unit, and a real number $\alpha > 0$, such that for every $x \in k$ we have $\|x\| = |j(x)|^\alpha$.

A. We propose to replace $\|\ \|$ by a triangular absolute value.

(1) Consider the special case $k = \mathbb{Q}(\zeta)$, where ζ is a root of $X^2 + X + 1$, and for a and $b \in \mathbb{Q}$, set

$$\|a + b\zeta\| = a^2 - ab + b^2.$$

Show that $\|\ \|$ is an Archimedean absolute value on k; find an isomorphism j and the value of α.

(2) Let us return to the general case.

Show that the prime subfield of k is the field $\mathbb{Q}$ of the rationals, that there exists a real number $\alpha > 0$ such that $\|x\| = |x|_\infty^\alpha$ for every $x \in \mathbb{Q}$, and that the completion $\hat{k}$ of k for $\|\ \|^{1/\alpha}$ contains $\mathbb{R}$.

B. Replace $\|\ \|$ by $\|\ \|^{1/\alpha}$, i.e. assume now that $\|\ \|$ is a triangular absolute value. Take $x \in k$; let us show that x is a root of an equation of degree two over $\mathbb{R}$.

(1) Let $x \in k$. If $z \in \mathbb{C}$, the numbers $z + \bar{z}$ and $z\bar{z}$ are in $\mathbb{R}$, so we can consider the map

$$h_x \begin{cases} z \longmapsto \|x^2 - (z + \bar{z})x + z\bar{z}\| \\ \mathbb{C} \longrightarrow \mathbb{R}_+. \end{cases}$$

Show that h_x is a continuous function of $\mathbb{C}$.

(2) Show that $h_x(z)$ tends to infinity when $|z|$ tends to infinity. Deduce that h_x admits a minimum $m = h_x(z_0)$ in $\mathbb{C}$.

(3) Show that $h_x^{-1}(m)$ is compact and that there exists $z_1 \in h_x^{-1}(m)$ such that $|z_1|$ is maximal in $h_x^{-1}(m)$.

We now propose to show that $m = 0$.

(4) Set

$$g(X) = X^2 - (z_1 + \bar{z}_1)X + z_1\bar{z}_1 \in \mathbb{R}[X]$$
$$h(X) = g(X) + \frac{m}{2}.$$

Show that if $m > 0$, then $h(X)$ has at least one root $z_2 \in \mathbb{C}$ such that $|z_2| > |z_1|$. Deduce that $z_2 \notin h_x^{-1}(m)$.

(5) Set

$$G(X) = g(X)^n - \left(\frac{-m}{2}\right)^n \in \mathbb{R}[X].$$

Show that $G(z_2) = 0$.

Let t_i $(i = 1, \ldots, 2n)$ be the roots of G in $\mathbb{C}$. Writing

$$G^2(X) = \prod_{i=1}^{2n} (X^2 - (t_i + \bar{t}_i)X + t_i\bar{t}_i)$$

and replacing X by x, show that

$$\|G^2(x)\| \geq m^{2n-1} h_x(z_2).$$

(6) Show that we also have

$$\|G^2(x)\| = \|G(x)\|^2 \leq m^{2n}\left(1 + \frac{1}{2^n}\right)^2.$$

(7) Deduce an upper bound for $h_x(z_2)$ and obtain a contradiction by letting n tend to infinity.

C. We now want to prove the existence of $j : k \to \mathbb{C}$ such that

$$\|x\| = |j(x)|.$$

(1) Show that $[\hat{k} : \mathbb{R}] \leq 2$.
(2) Construct j when $\hat{k} = \mathbb{R}$.
(3) Using Steinitz' theorem, show that j exists when $[\hat{k} : \mathbb{R}] = 2$.
How many possible isomorphisms j are there?

Problem 3

(Hermite's Theorem)

We want to show that π does not satisfy any non-trivial algebraic equation with coefficients in $\mathbb{Z}$. To show this, we use Euler's relation

$$e^{i\pi} = -1.$$

(1) Let $f(x) = \sum_{m=0}^{n} a_m x^m \in \mathbb{C}[X]$.
Set

$$\begin{cases} F(x) = \sum_{k=0}^{n} f^{(k)}(x) \\ Q(x) = F(0)e^x - F(x) \in \mathbb{C}[[x]]. \end{cases}$$

Show that for every $x \in \mathbb{C}$, we have

$$|Q(x)| \leq e^{|x|} \sum_{m=0}^{n} |a_m||x|^m.$$

One can start by considering the case $f(x) = x^n$.

(2) We will now obtain a contradiction, by assuming that $i\pi$ is the root α_1 of the polynomial

$$E(x) = ax^m + a_1x^{m-1} + \cdots + a_m \in \mathbb{Z}[x]$$

such that $a > 0$.

Show that ax is then a root of a monic polynomial $P(x)$ with integral coefficients of degree m.

(3) Set

$$R = \prod_{h=1}^{m}(1 + e^{\alpha_h}) = \prod_{h=1}^{m}(e^0 + e^{\alpha_h}),$$

where the α_h are the roots of $E(x)$.

Show that $R = c + e^{\beta_1} + \cdots + e^{\beta_r}$, where c represents the number of terms of the expanded product whose exponent is zero and where the β_i are (not necessarily distinct) sums of α_h.

(4) Let p be an **arbitrary** prime.

Show that

$$f(x) := \frac{1}{(p-1)!}(ax)^{p-1}\prod_{h=1}^{r}(ax - a\beta_h)^p$$

can be written $(A_{p-1}x^{p-1} + A_px^p + \cdots)/(p-1)!$ with $A_i \in \mathbb{Z}$.

(5) Show that for every $h \in \{1, \ldots, r\}$, we can also write

$$f(x) = \frac{B_{p,h}(x - \beta_h)^p + B_{p+1,h}(x - \beta_h)^{p+1} + \cdots}{(p-1)!},$$

where the $B_{\ell,h}$ are polynomials with integral coefficients in the roots of $P(x)$.

(6) Assume now that p is strictly greater than $\sup\{c, a, \prod_{h=1}^{r}(a|\beta_h|)\}$. Writing

$$cF(0) + \sum_{h=1}^{r} F(\beta_h) + \sum_{h=1}^{r} Q(\beta_h) = F(0)\left(c + \sum_{h=1}^{r} e^{\beta_h}\right) = 0,$$

show that $cF(0) \equiv cA_{p-1} \not\equiv 0 \pmod p$.

(7) Show that $\sum_{h=1}^{r} F(\beta_h) = pc_p + p(p+1)c_{p+1} + \cdots$ where the c_s are symmetric functions of $a\beta_1, \ldots, a\beta_h$. Deduce that the c_s lie in $\mathbb{Z}$ and that $\sum_{\ell=1}^{r} F(\beta_h) \in p\mathbb{Z}$.

(8) Deduce from (6) and (7) that $|cF(0) + \sum_{h=1}^{r} F(\beta_h)| \geq 1$, and obtain a contradiction using (1) when p is large enough.

Problem 4

The goal of this problem is the construction of an infinity of elements in $\mathrm{Aut}(\mathbb{C})$; such elements are called **involutions** of $\mathbb{C}$ when their order in $\mathrm{Aut}(\mathbb{C})$ is equal to 2.

(1) Let τ be an involution of $\mathbb{C}$. Show that if $\tau(\mathbb{R}) \subset \mathbb{R}$, then τ is the complex conjugation σ.

(2) Let τ be an involution of $\mathbb{C}$. Show that if τ is continuous (for the usual topology on $\mathbb{C}$), then $\tau = \sigma$.

(3) Recall that an ordered set S is said to be *inductive* if every totally ordered subset of S is bounded. We will admit Zorn's lemma:

Every non-empty inductive set admits (at least) one maximal element.

Let S be the set of subfields of $\mathbb{R}$ which do not contain $\sqrt{2}$. Show that S contains a maximal element $K_{\sqrt{2}}$.

(4) Show that for every $x \in \mathbb{R} \setminus K_{\sqrt{2}}$, we have $\sqrt{2} \in K(x)$. Deduce that $\mathbb{R}$ is algebraic over K.
(5) Show that $\mathbb{C}$ is an algebraic closure of $K_{\sqrt{2}}$.
(6) Set $L = K_{\sqrt{2}}(\sqrt{2})$, and let θ denote the automorphism of L such that

$$\theta \mid_{K_{\sqrt{2}}} = \mathrm{id}_{K_{\sqrt{2}}}, \qquad \theta(\sqrt{2}) = -\sqrt{2}.$$

Show that there exists an automorphism τ of $\mathbb{C}$ such that $\tau \mid_L = \theta$.
(7) Considering $\tau(\sqrt[4]{2})$, show that τ does not commute with the complex conjugation involution σ.
(8) Deduce that $\sigma_{\sqrt{2}} := \tau\sigma\tau^{-1}$ is an involution of $\mathrm{Aut}(\mathbb{C})$ *different* from σ, and that $\sigma_{\sqrt{2}}(\sqrt{2}) = \sqrt{2}$.
(9) Let R denote the field of invariants of $\{\mathrm{id}_{\mathbb{C}}, \sigma_{\sqrt{2}}\}$ (i.e. $R := \psi(\{\mathrm{id}_{\mathbb{C}}, \sigma_{\sqrt{2}}\})$ in $\mathbb{C}/\mathbb{Q}$). Show that $[\mathbb{C} : R] = 2$ and that $R \supset K_{\sqrt{2}}(\sqrt{2})$.
(10) Let p be an odd prime. Show that there exists an automorphism τ_p of $\mathbb{C}$ such that

$$\tau_p(\sqrt{2}) = -\sqrt{2}, \qquad \tau_p(\sqrt{p}) = \sqrt{p}.$$

(11) Show that $\mathrm{Aut}(\mathbb{C})$ is infinite.
(12) Let e and π denote two transcendental elements of $\mathbb{C}$. Show that $\mathbb{Q}(e)$ is isomorphic to $\mathbb{Q}(\pi)$.
(13) Let S_e (resp. S_π) be the set of subfields of $\mathbb{C}$ which do not contain e (resp. π). Show that S_e (resp. S_π) contains a maximal element E (resp. P), and $\overline{E(e)} = \mathbb{C} = \overline{P(\pi)}$. Deduce the existence of an automorphism of $\mathbb{C}$ sending e to π. What can one say about the orbit of a transcendental element of $\mathbb{C}$ under the action of $\mathrm{Aut}(\mathbb{C})$?

Problem 5

(Newton–Puiseux theorem)

Let K be a field, and let $K^{\mathbb{Q}}$ denote the set of maps from $\mathbb{Q}$ to K. If $f \in K^{\mathbb{Q}}$, we set

$$Supp(f) = \{\alpha \in \mathbb{Q}; f(\alpha) \neq 0\}.$$

Let $N(K)$ be the subset of $K^{\mathbb{Q}}$ consisting of the elements f such that

(i) $Supp(f) \subset \mathbb{Z}n_f^{-1}$ for a certain $n_f \in \mathbb{N}^*$
(ii) There exists $\beta_f \in \mathbb{Q}$ such that $Supp(f) \subset [\beta_f, \infty[$.
Write

$$f = \sum_{\alpha \in Supp(f)} f(\alpha)t^\alpha,$$

and set

$$(f+g)(\alpha) = f(\alpha) + g(\alpha), \quad (fg)(\alpha) = \sum_{\beta+\gamma=\alpha} f(\beta)g(\gamma).$$

A. (1) Show that $N(K)$ is a field.
(2) We define a map $N(K) \xrightarrow{v} \mathbb{Q} \cup \{\infty\}$ by

$$v(0) = +\infty, \qquad v(f) = \text{ smallest element of } Supp(f) \text{ if } f \neq 0.$$

Show that v is a valuation on $N(K)$.

(3) Let $\lambda \in \mathbb{Q}_+^*$, show that the map

$$\eta_\lambda : N(K) \longrightarrow N(K)$$

defined by $\eta_\lambda(f)(\alpha) = f(\lambda^{-1}\alpha)$ is an automorphism of $N(K)$, and that

$$v(\eta_\lambda(f)) = \lambda v(f).$$

B. Assume now that K is an algebraically closed field of characteristic zero. The goal of this part is to show that $N(K)$ is also algebraically closed.

(1) Let $G(X) \in N(K)[X]$ be a monic polynomial of degree $n > 1$. Show that using a suitable automorphism of $N(K)$, we can assume that

$$G(X) \in K((t))[X],$$

and that we even have $G(X) = X^n + f_2X^{n-2} + \cdots + f_n$ with $f_i \in K((t))$.

(2) Assume that the f_i are in $K[[t]]$ and that there exists an index i such that $f_i(0) \neq 0$.
Set $\bar{G}(X) = X^n + f_2(0)X^{n-2} + \cdots + f_n(0) \in K[X]$.
Show that $\bar{G}(X)$ is not a power of a polynomial of degree 1 in X.

(3) Assume that $F(X) \in K[[t]][X]$ and that $\bar{F}(X) = \rho(X)\sigma(X)$ with $\rho(X), \sigma(X) \in K[X]$, $(\rho(X), \sigma(X)) = 1$ and $\deg \rho(X) > 0$. Show that $F(X) = R(X)S(X)$, with $R(X), S(X) \in K[[t]][X]$, $\bar{R}(X) = \rho(X)$, $\bar{S}(X) = \sigma(X)$ and $\deg R(X) = \deg(\rho(X))$ (another form of **Hensel's lemma**).

(4) Deduce from (2) and (3) that $G(X)$ has a factor of degree m in $K[[t]][X]$ with $1 \leq m < n$.

(5) Our goal is to generalise the result of (4) to the case where $G(X) = X^n + f_2X^{n-2} + \cdots + f_n$ is in $K((t))[X]$ but not necessarily in $K[[t]][X]$. We assume that the f_i are not all zero, and we choose $r \in \{2, \ldots, n\}$ such that

$$\frac{v(f_r)}{r} = \inf\left\{\frac{v(f_i)}{i};\ 2 \leq i \leq n\right\}.$$

(5.1) Show that applying the automorphism η_r to the coefficients of $G(X)$ and replacing X by $t^{v(f_r)}X$, we obtain a polynomial

$$t^{nv(f_r)}[X^n + g_2X^{n-2} + \cdots + g_n]$$

with $g_i \in K[[t]]$ for $2 \leq i \leq n$.

(5.2) Applying question (4) to the polynomial

$$H(X) := X^n + g_2X^{n-2} + \cdots + g_n,$$

obtain the desired result.

C. Now take an arbitrary prime number p, and assume that K is a field of characteristic zero having the property

($\mathcal{P}$) *Every non-zero polynomial of $K[X]$ whose degree is not divisible by p has a root in K.*

Applying the method of B, show that if K has the property $\mathcal{P}$ then $N(K)$ also has the property $\mathcal{P}$.

Problem 6

(Casus irreducibilis)

When an equation of degree three with rational coefficients admits three real roots, its discriminant is negative. It follows that the first of the equations (7) of section 3.7 contains the cube root of an "imaginary" quantity. Is it possible to avoid having recourse to "imaginary" quantities in resolving equations by radicals? We will see that the answer is no when the equation has no rational roots.

(1) Let K be a field of characteristic zero. Take $a \in K$, and show the equivalence of the conditions (i) and (ii):

(i) a is a cube in K

(ii) The polynomial $X^3 - a$ is reducible in $K[X]$.

(2) Consider the equation

$$X^3 - 6X + 2 = 0. \qquad (E)$$

Does this equation have a root in $\mathbb{Q}$?

(3) Show that (E) has three distinct real roots α, β and γ.

(4) Do we have $\mathbb{Q}(\alpha) \simeq \mathbb{Q}(\beta)$?

(5) Set $K = \mathbb{Q}(\sqrt{21})$. Show that

$$K(\alpha) = K(\beta) = K(\gamma).$$

Letting L denote this field, show that $[L : K] = 3$ and determine $\mathrm{Gal}(L/K)$.

(6) Suppose that there exists a sequence of extensions

$$K = K_0 \subset K_1 \subset K_2 \subset \cdots \subset K_n \subset \mathbb{R}$$

such that $K_{i+1} = K_i(\sqrt[p_i]{a_i})$ with p_i prime for $i \in \{0, \ldots, n-1\}$, $a_i \in K$ and $a_i > 0$, $\alpha \notin K_i$ for $i < n$ and $\alpha \in K_n$.

Show that then we have $p_{n-1} = 3$.

(7) Show that K_n/K_{n-1} is a Galois extension, and deduce that j and j^2 belong to K_n.

(8) Conclude.

Problem 7

("Galois' Last Theorem")

A. Let F be a commutative field, and let GA be the set

$$GA = \{ax + b;\ a \in F^*,\ b \in F\}$$

of polynomials of degree 1, equipped with the composition law

$$(ax + b) \circ (a'x + b') = a(a'x + b') + b = aa'x + (ab' + b).$$

(1) Show that $(GA, \circ)$ is a group, and that the map

$$\det \begin{cases} GA \longrightarrow F^* \\ ax + b \longmapsto a \end{cases}$$

is a group homomorphism.

The group G is called the group of **affine transformations** of F, and $T = \text{Ker}(\det)$ is the group of **translations** of F.

(2) Let $\lambda \in F^*$, and let $\theta(\lambda)$ denote the automorphism of $(F, +)$ which sends b to λb.

Show that $\theta : F^* \to \text{Aut}(F, +)$ is a group homomorphism, and that GA is isomorphic to the **semi-direct product** $F \underset{\theta}{\rtimes} F^* = \{(b, a) \in F \times F^*; *\}$ with

$$(b, a) * (b', a') = (b + \theta(a)(b'), aa')$$

(3) Deduce that in order for GA to be commutative, it is necessary and sufficient that $F = \mathbb{F}_2$.

(4) Show that the isomorphism

$$GA \xrightarrow{\rho} \begin{pmatrix} a & b \\ 0 & 1 \end{pmatrix} \in M^*_{2\times 2}(F)$$

is the matrix given by a representation of degree 2 of GA over F. Is it irreducible?
Can we apply Maschke's theorem when F is finite?

(5) Show that every subgroup of GA which contains T is solvable.

B. Let K be an arbitrary field, and $P(X)$ an **irreducible separable** polynomial in $K[X]$ whose degree is a **prime** number p. Let $\bar{K}$ denote an algebraic closure of K, and L the smallest subfield of $\bar{K}$ containing K and all the roots of $P(X)$ in $\bar{K}$.

(1) Show that the extension L/K is Galois.

(2) Let $\alpha_1, \ldots, \alpha_p$ denote the roots of $P(X)$ in $\bar{K}$, and G the Galois group of L/K. Recall that G acts transitively on these roots, i.e. for every α_i, there exists an element $\sigma \in G$ such that $\sigma(\alpha_1) = \alpha_i$. Let $H = Stab_G(\alpha_1) := \{\sigma \in G; \sigma(\alpha_1) = \alpha_1\}$. Show that $(G : H) = p$ and deduce that p divides $\#G$.

(3) Let N be a normal subgroup of G, and let $S_1 = \{\alpha_1, \alpha_2, \ldots, \alpha_s\}$ the orbit of α_1 under the action of N. Show that every other orbit of $S = \{\alpha_1, \ldots, \alpha_p\}$ under the action of N is of the form $S_i = \{\sigma(\alpha_1), \ldots, \sigma(\alpha_s)\}$ for $\sigma \in G$. Deduce that $s = 1$ or $s = p$, and that N acts transitively on the roots of $P(X)$ if $N \neq \{e\}$.

(4) Choose indices $1, 2, \ldots, p$ for the roots α_i in the field $\mathbb{F}_p$. Following Galois, we see that G is isomorphic to a group of permutations of the elements of the field $\mathbb{F}_p$, where a permutation σ acts via

$$\sigma(\alpha_i) = \alpha_{\hat{\sigma}(i)}.$$

Show that if $\#G = p$, then $G \cong T$.

(5) Show that if G is Abelian and $p > 2$, then $G \cong T$.

(6) By induction on $\#G$, show that if G is solvable, then G is isomorphic to a subgroup of GA containing T.

Assume that G is not Abelian, and consider a non-trivial normal subgroup N of G such that $N \neq G$. Show that if σ is given by $x \mapsto x + 1$, then for every $\tau \in G$, $\tau\sigma\tau^{-1}$ is a cycle of order p contained in N; determine its expression in GA by using the induction hypothesis. Finally, show that $\tau \in GA$.

4

ELLIPTIC CURVES

Over the field of complex numbers, an elliptic curve is a curve which is isomorphic to a Weierstrass cubic (this will be proved in Chapter 5). Such a curve is thus the quotient of $\mathbb{C}$ by a lattice in $\mathbb{C}$.

Now, however, we wish to define the notion of an elliptic curve over an arbitrary field, for example over the field $\mathbb{Q}$ of rational numbers or over a finite field (of positive characteristic).

In such a context, it would be illusory to try to define elliptic, or even loxodromic functions, and we are reduced to the resources afforded by algebraic geometry, which are fortunately considerable.

It so happens, however, that in recent years algebraic geometry has developed into a vast territory whose language takes years of study to master, and we do not wish to assume here that the reader is familiar with it. Thus, we will adopt the more naive points of view of the "analytic geometry" due to Descartes[42] and Fermat, and the projective geometry of Chasles[43] and Poncelet[44]. We will take advantage of the fact that, in the projective plane, an elliptic curve can be considered as a hypersurface, i.e. as an object given by a non-zero homogeneous polynomial $F(X, Y, Z) \in k[X, Y, Z]$ of a certain type (modulo the constants of k^*), and we will study these hypersurfaces using the notion of an intersection product.

The main results of this chapter can be found in Sections 4.4–4.7, then Sections 4.9 and 4.15. The theorems in Sections 4.4, 4.5 and the first theorem of Section 4.7 are particularly interesting. As for Mazur's theorem (Section 4.10), it will play a crucial role in Section 6.6 of Chapter 6 (irreducibility of the representation using the p-division points of the Hellegouarch–Frey curve).

4.1 CUBICS AND ELLIPTIC CURVES

We have already considered cubics and elliptic curves at length in the preceding chapters; recall that in Chapter 1, we studied the third-degree Fermat equation, and in Chapter 2, the

[42] R. Descartes (1596–1650).

[43] M. Chasles (1793–1880).

[44] J.V. Poncelet (1788–1867).

Weierstrass cubic. In the first situation, we considered the curve as being defined over $\mathbb{Q}$, and in the second case the Weierstrass cubics were defined over $\mathbb{C}$: in fact, what we want is to consider a general situation in which curves will be defined over an arbitrary field. Recall that we saw that the most natural framework for studying the geometry of these curves was the framework of projective geometry, which arises naturally in Diophantine problems, and which is motivated by internal reasons in the study of Weierstrass cubics (in order to ensure that the set of points of the curve forms an additive group, which in particular contains an identity element).

In order to develop a precise vocabulary, we need to return to the notion of a plane algebraic curve. This notion goes back to the invention of analytic geometry by Descartes and Fermat, in the first half of the 17th century. For example, consider a conic Γ having a focus at the origin and for associated directrix the line given by the following homogeneous equation in $(X, Y, Z) : uX + vY + wZ = 0$. This conic is the set of points of $\mathbb{R}^2$ satisfying an equation of the form:

$$\Gamma(X, Y, Z) \equiv \lambda(X^2 + Y^2) + \mu(uX + vY + wZ)^2 = 0,$$

with $(\lambda, \mu) \in \mathbb{R}^2 \setminus \{(0, 0)\}$.

When $(\lambda, \mu) = (1, 0)$ and $(0, 1)$, we obtain two limit cases of curves of the preceding type: the "real point" of equation $X^2 + Y^2 = 0$ and the "double line" of equation $(uX + vY + wZ)^2 = 0$.

At the beginning of the 19th century, a certain number of geometers considered that these two limit cases contained the very essence of the idea of a conic defined by focus and directrix, but that in order to understand this idea, one should consider these conics as objects inside $\mathbb{P}_2(\mathbb{C})$. In other words, one should **extend** the field of scalars from $\mathbb{R}$ to $\mathbb{C}$.

If Γ_0 denotes the conic of equation $X^2 + Y^2 = 0$, we see that this curve is such that

$$\begin{cases} \Gamma_0(\mathbb{R}) := \{(a, b, c) \in \mathbb{P}_2(\mathbb{R}); a^2 + b^2 = 0\} = \{(0, 0, 1)\} \\ \Gamma_0(\mathbb{C}) := \{(a, b, c) \in \mathbb{P}_2(\mathbb{C}); a^2 + b^2 = 0\} = \Delta(\mathbb{C}) \cup \bar{\Delta}(\mathbb{C}), \end{cases}$$

where Δ denotes the line $X - iY = 0$ and $\bar{\Delta}$ the "conjugate line" $X + iY = 0$, and where we set

$$\begin{cases} \Delta(\mathbb{C}) := \{(a, b, c) \in P_2(\mathbb{C}); a + ib = 0\} \\ \bar{\Delta}(\mathbb{C}) := \{(a, b, c) \in P_2(\mathbb{C}); a - ib = 0\}. \end{cases}$$

The extension of our field of vision from $\mathbb{R}^2$ to $\mathbb{C}^2$ makes it possible to "see" that the idea of a conic with focus at the origin and associated directrix D can be identified with that of a conic tangent to the "isotropic lines" (here Δ and $\bar{\Delta}$ are called the isotropic lines at the origin) at points situated on the line D (this definition is due to Plücker[45]).

The main idea which should be retained from this brief historical overview is that the equation of a curve is a more fundamental notion that the set of its points (in $\mathbb{R}^2$, $\mathbb{P}_2(\mathbb{R})$, $\mathbb{C}^2$, $\mathbb{P}_2(\mathbb{C})$, etc.).

[45] J. Plücker, 1801–1868.

Definition 4.1.1 *Let d be an integer ≥ 1. A plane projective curve of degree d* **defined over** k *is just an element of* $\mathbb{P}(k[X, Y, Z]_d)$, *where $k[X, Y, Z]_d$ denotes the k-vector space of homogeneous polynomials of degree d in X, Y, Z and $\mathbb{P}(k[X, Y, Z]_d)$ the associated projective space.*

Thus, a curve is given by a non-zero polynomial $F \in k[X, Y, Z]_d$, of degree $d > 0$, determined up to a multiplicative constant. We will often write it as $F(x, y, z) = 0$.

If K is a field containing k, we set

$$F(K) := \{(a, b, c) \in \mathbb{P}_2(K); F(a, b, c) = 0\};$$

this makes sense since F is homogeneous.

Thus, we see that the curve F defines a functor

$$K \longmapsto F(K) \in \mathcal{P}(\mathbb{P}_2(K)),$$

where $\mathcal{P}(\mathbb{P}_2(K))$ denotes the set of subsets of $\mathbb{P}_2(K)$.

Example 4.1.1

(1) Curves of degree 1 are called **lines**.

(2) Curves of degree 2 are often called *conics*. A conic can sometimes be **decomposed** as the union of two distinct or identical lines; this can be seen by taking

$$F(X, Y, Z) := (u_1 X + v_1 Y + w_1 Z)(u_2 X + v_2 Y + w_2 Z).$$

(3) Curves of degree 3 are often called *cubics*. A cubic can sometimes be *decomposed* as the union of a line and a conic; this can be seen by taking

$$F(X, Y, Z) := (uX + vY + wZ)\Gamma(X, Y, Z)$$

where $\Gamma(X, Y, Z)$ is a conic.

Definition 4.1.2 *Let $\bar{k}$ be an algebraic closure of k. If the homogeneous polynomial $F \in k[X, Y, Z]$ is irreducible in $\bar{k}[X, Y, Z]$, we say that the curve F is* **absolutely irreducible**.

Let us now recall the definition of a **singular point** (or **multiple point**) of a curve F.

Definition 4.1.3 *Let $P = (a, b, c) \in F(\bar{K})$. We say that P is a* **singular point** *(or a* **multiple point***) of F if*

$$\frac{\partial F}{\partial X}(a, b, c) = \frac{\partial F}{\partial Y}(a, b, c) = \frac{\partial F}{\partial Z}(a, b, c) = 0.$$

Otherwise we say that P is **non-singular** *or* **simple**.

Remark 4.1.1

(1) This definition is meaningful since F is homogeneous.

(2) When the degree n of F is not divisible by the characteristic of k, the conditions of the definition are redundant since Euler's identity says that

$$nF = X\frac{\partial F}{\partial X} + Y\frac{\partial F}{\partial Y} + Z\frac{\partial F}{\partial Z}.$$

(3) A curve having no singular points in $\mathbb{P}_2(\bar{K})$ is said to be **non-singular**, or **smooth**.

Example 4.1.2 **(1)** The curves

$$X^2, \qquad X^2 + Y^2, \qquad Y^2Z - X^3, \qquad Y^2Z - X^2(X - Z)$$

are singular.

(2) The cubic

$$L: \quad Y^2Z - X(X - Z)(X - \lambda Z), \quad \lambda \in k \setminus \{0, 1\}$$

is non-singular when the characteristic of k is not equal to 2. But if $k = \mathbb{F}_4 = \{0, 1, \lambda, \mu\}$, we see that the point $P = (\mu, 1, 1)$ is a singular point of L.

(3) The cubic

$$F: \quad X^3 + Y^3 + Z^3$$

is non-singular when the characteristic of k is not equal to 3, but if it is equal to 3 we have

$$F(X, Y, Z) = (X + Y + Z)^3$$

and all its points are singular (it is a triple line).

(4) If the line D does not intersect the smooth conic C in $\mathbb{P}_2(\mathbb{Q})$, then the cubic $F = DC$ has no singular point in $\mathbb{P}_2(\mathbb{Q})$. However, it admits singular points in $\mathbb{P}_2(\bar{\mathbb{Q}})$, as shown by the following property.

Property 4.1.1 *Let F and G be two plane curves defined over k, and let (a, b, c) be a point of $F(k) \cap G(k)$. Then (a, b, c) is a multiple point of the curve FG.*

Proof. Let u denote one of the variables X, Y, Z, and P the point $(a, b, c) \in k^3$. We have

$$\frac{\partial FG}{\partial u}(P) = \frac{\partial F}{\partial u}(P)G(P) + F(P)\frac{\partial G}{\partial u}(P),$$

so $(\partial FG/\partial u)(P) = 0$ for every choice of u. □

Corollary 4.1.1 *If $F(\bar{k})$ does not contain any singular points, then F is absolutely irreducible.*

We can now give the precise conditions to be satisfied by a cubic in order for it to qualify as an **elliptic curve**.

> **Definition 4.1.4** *A cubic F defined over k is an* **elliptic curve** *defined over k if and only if $F(k) \neq \phi$ and it is non-singular.*

Example 4.1.3

(1) $Y^2Z - X^3$ and $Y^2Z - X^2(X - Z)$ are not elliptic curves over any field.
(2) $Y^2Z - X(X - Z)(X - 2Z)$ is an elliptic curve over every field of characteristic not equal to 2.
(3) $X^3 + Y^3 + Z^3$ is an elliptic curve over every field of characteristic not equal to 3.
(4) $X^3 + pY^3 + p^2Z^3$, where p is a prime number, is not an elliptic curve over $\mathbb{Q}$, but is an elliptic curve over $\bar{\mathbb{Q}}$ and even over $\mathbb{Q}(\sqrt[3]{p})$.

When the projective plane curve $F \in k[X, Y, Z]_d$ is not absolutely irreducible, there exist G and H in $\bar{k}[X, Y, Z] \setminus \bar{k}$ such that $F = GH$. One can then ask whether they are homogeneous.

Proposition 4.1.1 *Let $d \geq 1$ and $F \in k[X, Y, Z]_d \setminus \{0\}$. If $F = GH$ with G and $H \in k[X, Y, Z] \setminus k$, then G and H are homogeneous, i.e. there exist integers g and $h \geq 0$ such that $G \in k[X, Y, Z]_g$ and $H \in k[X, Y, Z]_h$ with $F = GH$.*

Proof. Set $K = k(X, Y, Z)$ and consider a new indeterminate T. We have

$$\begin{cases} F(TX, TY, TZ) = \ F(X, Y, Z)T^d \in K[T] \\ G(TX, TY, TZ) = \sum_i a_i(X, Y, Z)T^i \in K[T] \\ H(TX, TY, TZ) = \sum_j b_j(X, Y, Z)T^j \in K[T]. \end{cases}$$

From the relation

$$T^d = \left(\sum_i \frac{a_i(X, Y, Z)}{F(X, Y, Z)} T^i\right)\left(\sum_j \frac{b_j(X, Y, Z)}{F(X, Y, Z)} T^j\right)$$

in $K[T]$, we deduce that the first factor is associated to T^g for a certain $g \geq 1$ and the second factor is associated to T^h, for $h \geq 1$. Thus all the a_i except for a_g are zero and all the b_j except for b_h are zero. □

Corollary 4.1.2 *Let F be a projective plane curve. If*

$$F = \prod_{i=1}^{r} F_i^{e_i}$$

is the decomposition of F as a product of irreducible factors in $\bar{k}[X, Y, Z]$, then all the polynomials F_i are homogeneous, and we have

$$F(\bar{k}) = \bigcup_{i=1}^{r} F_i(\bar{k}).$$

Definition 4.1.5 *Under the conditions of the corollary, the F_i are called the irreducible components of the curve F, and we say that e_i is the multiplicity of the component F_i.*

Let $P = (a, b, c) \in F(k)$ be a point on the plane curve $F(X, Y, Z) \in k[X, Y, Z]_d \setminus \{0\}$. As one of the three numbers a, b, c is non-zero, we can assume, for example, that $c = 1$ and $F(a, b, 1) = 0$.

Definition 4.1.6 *The* **dehomogenisation** *of F in Z is the affine curve $F_\flat(x, y)$ defined by*

$$F_\flat(x, y) = F(x, y, 1).$$

We then have

$$F_\flat(x, y) = F_1(x - a, x - b) + \cdots + F_d(x - a, x - b),$$

where each $F_i(X, Y) \in k[X, Y]_i$.

We know that

(1) If P is simple, then $F_1(X, Y) \neq 0$ and the equation of the tangent to $F_\flat$ at (a, b) is

$$F_1(x - a, y - b) = 0.$$

(2) If P is multiple, then $F_1(X, Y) \equiv 0$. In fact, we easily see that a necessary and sufficient condition for P to be singular on F is that $F_1(X, Y) \equiv 0$.

Definition 4.1.7 *The* **order of multiplicity** *of P on F, denoted by $m_p(F)$, is the smallest index i such that $F_i(X, Y) \not\equiv 0$.*

Remark 4.1.2 The number $m_p(F)$ does not depend on the choice of a system of coordinates in $\mathbb{P}_2(k)$.

Definition 4.1.8 *Consider two curves F and G defined over k, and take three homogeneous polynomials of the same degree $d > 0$ in $k[X, Y, Z]$:*

$$A(X, Y, Z), \quad B(X, Y, Z) \quad \text{and} \quad C(X, Y, Z).$$

We say that A, B and C define a **rational map** *φ from F to G, defined over k, if for all but a finite number of points $(x, y, z) \in F(\bar{k})$, $\varphi(x, y, z) = (A(x, y, z), B(x, y, z), C(x, y, z))$ is defined and lies in $G(\bar{k})$.*

Example 4.1.4 **(1)** Let F be the line given by

$$F(X, Y, Z) = Z,$$

and consider the conic

$$G(X, Y, Z) = Y^2 - XZ.$$

Then

$$\varphi(x, y, z) = (x^2, xy, y^2)$$

is a rational map from F to G defined over every field (it is a rational parametrisation of the conic).

(2) Take the cubics

$$F(X, Y, Z) = X^3 + Y^3 + dZ^3 \in \mathbb{Q}[X, Y, Z]_3$$

and

$$G(X, Y, Z) = Y^2Z - (X^3 - 2^4.3^3d^2Z^3) \in \mathbb{Q}[X, Y, Z]_3.$$

Then

$$\varphi(x, y, z) = (-2^2 \cdot 3dz, 2^2 \cdot 3^2d(x - y), x + y)$$

is a rational map from F to G defined over $\mathbb{Q}$. Show that $\varphi(x, y, z)$ is defined for every $(x, y, z) \in F(\bar{\mathbb{Q}})$ when $d \neq 0$.

(3) Consider the following cubics defined over $\mathbb{Q}$:

$$F(X, Y, Z) \equiv Y^2Z - X(X^2 + aXZ + bZ^2) \qquad \text{with } b \neq 0 \quad \text{and} \quad a^2 - 4b \neq 0,$$
$$G(X, Y, Z) \equiv Y^2Z - X(X^2 - 2aXZ + (a^2 - 4b)Z^2).$$

Then

$$\varphi(x, y, z) = (y^2z, y(x^2 - bz^2), x^2z)$$

is a rational map from F to G such that $\varphi(x, y, z)$ is defined for every $(x, y, z) \in F(\bar{\mathbb{Q}})$.

The properties of the two last examples generalise, and we have the following result ([Fu] p. 160).

Theorem 4.1.1 *If the curve F is smooth, a rational map $F \xrightarrow{\varphi} G$ is defined at every point of $F(\bar{k})$, where naturally, k denotes a field of definition of F, G and φ.*

Remark 4.1.3 If F is singular, it can be shown that φ is defined at every non-singular point of $F(\bar{k})$.

Definition 4.1.9 *Consider two curves F and G defined over k, and let $\varphi : F \to G$ be a rational map defined over k.*

We say that φ is **birational** *if there exists a rational map $\psi : G \to F$, defined over k, such that for "almost all" the points of $F(\bar{k})$ and $G(\bar{k})$, the maps $\psi \circ \varphi$ and $\varphi \circ \psi$ are defined and equal to the identity.*

We say that F and G are **birationally equivalent over** *k, if there exists a birational map φ from F to G defined over k.*

Remark 4.1.4

(1) The phrase "almost all" means "all but a finite number".
(2) A birational map is bijective almost everywhere on $F(\bar{k})$ and $G(\bar{k})$.

Definition 4.1.10 *Two elliptic curves defined over k are said to be* **equivalent** *over k if there exists a birational map, defined over k, from one to the other.*

4.2 BÉZOUT'S THEOREM

A famous result usually attributed to Etienne Bézout[46], although it was already partially known to Maclaurin[47] and Cramer[48], states that two algebraic plane curves of degrees m and n respectively, having **no common component**, "intersect" in exactly mn points.

In one direction, this result can be usefully restated as follows: if two curves of degrees m and n have more than mn distinct intersection points, then they have a common component.

However, in the other direction, it needs to be made precise (who doesn't know about parallel lines or lines which do not meet a circle?). As we already noted in Section 4.1, the adequate framework for this situation is that of projective geometry over an algebraically closed field.

We will first define the intersection multiplicity of the curves F and G at a point $P \in \mathbb{P}_2(\bar{k})$, where $\bar{k}$ denotes an algebraic closure of a field of definition common to both F and G.

Since we are working here with a local notion, we can choose a system of projective coordinates in such a way that the homogeneous coordinates of P are $(0, 0, 1)$. Let $F_\flat$ and $G_\flat$ be the dehomogenised polynomials of F and G in Z (i.e., we set $Z = 1$), and let $\mathcal{O}_P = \{U(x, y)/V(x, y)$, with U and $V \in \bar{k}[X, Y]$ and $V(0, 0) \neq 0\}$ be the local ring of P in $\bar{k}^2$. Then we know ([Fu], ch. VI) that the dimension of the quotient ring $\mathcal{O}_P/(F_\flat, G_\flat)$ over $\bar{k}$ depends only on P and on the curves F and G (here, $(F_\flat, G_\flat)$ denotes the ideal

[46] E. Bézout (1730–1783).
[47] C. Maclaurin, 1698–1746.
[48] G. Cramer, 1704–1752.

generated by $F_\flat$ and $G_\flat$ in $\mathcal{O}_P$), and not on the choice of a particular system of homogeneous coordinates.

Definition 4.2.1 *The* **intersection multiplicity** *of F and G at P, denoted by $\mu_P(F, G)$, is the integer defined by*

$$\mu_P(F, G) = \dim_{\bar{k}} \mathcal{O}_P/(F_\flat, G_\flat).$$

Using this definition, we can state our desired result as follows.

Bézout's Theorem 4.2.1 *Let F and G be two algebraic plane curves in $\mathbb{P}_2(\bar{k})$, of degree m and n,* **having no common component** *in $\mathbb{P}_2(\bar{k})$. Then we have*

$$\sum_{P \in \mathbb{P}_2(\bar{k})} \mu_P(F, G) = mn.$$

Example 4.2.1 Let F be a cubic defined by a homogeneous polynomial of degree 3 of $k[X, Y, Z]$. Assume that $A \in \mathbb{P}_2(\bar{k})$ is a multiple point of $F(\bar{k})$. Then, choosing the coordinates as above, we have

$$F_\flat(x, y) = a_2(x, y) + a_3(x, y),$$

where the polynomials $a_i(x, y)$ are homogeneous of degree i in $\bar{k}[x, y]$.

Up to making a projective coordinate change, we can assume that the equation of an arbitrary line D passing through A is $Y = 0$. Thus we have $D_\flat = y$.

We have three possibilities for $I := (F_\flat, D_\flat)$ namely

$$\begin{cases} (y) & \text{(i)} \\ (x^2, y) & \text{(ii)} \\ (x^3, y) & \text{(iii)}. \end{cases}$$

In the first case, we see that D divides F, so D is a common component of D and F. We have

$$\mathcal{O}_P/I = R = \left\{ \frac{u(x)}{v(x)} \in \bar{k}(x);\ v(0) \neq 0 \right\},$$

where $u(x)$ and $v(x)$ are polynomials in $\bar{k}[x]$. Thus $\dim_{\bar{k}}(\mathcal{O}_P/I) = \infty$.

In cases (ii) and (iii), we have

$$\mathcal{O}_P/I \cong R/(x^2) \ \ (\text{resp. } R/(x^3)) \quad \text{and} \quad \dim_{\bar{k}}(\mathcal{O}_P/I) = 2 \ \ (\text{resp. } 3).$$

We can now see that if the cubic F has two distinct multiple points, then F is reducible.

Indeed if this were not the case, and if D denoted the line joining these multiple points A and B, we would have $\mu_A(F, D) \geq 2$ and $\mu_B(F, D) \geq 2$. Thus we would have

$$3 = \sum_{P \in \mathbb{P}_2(\bar{k})} \mu_P(F, D) \geq 2 + 2 = 4,$$

which is absurd. □

The preceding example shows that the direct computation of $\mu_P(F, G)$ is quite difficult. In practice, it is easier to use the following seven properties and a simple algorithm to compute $\mu_P(F, G)$.

List of the seven properties characterising μ_P.

(1) *$\mu_P(F, G) \in \mathbb{N} \cup \{\infty\}$ and $\mu_P(F, G) \neq \infty$ if and only if F and G have no common component containing P.*

(2) *$\mu_P(F, G) = 0$ if and only if P is not a point lying on both F and G. Moreover, $\mu_P(F, G)$ does not depend on the components of F and G passing through P.*

(3) *$\mu_P(F, G)$ is invariant under change of coordinates.*

(4) *$\mu_P(F, G) = \mu_P(G, F)$.*

(5) *$\mu_P(F, G) \geq m_P(F)m_P(G)$, and equality holds if and only if F and G have no common tangents at P.*

(6) *If $F = \prod F_i^{r_i}$ and $G = \prod G_j^{s_j}$, then*

$$\mu_P(F, G) = \sum_{i,j} r_i s_j \mu_P(F_i, G_j).$$

(7) *For every $A \in \bar{k}[x, y]$, we have*

$$\mu_P(F, G) = \mu_P(F, G + AF).$$

Algorithm 4.2.1 Assume that $P = (0, 0)$ (this is legitimate by property 3), and $\mu_P(F, G) < \infty$ (which can be seen using property 1).

Property 2 tells us if $\mu_P(F, G) = 0$.

Let us use induction on $n = \mu_P(F, G)$, assuming that we know how to compute $\mu_P(A, B)$ if $\mu_P(A, B) < n$.

Let r and s be the degrees of $F(x, 0)$ and $G(x, 0)$. By property 4, we may assume that $r \leq s$.

First case: $r = -\infty$.

Then $F = yH$ and $\mu_P(F, G) = \mu_P(y, G) + \mu_P(H, G)$ by property 6.

But

$$\mu_P(y, G) \overset{\text{property 7}}{=} \mu_P(y, G(x, 0)) = \mu_P(y, x^m) \overset{\text{properties 5 and 6}}{=} m.$$

Since $P \in G(\bar{k})$, we have $m > 0$, so $\mu_P(H, G) < n$, and we finish by induction.

Second case: $r \geq 0$.

We can take $F(X, 0)$ and $G(X, 0)$ monic, and set

$$H = G - X^{s-r}F.$$

We then have

$$\mu_P(F, G) = \mu_P(F, H),$$

and $\deg(H(x, 0)) = t < s$.

Repeating this procedure a finite number of times, and possibly inverting the curves, we reduce to the first case. □

Example 4.2.2 Computation of $\mu_P(F, G)$ when

$$\begin{aligned} P &= (0,0), \\ F &= (x^2+y^2)^2 + 3x^2y - y^3, \\ G &= (x^2+y^2)^3 - 4x^2y^2. \end{aligned}$$

In characteristic zero, we can draw the curves $F(\mathbb{R})$ and $G(\mathbb{R})$ as in the following figure.

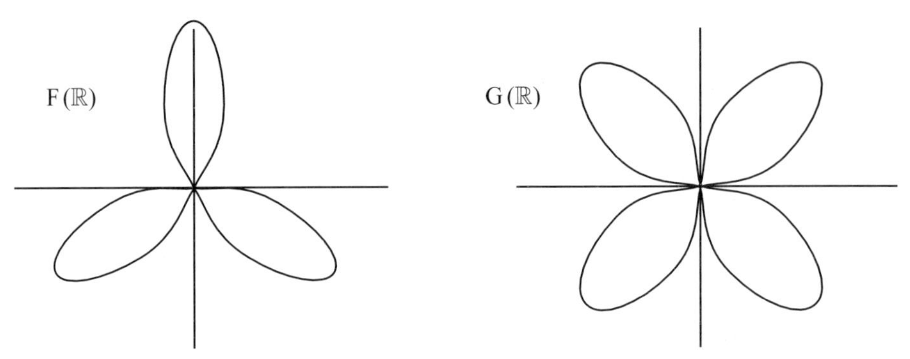

Three-leaved clover Four-leaved clover

We have

$$\mu_P(F, G) = \mu_P(F, G - (x^2+y^2)F) = \mu_P(F, yH),$$

with $H = (x^2+y^2)(y^2-3x^2) - 4x^2y$. Let us compute $\mu_P(F, H)$. We have

$$F(x,0) = x^4, \qquad H(x,0) = -3x^4,$$

so, if the characteristic of k is different from 3, $r = s = 4$. We find that

$$3F + H = y(5x^2 - 3y^2 + 4y^3 + 4x^2y) = yE.$$

Hence

$$\mu_P(F, G) = 2\mu_P(F, y) + \mu_P(F, E).$$

But $\mu_P(F, y) = \mu_P(x^4, y) = 4$ and $\mu_P(F, E) = m_P(F)m_P(E) = 6$ if the characteristic of k does not divide 10, so $\mu_P(F, G) = 14$.

If the characteristic of k is equal to 5, we have

$$\begin{aligned} \mu_P(F, E) &= \mu_P(F, -3y^2 + 4y^3 + 4x^2y) \\ &= \mu_P(F, y) + \mu_P(F, -3y + 4y^2 + 4x^2) \\ &= 4 + \mu_P(-3y + 4y^2 + 4x^2, F). \end{aligned}$$

For the last multiplicity $r = 2$ and $s = 4$, we consider

$$4F - x^2(4x^2 + 4y^2 - 3y) = y^2[(4x^2 + 4y^2) - 4y],$$

so

$$\begin{aligned}\mu_P(-3y + 4y^2 + 4x^2, F) &= \mu_P(F, y^2) + \mu_P(F, 4x^2 + 4y^2 - 4y) \\ &= 8 + \mu_P(F, x^2 + y^2 - y).\end{aligned}$$

Now,

$$F - (x^2 + y^2)(x^2 + y^2 - y) = 3x^2y - y^3 + y(x^2 + y^2) = 4x^2y,$$

so

$$\begin{aligned}\mu_P(F, x^2 + y^2 - y) = \mu_P(F, x^2y) &= \mu_P(F, y) + 2\mu_P(F, x) \\ &= 4 + 2\mu_P(y^3, x) = 4 + 6 = 10.\end{aligned}$$

Thus, $\mu_P(F, E) = 22$ and $\mu_P(F, G) = 30$.

If the characteristic of k is equal to 2, we have

$$\mu_P(F, E) = \mu_P(F, x^2 + y^2) = \mu_P(x^2y + y^3, x^2 + y^2) = \infty,$$

so $\mu_P(F, G) = \infty$.

If the characteristic of k is equal to 3, we have

$$\mu_p(F, G) = \mu_p(F, y^2H) = 8 + \mu_p(F, H)$$

with $H = y(x^2 + y^2) - x^2$. Now,

$$\mu_P(H, F) = \mu_P(H, yK) = \mu_P(H, y) + \mu_P(H, K)$$

with $K = x^2(x^2 + y^2) - yx^2 + y^3 - y^2$. We also have

$$\mu_P(H, K) = \mu_P(H, K + x^2H) = \mu_P(H, y) + \mu_P(H, L),$$

with $L = x^4 - x^2 + (y^2 + y)x^2 + y^2 - y$. Furthermore, $\mu_P(H, L) = \mu_P(H, L + (x^2 - 1)H) = \mu_P(H, y)$. It follows that $\mu_P(F, G) = 14$. □

4.3 NINE-POINT THEOREM

Bézout's theorem can be expressed in a particularly agreeable way if we introduce the Abelian group of (dimension zero) cycles of $\mathbb{P}_2(k)$. This is the free $\mathbb{Z}$-module generated by the points of $\mathbb{P}_2(\bar{k})$, where $\bar{k}$ denotes an algebraic closure of k:

$$\mathcal{Z} := \bigoplus_{P \in \mathbb{P}_2(\bar{k})} \mathbb{Z}\overrightarrow{P} \cong \mathbb{Z}^{(\mathbb{P}_2(\bar{k}))}.$$

A *cycle* is thus a sum $\sum_{P \in \mathbb{P}_2(\bar{k})} n_P \overrightarrow{P}$, where $n_p \in \mathbb{Z}$ and where all but a finite number of the n_p are zero.

Definition 4.3.1 *The homomorphism*

$$\deg \begin{cases} \mathcal{Z} \longrightarrow \mathbb{Z} \\ \sum n_P \overrightarrow{P} \longmapsto \sum n_P \end{cases}$$

is known as the **degree**, *and* $\sum n_P$ *is the degree of the cycle* $\sum n_P \overrightarrow{P}$.

Let us define a partial ordering on $\mathcal{Z}$ by setting

$$\left(\sum m_P \overrightarrow{P} \geq \sum n_P \overrightarrow{P}\right) \iff \text{for every } P \in \mathbb{P}_2(\bar{k}),\ m_P \geq n_P.$$

A **cycle** is said to be **positive** if it is greater than or equal to the zero cycle.

Now, consider two plane curves F and G having no common component.

Definition 4.3.2 *The* **intersection cycle** *of F and G, denoted by $F \cdot G$, is the cycle*

$$F \cdot G := \sum_{P \in \mathbb{P}^2(\bar{k})} \mu_P(F, G) \overrightarrow{P}.$$

Remark 4.3.1 With this notation, Bézout's theorem (theorem 4.2.1) can be stated as

$$\deg(F \cdot G) = \deg F \cdot \deg G.$$

Given this, we can now state the following consequence of "Max Noether's fundamental theorem"[49] ([Fu] p. 120); all curves here are assumed to be defined over k.

Theorem 4.3.1 *Consider two curves F and G having no common component, and such that all the points of $F(\bar{k}) \cap G(\bar{k})$ are simple over F.*

Then if H is a third curve such that

$$H \cdot F \geq G \cdot F,$$

there exists a homogeneous polynomial A such that $A \cdot F = H \cdot F - G \cdot F$.

Corollary (Nine-point Theorem) 4.3.1 *Let C be an absolutely irreducible cubic and let C' and C'' be two arbitrary cubics.*

Assume that $C \cdot C' = \sum_{i=1}^{9} \overrightarrow{P}_i$ and that the points P_i are **simple** *(but not necessarily distinct) on C.*

Then if $C \cdot C'' = \sum_{i=1}^{8} \overrightarrow{P}_i + \overrightarrow{Q}$, we have $Q = P_9$.

[49] M. Noether, 1844–1921.

Proof. We will reason by the absurd, by assuming that $Q \neq P_9$.

Let D be a line passing through P_9 but not through Q. We have

$$D \cdot C = \overrightarrow{P}_9 + \overrightarrow{R} + \overrightarrow{S}.$$

Thus,

$$(DC'') \cdot C = C' \cdot C + (\overrightarrow{Q} + \overrightarrow{R} + \overrightarrow{S}) > C' \cdot C.$$

The preceding theorem ensures the existence of a line D' such that

$$D' \cdot C = \overrightarrow{Q} + \overrightarrow{R} + \overrightarrow{S},$$

but this is absurd since the line RS intersects C at P_9! □

Remark 4.3.2 The nine-point theorem illustrates a situation which was remarked around the middle of the 18th century by Cramer[50] and which gave rise to the theory of determinants.

Let $k[X, Y, Z]_n$ be the space of homogeneous polynomials of degree n. It is not difficult to see that

$$\dim_k k[X, Y, Z]_n = \frac{(n+2)(n+1)}{2}.$$

"Cramer's paradox" consists in noting that, in general, an algebraic plane curve of degree n is determined by

$$\frac{(n+2)(n+1)}{2} - 1 = \frac{n(n+3)}{2}$$

points ("in general position"); however Bézout's theorem (theorem 4.2.1) implies that two distinct curves of degree n intersect, in general, in n^2 points. But $n^2 \geq n(n+3)/2$ if $n \geq 3$.

The solution of this paradox for plane cubics comes naturally from the fact that the nine points of intersection of two distinct cubics are not "in general position".

4.4 GROUP LAWS ON AN ELLIPTIC CURVE

Let us choose our elliptic curve in the form of a plane cubic C defined over k having no multiple point in $\mathbb{P}_2(\bar{k})$.

When k is the field of complex numbers $\mathbb{C}$ and the curve C is given in Weierstrass form, the theory of elliptic functions gives us a key to equip $C(k)$ with a group structure (see Section 2.8 of Chapter 2).

We will generalise this construction in two different directions: k will be an **arbitrary** field, and the zero of the group will be an **arbitrary** point O of $C(k)$ (rather than an inflection point).

Rule 4.4.1 *Let A and B be in $C(k)$ and let R be the third point of intersection of the line AB with C. Then the sum $A \oplus B$ relative to the origin O will be the third point of intersection of the line OR with C.*

[50] G. Cramer (1704–1752).

Remark 4.4.1 We easily check that R and $A \oplus B$ lie in $C(k)$.

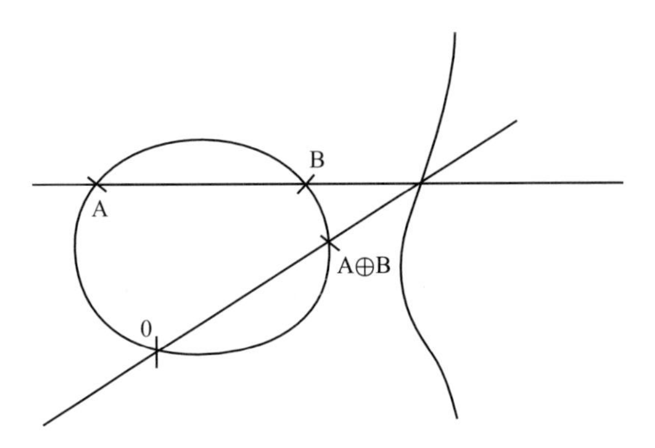

We then check that for every point $A \in C(k)$, we have $A \oplus O = A$. In particular we compute $O \oplus O$ by drawing the tangent to C at the point O. It intersects C at the point T, and the line TO gives back the point O. Thus, we have $O \oplus O = O$.

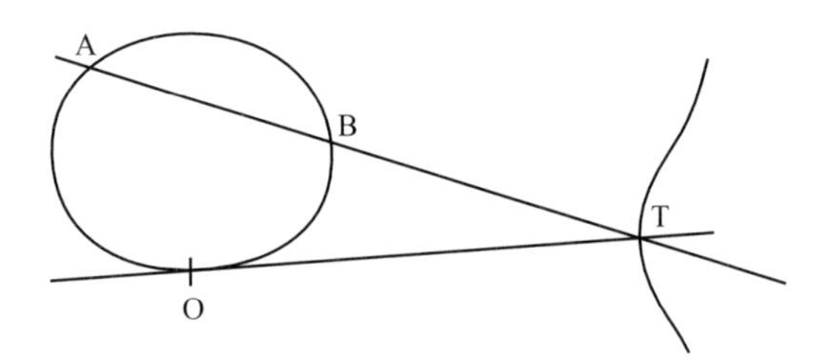

Finally, A and B will be opposites in the group if and only if the line AB passes through T.

Theorem 4.4.1

(1) *The addition law associated to the choice of an origin O is an Abelian group law on $C(k)$.*
(2) *If O and O' are two points of $C(k)$ and if $\oplus$ and $\oplus'$ are the corresponding laws, the groups $(C(k), \oplus)$ and $(C(k), \oplus')$ are isomorphic via a birational isomorphism defined over k.*

Proof. **(1)** In order to show that $(C(k), \oplus)$ is a group, it remains to show that the addition law is associative. Let $P, Q, R \in C(k)$; we want to show that

$$(P \oplus Q) \oplus R = P \oplus (Q \oplus R).$$

To construct the point $S := -[(P \oplus Q) \oplus R]$, consider the following collinearities:

$$L_1 \cdot C = \vec{P} + \vec{Q} + \vec{U'}, \quad M_1 \cdot C = \vec{O} + \vec{U'} + \vec{U}, \quad L_2 \cdot C = \vec{U} + \vec{R} + \vec{S}.$$

To construct the point $T := -[P \oplus (Q \oplus R)]$, we consider the following collinearities:

$$M_2 \cdot C = \overrightarrow{Q} + \overrightarrow{R} + \overrightarrow{V'}, \qquad L_3 \cdot C = \overrightarrow{O} + \overrightarrow{V'} + \overrightarrow{V}, \qquad M_3 \cdot C = \overrightarrow{P} + \overrightarrow{V} + \overrightarrow{T}.$$

It remains to show that $S = T$. For this, we apply the nine-point theorem to the smooth cubic C and to the decomposed cubics

$$C' = L_1L_2L_3, \qquad C'' = M_1M_2M_3.$$

(2) Let P and $Q \in C(k)$, and let $P * Q = R$ denote the third intersection point of the line PQ with C. We propose to show that

$$\varphi : P \in (C(k), \oplus) \longmapsto O * (O' * P) \in (C(k), \oplus')$$

is a birational isomorphism of groups defined over k.

If we set $\psi : Q \mapsto O' * (O * Q)$, it is clear that $\psi \circ \varphi = id$ and $\varphi \circ \psi = id$, so φ is birational and bijective.

Moreover, it is clear that $\varphi(O) = O'$ and the nine-point theorem shows that $\varphi(\ominus P) = \ominus'\varphi(P)$. Thus, it remains to show that $\varphi(P \oplus Q) = \varphi(P) \oplus' \varphi(Q)$, and to see this, it suffices to check that

$$O * (O' * (O * (P * Q))) = O' * ((O(O'P)) * (O(O'Q)))$$

(we have omitted the $*$ on the right for simplicity).

Let us temporarily admit the following lemma.

Lemma 4.4.1 *For every quadruple L, Ω, M, N of points of $C(k)$, we have*

$$L * (\Omega * (M * N)) = M * (\Omega * (L * N)).$$

For the right-hand term we have

$$\begin{aligned} O' * & ((O(O'P)) * (O(O'Q))) \\ &= O * ((O(O'P)) * (O' * (O'Q)) \quad \text{(here } \Omega = O(O'P)) \\ &= O * ((O(O'P)) * Q) \\ &= O * [(Q * (O(O'P))] \\ &= O * [O' * (O(QP)] \quad \text{(here we take } \Omega = O \text{ inside the square bracket),} \end{aligned}$$

which is indeed the left-hand term. □

Proof of the lemma. Since the law of origin Ω is associative, we have

$$\Omega * (L * (\Omega * (M * N))) = \Omega * ((\Omega * (L * M)) * N),$$

and simplifying by Ω (which is legitimate), we obtain the result. □

Remark 4.4.2 When C is singular, let $C_0(k)$ denote the set of non-singular points of C in $\mathbb{P}_2(k)$. Then if $O \in C_0(k)$, the above construction makes it possible to equip $C_0(k)$ with an Abelian group law and the theorem continues to hold *mutatis mutandis.*

When the singular point of C is a cusp, $C_0(k)$ is isomorphic to the group $(k, +)$, see Exercise 4.11.

When the singular point of C is a point with distinct tangents in $\mathbb{P}_2(k)$, $C_0(k)$ is isomorphic to $(k^*, \times)$, see Exercise 4.12.

When the singular point of $C_0(k)$ is an "isolated point" whose tangents are in an extension of k (which we can assume to be quadratic), then the group law group of $C_0(k)$ is a little more complicated; see Exercise 4.13.

Formulae 4.4.1 We restrict ourselves to the case where the curve is given by the long Weierstrass equation

$$W: \qquad y^2 + a_1xy + a_3y - (x^3 + a_2x^2 + a_4x + a_6) = 0,$$

and where we choose $O = (0, 1, 0)$. We say that W is a *Weierstrass cubic* if this cubic is *smooth*.

Then if $P = (x, y)$, we have

$$-P = (x, -y - a_1x - a_3).$$

If $P_1 \oplus P_2 = P_3$ with $P_i = (x_i, y_i)$, then

(i) If $x_1 \neq x_2$, set

$$\lambda = \frac{y_2 - y_1}{x_2 - x_1}, \qquad \mu = \frac{y_1x_2 - y_2x_1}{x_2 - x_1}$$

(ii) If $x_1 = x_2$ and $P_1 = P_2$, set

$$\lambda = \frac{3x_1^2 + 2a_2x_1 + a_4 - a_1y_1}{2y_1 + a_1x_1 + a_3} \qquad \mu = \frac{-x_1^3 + a_4x_1 + 2a_6 - a_3y_1}{2y_1 + a_1x_1 + a_3};$$

in both cases, we have

$$\begin{cases} x_3 = \lambda^2 + a_1\lambda - a_2 - x_1 - x_2 \\ y_3 = -(\lambda + a_1)x_3 - \mu - a_3. \end{cases}$$

Remark 4.4.3

(1) The line $y = \lambda x + \mu$ is the line P_1P_2.

(2) If $a_1 = a_3 = 0$ (short Weierstrass form), the formulae simplify, and we have

$$\begin{cases} x_3 = \lambda^2 - a_2 - x_1 - x_2 \\ y_3 = -\lambda x_3 - \mu. \end{cases}$$

4.5 REDUCTION MODULO *P*

It is clear that when we want to study the rational points of a projective plane curve of equation

$$F(X, Y, Z) \in \mathbb{Q}[X, Y, Z]_n,$$

we can assume that the polynomial F has integral coefficients whose greatest common divisor is 1, and that the points of $F(\mathbb{Q})$ are represented by triples $(a, b, c) \in \mathbb{Z}^3$ whose greatest common divisor is also 1 ("primitive solutions"). We can now consider the equation modulo p:

$$\bar{F}(\bar{a}, \bar{b}, \bar{c}) = O \in \mathbb{F}_p.$$

Example 4.5.1 Consider the curve

$$X^{2n} + Y^{2n} - Z^{2n} \in \mathbb{Q}[X, Y, Z]_{2n}.$$

If this equation admits a primitive solution $(a, b, c) \in \mathbb{Z}^3$, we have

$$a^{2n} + b^{2n} \equiv c^{2n} \bmod 3,$$

and as $u^{2n} \in \{\bar{0}, \bar{1}\}$ in $\mathbb{F}_3$, we see that $abc \equiv 0 \bmod 3$.

Similarly, if $2n + 1$ is a prime number p ($n = 5$ for example), we have $u^{2n} \in \{\bar{0}, \bar{1}\}$ in $\mathbb{F}_p$ and $abc \equiv 0 \bmod p$.

Now consider a plane cubic which is smooth and rational over $\mathbb{Q}$, given by the long Weierstrass equation

$$W = Y^2Z + a_1XYZ + a_3YZ^2 - (X^3 + a_2XZ^2 + a_4X^2Z + a_6Z^3) = 0,$$

where a_1, a_3, a_2, a_4, a_6 are rational numbers. When we make a projective transformation of the type

$$X = \frac{\xi}{\lambda^2}, \qquad Y = \frac{\eta}{\lambda^3}, \qquad Z = \zeta$$

with $\lambda \in \mathbb{Q}^*$, we find a new equation (in (ξ, η, ζ)) which is of the same type but where a_i becomes $\lambda^i a_i$.

Thus, by choosing λ suitably, we reduce to an equation with integral coefficients which are "not too large" (in order to avoid systematically finding the uninteresting reduced equation $Y^2Z - X^3$).

We now consider an arbitrary prime number p, and the following reduced equation:

$$\bar{W} = Y^2Z + \bar{a}_1XYZ + \bar{a}_3YZ^2 - (X^3 + \bar{a}_2XZ^2 + \bar{a}_4X^2Z + \bar{a}_6Z^3) \in \mathbb{F}_p[X, Y, Z].$$

Our goal is to study the reduction map

$$\begin{cases} \mathbb{P}_2(\mathbb{Q}) \xrightarrow{\pi} \mathbb{P}_2(\mathbb{F}_p) \\ (a, b, c) \longmapsto (\bar{a}, \bar{b}, \bar{c}), \end{cases}$$

where a, b, c are chosen relatively prime, and to show that it induces a homomorphism from $W(\mathbb{Q})$ to $\bar{W}(\mathbb{F}_p)$ *when* $\bar{W}$ is **smooth.**

Let us first note, by duality, that a line D of $\mathbb{P}_2(\mathbb{Q})$ has an equation of the form

$$D: \quad uX + vY + wZ = 0$$

such that the greatest common divisor of $(u, v, w) \in \mathbb{Z}^3$ is 1. Thus if P_1, P_2, P_3 are collinear on D, the points $\pi(P_1), \pi(P_2), \pi(P_3)$ are collinear on the line

$$D: \quad \bar{u}X + \bar{v}Y + \bar{w}Z = \bar{0}.$$

Let us state the following special case of a much more general result.

Proposition 4.5.1 *Let C be a plane cubic defined over $\mathbb{Q}$, and D a line of $\mathbb{P}_2(\mathbb{Q})$. Assume that $C(\mathbb{Q}) \cap D(\mathbb{Q}) = \{P_1, P_2, P_3\}$, where P_i is repeated as many times as its intersection multiplicity (in $\mathbb{P}_2(\mathbb{Q})$). Let $\bar{C}$, $\bar{D}$ and $\bar{P}_i$ denote the objects deduced from C, D and P_i by reduction. Then if $\bar{D}$ is not a component of $\bar{C}$, we have*

$$\bar{C}(\mathbb{F}_p) \cap \bar{D}(\mathbb{F}_p) = \{\bar{P}_1, \bar{P}_2, \bar{P}_3\}$$

with the same convention on the repetition of points.

Proof. **(1)** First consider the special case where the equation of D is $Z = 0$. Set $P_i = (\ell_i, m_i, 0)$ with $(\ell_i, m_i) \in \mathbb{Z}^2$ relatively prime. Thus, we have

$$C(X, Y, O) = c \prod_{i=1}^{3} (m_i X - \ell_i Y) = c\, F(X, Y),$$

with $c \in \mathbb{Q}^*$.

Since $\bar{m}_i X - \bar{\ell}_i Y \neq 0$ in $\mathbb{F}_p[X, Y]$, we see that $\bar{F}(X, Y) \neq 0$, so that p does not divide all the coefficients of $F(X, Y)$.

Since $\bar{D}$ is not a component of $\bar{C}(X, Y, Z)$, we see that $\bar{C}(X, Y, O) \neq 0$ and that p cannot divide either the numerator or the denominator of c (written as an irreducible fraction). Thus, we have

$$\bar{C}(X, Y, O) = \bar{c}.\bar{F}(X, Y) = \bar{c} \prod_{i=1}^{3} \left(\bar{m}_i X - \bar{\ell}_i Y\right),$$

which gives the result in this case.

(2) When the line D is arbitrary, we can reduce to the preceding case by a suitable change of coordinates in $\mathbb{P}_2(\mathbb{Q})$. But in order for this change of coordinates to "pass to the quotient modulo p", we need to make sure to choose it in $SL_3(\mathbb{Z})$.

Let $D = uX + vY + wZ,$, where the greatest common divisor of $(u, v, w) \in \mathbb{Z}^3$ is 1. We want a matrix of the form

$$\begin{pmatrix} u_1 & v_1 & w_1 \\ u_2 & v_2 & w_2 \\ u & v & w \end{pmatrix} \in SL_3(\mathbb{Z}),$$

so that if we take

$$\begin{pmatrix} X' \\ Y' \\ Z' \end{pmatrix} = \begin{pmatrix} u_1 & v_1 & w_1 \\ u_2 & v_2 & w_2 \\ u & v & w \end{pmatrix} \begin{pmatrix} X \\ Y \\ Z \end{pmatrix},$$

the equation of D is $Z' = 0$.

Let d be the greatest common divisor of v and w. We know that there exists v_2 and w_2 such that $v_2 w - w_2 v = d$, and v_2 and w_2 are relatively prime. Now, we know that the greatest common divisor $(u, d) = 1$, so there exists u_1 and x such that $u_1 d + xu = 1$. Finally the greatest common divisor $(v_2, w_2) = 1$, and there exist v_1 and w_1 such that $v_1 w_2 - v_2 w_1 = x$. We check that the determinant of the matrix

$$\begin{pmatrix} u_1 & v_1 & w_1 \\ 0 & v_2 & w_2 \\ u & v & w \end{pmatrix}$$

is indeed equal to 1. □

Theorem 4.5.1 *Let C be a smooth cubic in $\mathbb{P}_2((\mathbb{Q}))$, and let $O \in C(\mathbb{Q})$. Let $\bar{C}$ be the cubic reduced modulo p, and let $\bar{O}$ be the reduction of O.*

(1) *If the cubic $\bar{C}$ is smooth, then the reduction map*

$$\mathbb{P}_2(\mathbb{Q}) \xrightarrow{\pi} \mathbb{P}_2(\mathbb{F}_p)$$

induces a homomorphism from $C(\mathbb{Q})$ equipped with the addition law of origin O to $\bar{C}(\mathbb{F}_p)$ equipped with the addition law of origin $\bar{O}$.

(2) *If the cubic $\bar{C}$ admits a double point S in $\mathbb{P}_2(\mathbb{F}_p)$, let $C_0(\mathbb{Q})$ denote the set $C(Q)\backslash\pi^{-1}(S)$.*

Then if $O \in C_0(\mathbb{Q})$, the set of points of $C_0(\mathbb{Q})$ forms a subgroup of $C(\mathbb{Q})$ and the reduction map induces a homomorphism of this subgroup on the set of non-singular points of $\bar{C}(\mathbb{F}_p)$ equipped with the addition law of origin $\bar{O}$.

Proof. Saying that $P_1 \oplus P_2 = P_3$ is equivalent to stating that each of the triples P_1, P_2, R and R, O, P_3 is collinear. The preceding proposition then shows that these collinearities are preserved by π. □

Vocabulary 4.5.1

- When $\bar{W}$ is smooth, we say that W has **good reduction** at p.
- When $\bar{W}$ admits a double point with distinct tangents in $\mathbb{P}_2(\bar{\mathbb{F}}_p)$, we say that W has **multiplicative reduction** at p.
- When $\bar{W}$ has a cusp in $\mathbb{P}_2(\bar{\mathbb{F}}_p)$, we say that W has an **additive reduction** at p.

Remark 4.5.1 It is clear that the whole of this theory continues to hold if $\mathbb{Z}$ is replaced by an arbitrary principal ideal domain A, and $\mathbb{Q}$ by the fraction field of A.

4.6 n-DIVISION POINTS OF AN ELLIPTIC CURVE

Let E be an elliptic curve defined over k, and let Ω be an arbitrary field containing k. It is clear that $E(k)$ is a subgroup of $E(\Omega)$.

Definition 4.6.1 *Let $n \in \mathbb{N}$ and $P \in E(\Omega)$. We say that P is an n-**division point** of E if*

$$nP := \underbrace{P \oplus P \oplus \cdots \oplus P}_{n \text{ times}} = O.$$

*We say that P is of **order** n if $nP = O$ and if $mP \neq 0$ for every $m \in \mathbb{N}^*$ smaller than n.*

The goal of this paragraph is to show that all the n-division points of E lie in $\bar{k}$ (where $\bar{k}$ denotes an algebraic closure of k), and that if the field k is a field of characteristic zero, they form a group isomorphic to $\mathbb{Z}/n\mathbb{Z} \times \mathbb{Z}/n\mathbb{Z}$.

Since the isomorphism class of the group $(E(k), \oplus)$ is invariant under birational transformation defined over k (let us admit this point), we can replace E by a Weierstrass cubic.

4.6.1 2-Division Points

In order for $P \in E(\Omega)$ to be a 2-division point, it is necessary and sufficient that $2P = P \oplus P = O$.

It is clear that $P = O$ is a 2-division point.

Suppose now that $P \neq O$, and take E in long Weierstrass form:

$$E: \quad y^2 + a_1xy + a_3y - \left(x^3 + a_2x^2 + a_4x + a_6\right) = 0.$$

By Section 4.4, the abscissa ξ and the ordinate η of $2P$ are given by

$$\begin{cases} \xi = \lambda^2 + a_1\lambda - a_2 - 2x \\ \eta = -(\lambda + a_1)\xi - \mu - a_3. \end{cases}$$

But ξ is infinite only if λ is infinite, i.e. if $2y + a_1x + a_3 = 0$.

When the characteristic of k is equal to 2, we see that these points satisfy the equation $a_1x + a_3 = 0$, and are thus contained in the intersection of E and of this line. If $a_1 \neq 0$, we find a single point in $\bar{k}$.

If $a_1 = 0$ and $a_3 \neq 0$, then E has no points of order 2.

If $a_1 = a_3 = 0$, then E admits a multiple point: it is not an elliptic curve.

Example 4.6.1 Let $E : y^2 + y - (x^3 + 1)$. We have

$$\lambda = x^2, \qquad \mu = x^3 + y,$$

so (ξ, η) cannot be the point O. □

When the characteristic of k is not equal to 2, we can take E in short Weierstrass form (see Section 4.13):

$$E: \quad y^2 - \left(x^3 + a_2x^2 + a_4x + a_6\right).$$

The curve is smooth if and only if the polynomial $D(x) = x^3 + a_2x^2 + a_4x + a_6$ has no multiple roots. The points of order 2 of $E(\bar{k})$ are the three points $(e_1, 0)$, $(e_2, 0)$, $(e_3, 0)$ with ordinate equal to zero. Thus e_1, e_2, e_3 are the roots of $D(x)$ in $\bar{k}$. To summarize, we have the following theorem.

Theorem 4.6.1 *If E is given by a short Weierstrass equation, and if the characteristic of k is not equal to 2, the points of order 2 on $E(\bar{k})$ are the points with ordinate equal to zero. Together with the origin O, they form an Abelian group isomorphic to $\mathbb{Z}/2\mathbb{Z} \times \mathbb{Z}/2\mathbb{Z}$.*

Indeed, we know that there exist only two isomorphism classes of groups of order 4, namely $\mathbb{Z}/4\mathbb{Z}$ and $\mathbb{Z}/2\mathbb{Z} \times \mathbb{Z}/2\mathbb{Z}$. Since all the points of $\mathbb{Z}/4\mathbb{Z}$ are not 2-division points, only the latter group is possible.

4.6.2 3-Division Points

Here we restrict ourselves to the case where E can be given in short Weierstrass form, and where $\Omega = \bar{k}$. We set $D(x) = x^3 + a_2x^2 + a_4x + a_6$.

The condition $3P = O$ is equivalent to $2P = -P$.

If $P \neq O$, this is again equivalent to saying that P and $2P$ have the same abscissa x, which gives the equation

$$3x \cdot 4y^2 = \left(3x^2 + 2a_2x + a_4\right)^2 - 4a_2y^2,$$

which can be rewritten as

$$\psi_3(x) = 3x^4 + 4a_2x^3 + 6a_4x^2 + 12a_6x + \left(4a_2a_6 - a_4^2\right) = 0.$$

When the characteristic of k is equal to 3, we have

$$\psi_3(x) = a_2x^3 + a_2a_6 - a_4^2,$$

so if $a_2 = 0$ but $a_4 \neq 0$, $E(\Omega)$ has no points of order 3.

Now assume that the characteristic of k is either zero or strictly greater than 3. We check that

$$\psi_3(x) = 2D(x)D''(x) - [D'(x)]^2,$$

which gives

$$\psi_3'(x) = 12D(x).$$

Since $D(x)$ has no multiple roots, we see that the same holds for $\psi_3(x)$.

If $\beta_1, \beta_2, \beta_3, \beta_4$ are the roots of ψ_3 in $\bar{k}$, and if $\gamma_i = \pm\sqrt{D(\beta_i)}$ for $i \in \{1, 2, 3, 4\}$, it is clear that $\gamma_i \neq 0$ for every i (otherwise P would be of order 2), so we obtain 8 points of order 3, i.e. nine 3-division points.

Theorem 4.6.2 *If the characteristic of k is either zero or different from 2 and 3, the 3-division points of $E(\bar{k})$ form a group isomorphic to $\mathbb{Z}/3\mathbb{Z} \times \mathbb{Z}/3\mathbb{Z}$.*

Indeed, there are only two isomorphism classes of Abelian groups of order 9.

Remark 4.6.1 A moment of thought about the definition of $2P$ is enough to convince oneself that the 8 points of order 3 of $E(\bar{k})$ are the inflection points of E. As the point at infinity of E is also an inflection point, we see that E has exactly 9 inflection points in $\mathbb{P}_2(\bar{k})$. Note that the lines joining two of these points always contain a third such point (**Maclaurin's theorem**[51]).

4.6.3 n-division points of an elliptic curve defined over $\mathbb{Q}$

Clearly, it is prudent to limit oneself to the case of characteristic zero if one desires to keep to simple statements. Moreover, if $k = \mathbb{Q}$, we can take $\Omega = \mathbb{C}$ and make use of the theory of Chapter 2, as long as we admit that **every elliptic curve can be parametrised by elliptic functions attached to a period lattice** Λ (see theorem 5.4.2 of Section 5.4, Chapter 5).

Let $E[n]$ denote the set of n-division points of $E(\mathbb{C})$. We see that $E[n]$ is a group isomorphic to $(1/n)\Lambda/\Lambda \cong \mathbb{Z}/n\mathbb{Z} \times \mathbb{Z}/n\mathbb{Z}$.

Indeed, if ω_1 and ω_2 are generators of Λ, it is clear that

$$\frac{1}{n}\Lambda/\Lambda = \left\{\frac{a_1}{n}\omega_1 + \frac{a_2}{n}\omega_2;\ \frac{a_j}{n} \in \mathbb{Z}/n\mathbb{Z} \text{ for } j = 1 \text{ and } 2\right\}.$$

Set $m = n^2 - 1$, and let $(x_i, y_i) \in \mathbb{C}^2$, $1 \leq i \leq m$, denote the coordinates of the points of order n of $E(\mathbb{C})$. These coordinates generate a field extension $K_n = \mathbb{Q}(x_1, y_1, \ldots, x_m, y_m)$ of $\mathbb{Q}$, to which we now turn our attention.

Reminder of some notions of field theory

Let Aut($\mathbb{C}$) denote the group of all the automorphisms, continuous or discontinuous, of $\mathbb{C}$. In Chapter 3, Section 3.4, we noted that in order for a number $z \in \mathbb{C}$ to be **algebraic**, it suffices that its orbit under the action of the automorphism group of $\mathbb{C}$ be finite.

[51] C. Maclaurin 1698–1746.

We saw that an extension $L/\mathbb{Q}$ is **Galois** if L is an algebraic field and if for every $\sigma \in \mathrm{Aut}(\mathbb{C})$ we have $\sigma(L) \subset L$. The **Galois group** of $L/\mathbb{Q}$ is then the image of $\mathrm{Aut}(\mathbb{C})$ under the homomorphism $\sigma \mapsto \sigma/L$ of restriction to L.

Theorem 4.6.3 *Let E be an elliptic curve defined over $\mathbb{Q}$, and let n be an integer ≥ 1.*

(1) *The group $E[n](\mathbb{C})$ of n-division points of E is isomorphic to $\mathbb{Z}/n\mathbb{Z} \times \mathbb{Z}/n\mathbb{Z}$.*
(2) *The extension $K_n/\mathbb{Q}$ generated by the coordinates of the n-division points of $E(\mathbb{C})$ is a Galois extension of $\mathbb{Q}$.*

Proof.

(1) Let $E[n]$ denote the group of n-division points of $E(\mathbb{C})$.
If $P \in E[n]$, i.e. $[n]P = O$, it is clear that for every $\sigma \in \mathrm{Aut}(\mathbb{C})$, $n[\sigma(P)] = O$. Indeed, $\sigma|_{\mathbb{Q}} = 1\!\!1_{\mathbb{Q}}$, and since E is defined over $\mathbb{Q}$, the automorphism σ respects the addition law of the points on $E(\mathbb{C})$.
(2) Since $\#(E[n]) = n^2$, we see that the coordinates (x, y) of $P \in E[n]$ can only have a finite number of conjugates under the action of $\mathrm{Aut}(\mathbb{C})$; thus they are algebraic, so K_n is algebraic.
(3) For every $\sigma \in \mathrm{Aut}(\mathbb{C})$, we have $\sigma(E[n]) \subset E[n]$, so $\sigma(K_n) \subset K_n$. This shows that $K_n/\mathbb{Q}$ is Galois. □

Remark 4.6.2 Let E be an elliptic curve defined over $\mathbb{Q}$, and let p be a prime number. One can show ([Sil] p. 96) that $E[p]$ is equipped with a non-degenerate "bilinear" form

$$e_p : E[p] \times E[p] \longrightarrow \mu_p(\bar{\mathbb{Q}}),$$

where $\mu_p(\bar{\mathbb{Q}})$ denotes the multiplicative group (isomorphic to $(\mathbb{Z}/p\mathbb{Z}, +)$) of the p-torsion elements of $\bar{\mathbb{Q}}^*$. To every pair (P, Q) of points of $E[p](\bar{\mathbb{Q}})$, this form associates a root of unity $e_p(P, Q)$ by a construction which is **algebraic** (and nearly canonical), which shows that $\mu_p(\bar{\mathbb{Q}}) \subset K_p$. This **Weil form** is "bilinear alternating" in the sense that

$$\begin{cases} e(m_1P_1 + m_2P_2, Q) = e(P_1, Q)^{m_1} e(P_2, Q)^{m_2} \\ e(P, n_1Q_1 + n_2P_2) = e(P, Q_1)^{n_1} e(P, Q_2)^{n_2} \\ e(P, Q)e(Q, P) = 1. \end{cases}$$

Furthermore, it is non-degenerate, and commutes with the action of $G_Q = \mathrm{Gal}(\bar{\mathbb{Q}}/Q)$ in the sense that if $\sigma \in G_{\mathbb{Q}}$, then

$$e(\sigma(P), \sigma(Q)) = \sigma(e(P), e(Q)) = [e(P, Q)]^{\chi_p(\sigma)},$$

where χ_p denotes the cyclotomic character of $G_{\mathbb{Q}}$ (see Chapter 3, Section 3.6).

4.7 A MOST INTERESTING GALOIS REPRESENTATION

Let $\bar{\mathbb{Q}}$ denote the algebraic closure of $\mathbb{Q}$ in the field of complex numbers $\mathbb{C}$, i.e. the set of numbers $z \in \mathbb{C}$ which are algebraic over $\mathbb{Q}$.

We saw in Chapter 3 that $\bar{\mathbb{Q}}$ is a field and that $\bar{\mathbb{Q}}/\mathbb{Q}$ is a Galois extension.

We have also shown that the image of Aut($\mathbb{C}$) by restriction to $\bar{\mathbb{Q}}$ is the Galois group of $\bar{\mathbb{Q}}/\mathbb{Q}$; we denote it by $G_{\mathbb{Q}}$ and call it the **absolute Galois group**. One of the essential problems of number theory is the study of this group $G_{\mathbb{Q}}$, and we have seen that an important tool in this study is the notion of a representation of $G_{\mathbb{Q}}$ in a vector space (or a module) V, i.e. the notion of a homomorphism

$$G_{\mathbb{Q}} \longrightarrow GL(V).$$

Example 4.7.1 Let $n \geq 1$, and let $V = \mu_n(\mathbb{C}) \cong \mathbb{Z}/n\mathbb{Z}$ be the Abelian group of the n-division points of unity in $\mathbb{C}^*$.

Let ζ be a generator of V, and let $L_n = \mathbb{Q}(\zeta)$. For every $\sigma \in G_{\mathbb{Q}}$, it is clear that

$$\sigma(\zeta) = \zeta^{a(\sigma)} \quad \text{with } a(\sigma) \in \mathbb{Z}/n\mathbb{Z},$$

and in order for σ to preserve the order n of ζ, it is necessary and sufficient that $a(\sigma) \in (\mathbb{Z}/n\mathbb{Z})^\times$, in other words, it is necessary and sufficient that the endomorphism

$$\begin{cases} \mathbb{Z}/n\mathbb{Z} \longrightarrow \mathbb{Z}/n\mathbb{Z} \\ x \longmapsto a(\sigma) \cdot x \end{cases}$$

be an isomorphism.

Thus, we see that we have a **surjective homomorphism** (the surjectivity is non-trivial):

$$G_{\mathbb{Q}} \longrightarrow (\mathbb{Z}/n\mathbb{Z})^\times \cong GL_1(\mathbb{Z}/n\mathbb{Z}). \qquad \square$$

Now, let E be an elliptic curve defined over $\mathbb{Q}$, and let $n \geq 1$. Set $V = E[n](\bar{\mathbb{Q}})$; we will consider the action of G_Q on V.

Since V is a free $\mathbb{Z}/n\mathbb{Z}$-module of rank 2 (and even a vector space over $\mathbb{F}_p$ if n is a prime number p), we see that if P_1, P_2 is a $\mathbb{Z}/n\mathbb{Z}$-basis of V, we have

$$\begin{cases} \sigma(P_1) = a_\sigma P_1 + c_\sigma P_2 & a_\sigma, c_\sigma \in \mathbb{Z}/n\mathbb{Z} \\ \sigma(P_2) = b_\sigma P_1 + d_\sigma P_2 & b_\sigma, d_\sigma \in \mathbb{Z}/n\mathbb{Z} \end{cases}$$

i.e.

$$(\sigma(P_1), \sigma(P_2)) = (P_1, P_2) \begin{pmatrix} a_\sigma & b_\sigma \\ c_\sigma & d_\sigma \end{pmatrix}.$$

Now if we take an arbitrary pair Q_1, Q_2 of points of V such that

$$\begin{pmatrix} Q_1 \\ Q_2 \end{pmatrix} = \begin{pmatrix} \lambda & \nu \\ \mu & \delta \end{pmatrix} \begin{pmatrix} P_1 \\ P_2 \end{pmatrix},$$

then by linearity we obtain

$$\begin{pmatrix} \sigma(Q_1) \\ \sigma(Q_2) \end{pmatrix} = \begin{pmatrix} \lambda & \nu \\ \mu & \delta \end{pmatrix} \begin{pmatrix} \sigma(P_1) \\ \sigma(P_2) \end{pmatrix} = \begin{pmatrix} \lambda & \nu \\ \mu & \delta \end{pmatrix} \begin{pmatrix} a_\sigma & c_\sigma \\ b_\sigma & d_\sigma \end{pmatrix} \begin{pmatrix} P_1 \\ P_2 \end{pmatrix}.$$

Finally a simple computation shows that

$$\begin{pmatrix} a_{\sigma\circ\tau} & b_{\sigma\circ\tau} \\ c_{\sigma\circ\tau} & d_{\sigma\circ\tau} \end{pmatrix} = \begin{pmatrix} a_\sigma & b_\sigma \\ c_\sigma & d_\sigma \end{pmatrix} \begin{pmatrix} a_\tau & b_\tau \\ c_\tau & d_\tau \end{pmatrix},$$

and in particular,

$$\begin{pmatrix} 1 & 0 \\ 0 & 1 \end{pmatrix} = \begin{pmatrix} a_\sigma & b_\sigma \\ c_\sigma & d_\sigma \end{pmatrix} \begin{pmatrix} a_{\sigma^{-1}} & b_{\sigma^{-1}} \\ c_{\sigma^{-1}} & d_{\sigma^{-1}} \end{pmatrix}.$$

This whole situation is summarised in the following statement.

Theorem 4.7.1 *The action of $G_{\mathbb{Q}}$ on $V = E[n]$ defines a homomorphism*

$$G_{\mathbb{Q}} \xrightarrow{\rho_n} GL_2(\mathbb{Z}/n\mathbb{Z}),$$

so the image is isomorphic to the Galois group of the extension $K_n/\mathbb{Q}$ generated by the coordinates of the n-division points of E.

Proof. It remains to see that $\mathrm{Im}\rho_n \cong \mathrm{Gal}(K_n/\mathbb{Q})$. Now, it is clear that ρ_n factors through $\mathrm{Gal}(K_n/\mathbb{Q})$; thus it suffices to show that ρ_n restricted to $\mathrm{Gal}(K_n/\mathbb{Q})$ is injective. So let $\sigma \in G_{\mathbb{Q}}$ be such that $\rho_n(\sigma) = \left(\begin{smallmatrix} 1 & 0 \\ 0 & 1 \end{smallmatrix}\right)$. Then we have $\sigma(P_1) = P_1$ and $\sigma(P_2) = P_2$, which shows that $\sigma|_{K_n} = 1|_{K_n}$. □

Remark 4.7.1 Since $E[2]$ is formed by four points of $E(\bar{\mathbb{Q}})$, it may happen that $\mathrm{Im}\rho_n \neq GL_2(\mathbb{Z}/n\mathbb{Z})$. However, this does not happen frequently, as shown by the following theorem.

Serre's Theorem (1972) (Theorem 4.7.2) *Let E be an elliptic curve defined over $\mathbb{Q}$ which is not isomorphic over $\bar{\mathbb{Q}}$ to any curve having complex multiplications. Then there exists an integer $N \geq 1$, depending only on E, such that for every integer n prime to N, the representation ρ_n is surjective.*

Remark 4.7.2

(1) This result is far from elementary, and even its statement is not really elementary, since it uses the notion of an elliptic curve with complex multiplications, which will be defined in the next section.

(2) One can show that if ℓ is a prime number >163, then ρ_ℓ is irreducible; furthermore, if E is also semistable, then it suffices to take $\ell > 7$, see [D-D-T] p. 51.

4.8 RING OF ENDOMORPHISMS OF AN ELLIPTIC CURVE

Recall from Section 4.1 that two elliptic curves defined over k are **equivalent** over k if there exists a birational map of one onto the other defined over k. One can show that if this is the case, then the groups of points of these curves in $\mathbb{P}_2(k)$ are isomorphic.

In what follows, we will need a more general and more precise notion, which is analogous to the notion of a homomorphism of abstract groups.

Definition 4.8.1 *Consider two elliptic curves (E_1, O_1) and (E_2, O_2) defined over a field k.*

An **isogeny** $E_1 \xrightarrow{\varphi} E_2$ **defined over** k *is a rational map from E_1 to E_2,* **defined over** k, *such that $\varphi(O_1) = O_2$.*

One can show ([Sil] p. 75) that the isogenies are **homomorphisms** of $(E_1(\Omega), \oplus_1)$ in $(E_2(\Omega), \oplus_2)$, whatever the choice of the extension Ω of k. The theorem 4.9.2 shows the importance to be accorded to the field of definition of an isogeny.

Example 4.8.1 **(1)** Let n be an integer ≥ 0. The multiplication by n:

$$P \longmapsto [n](P) = \underbrace{P \oplus \cdots \oplus P}_{n \text{ times}}$$

is an isogeny $(E, \oplus) \to (E, \oplus)$.

We say that this isogeny is an **endomorphism** of E; when $n = 0$, it is the constant map $P \mapsto O$, and when $n = 1$, it is the identity.

The kernel $E[n]$ of this endomorphism is the group of n-division points studied in Section 3.6. When $n > 0$ and the characteristic of k is zero or large enough (i.e. if it does not divide n), we have

$$\#E[n] = n^2.$$

(2) Assume that the characteristic of k is not equal to 2, and consider the curves

$$\begin{cases} E_1 = y^2 - (x^3 + a_1x^2 + b_1x) \\ E_2 = y^2 - (x^3 + a_2x^2 + b_2x) \end{cases}$$

with

$$a_2 = -2a_1, \quad b_2 = a_1^2 - 4b_1, \quad 2b_1b_2 \neq 0.$$

Then the map $\varphi : (E_1, O) \to (E_2, O)$ defined by

$$(x, y) \longmapsto \left(\frac{y^2}{x^2}, \frac{y(x^2 - b_1)}{x^2}\right)$$

is an isogeny.

Indeed, one checks that for almost all the points P of $E_1(\Omega)$, we have $\varphi(P) \in E_2(\Omega)$. Moreover, we see that $O \mapsto O$, by writing φ in the form

$$(X, Y, Z) \longmapsto \big(Y^2X,\ Y\big(Y^2 - a_1X^2 - 2b_1XZ\big),\ \big(Y^2 - a_1X^2 - b_1XZ\big)Z\big).$$

If $x(P) = o$, we can use the form

$$(X, Y, Z) \longmapsto \big(Y\big(X^2 + a_1XZ + b_1Z^2\big),\ \big(X^2 - b_1Z^2\big)\big(X^2 + a_1XZ + b_1Z^2\big), XZ^2\big),$$

and we see that the point $A = (0, 0, 1)$ gives the point O. Thus, we have

$$\text{Ker}\,\varphi \supset \{0, A\},$$

so $\#\text{Ker}\,\varphi$ is even since $\{O, A\}$ is a subgroup of $\text{Ker}\,\varphi$.

(3) If we define $\psi : (E_2, O) \mapsto (E_1, O)$ by

$$(x, y) \longmapsto \left(\frac{y^2}{4x^2}, \frac{y(x^2 - b_2)}{8x^2}\right),$$

we see similarly that ψ is an isogeny, and that Ker $\psi \supset \{O, A\}$, so #Ker ψ is even.

A simple computation shows that

$$\begin{cases} \psi \circ \varphi = [2] : E_1 \longrightarrow E_1 \\ \varphi \circ \psi = [2] : E_2 \longrightarrow E_2. \end{cases}$$

(4) Assume that the curve E is **defined over a finite field** $\mathbb{F}_q$. Then we know that the map $z \mapsto z^q$ is an injective $\mathbb{F}_q$-endomorphism of the field $\mathbb{F}_q(x, y)$; we call it the **Frobenius endomorphism** of the field $\mathbb{F}_q(x, y)$, and denote it by Frob_q.

If O denotes a point of $E(\mathbb{F}_q)$, it is clear that Frob_q induces an isogeny of (E, O) to itself which we call the **Frobenius endomorphism**[52] **of** (E, O).

Remark 4.8.1 Let Ω be a field containing $\mathbb{F}_q$. Since Frob_q preserves collinearity in $\mathbb{P}_2(\Omega)$, it is clear that Frob_q preserves the group law of $E(\Omega)$ of origin O when $O \in E(\mathbb{F}_q)$. □

Definition 4.8.2 *Let (E, O) be an elliptic curve defined over k. An* **endomorphism of** (E, O) **defined over** $\bar{k}$ *is any isogeny $(E, O) \longrightarrow (E, O)$ defined over $\bar{k}$.*

Note that two isogenies $(E_1, O_1) \to (E_2, O_2)$ can be added using the addition law

$$(\varphi + \psi)(P) := \varphi(P) \oplus_2 \psi(P).$$

It is clear that if φ and ψ are two isogenies defined over $\bar{k}$, then $\varphi + \psi$ is again an isogeny defined over $\bar{k}$.

Finally, if φ and ψ are two endomorphisms of (E, O) defined over $\bar{k}$, the composition map $\varphi \circ \psi$ is again an endomorphism of (E, O) defined over $\bar{k}$.

Definition 4.8.3 *The* **ring of endomorphisms** *of E defined over $\bar{k}$, written $\text{End}_{\bar{k}}(E)$, is the set of endomorphisms of E defined over $\bar{k}$, equipped with the addition and multiplication laws defined above.*

In order to fully justify this definition, we still need to check that multiplication is distributive with respect to addition.

It is clear by definition that

$$(\varphi + \psi) \circ \theta = \varphi \circ \theta + \psi \circ \theta,$$

so it suffices to see that

$$\theta \circ (\varphi + \psi) = \theta \circ \varphi + \theta \circ \psi.$$

[52] G. Frobenius, 1849–1917.

Thus, for every $P \in E(\bar{k})$, we must show that

$$\theta(\varphi(P) \oplus \psi(P)) = \theta \circ \varphi(P) \oplus \theta \circ \psi(P),$$

but this is nothing more than the group homomorphism property of θ.

We need one last definition in order to formulate the essential properties of $\mathrm{End}_{\bar{k}}(E)$.

Definition 4.8.4 *Let φ be a* **non-constant** *isogeny defined over k : $(E_1, O_1) \to (E_2, O_2)$, and let $P \in \mathbb{P}_2(\Omega)$ be a point of $E(\Omega)$ whose coordinates are transcendental over $\bar{k}$ (we choose $\Omega \supset \bar{k}$ large enough for this).*

Set $Q = \varphi(P)$ and let $\bar{k}(P)$ (resp. $\bar{k}(Q)$) denote the extension of $\bar{k}$ obtained by adjoining the coordinates of P (resp. Q).

Then one can show that the degree $[\bar{k}(P) : \bar{k}(Q)]$ is independent of the choice of P, and we call this number the **degree of the isogeny** *φ (written $\deg(\varphi)$).*

Remark 4.8.2 **(1)** It is clear that if ψ is a second **non-constant** isogeny $(E_2, O_2) \to (E_3, O_3)$, then by the tower rule, we have

$$\deg(\psi \circ \varphi) = \deg \psi \cdot \deg \varphi.$$

(2) If the extension $\bar{k}(P)/\bar{k}(Q)$ is **separable** (in which case we say that the isogeny is separable), and if φ is not constant, one can prove that

$$\deg(\varphi) = \#\mathrm{Ker}\, \varphi.$$

This important relation will enable us, later on, to compute $\#E(\mathbb{F}_q)$ when E is defined over the field $\mathbb{F}_q$.

Example 4.8.2 In example 2 above, we saw that $\psi \circ \varphi = [2]$.

We saw earlier that if the characteristic is not equal to 2, we have

$$\#\mathrm{Ker}[2] = \#E[2] = 4,$$

which gives the relation $\deg([2]) = 4$.

We deduce from remark 1 above that

$$\deg \psi \cdot \deg \varphi = 4,$$

and since we noted that $\#\mathrm{Ker}\, \varphi = \deg \varphi$ and $\#\mathrm{Ker}\, \psi = \deg \psi$ are even, we have

$$\deg \varphi = \deg \psi = 2.$$

□

We refer to [Ca 3] or [Sil] for the proof of the following theorem.

Theorem 4.8.1 *Let (E, O) be an elliptic curve defined over k.*

(1) *The ring $End_{\bar{k}}(E)$ is an integral domain with unit, which is not necessarily commutative but is of characteristic zero.*
(2) *Set $\deg(\varphi) = 0$ if $\varphi = 0$. Then the map $\deg : End_{\bar{k}}(E) \to (\mathbb{N}, \times)$ is a morphism of multiplicative monoids.*
(3) *$End_{\bar{k}}(E)$ is equipped with an anti-involution $\varphi \mapsto \widehat{\varphi}$ such that*

$$\varphi \circ \widehat{\varphi} = \widehat{\varphi} \circ \varphi = [\deg \varphi] = [\deg \widehat{\varphi}].$$

(4) *The map $\varphi \mapsto \deg(\varphi)$ is a positive definite quadratic form on $End_{\bar{k}}(E)$.*

Remark 4.8.3 **(1)** Recall that an *anti-involution* σ of a (not necessarily commutative) ring A with unit is a map $A \to A$ such that

$$\begin{cases} \sigma^2 = \mathbb{1}_A & \text{(involution)} \\ \left.\begin{aligned} &\sigma(x+y) = \sigma(x) + \sigma(y) \\ &\sigma(1) = 1 \\ &\sigma(xy) = \sigma(y)\sigma(x) \end{aligned}\right\} & \text{(anti-morphism)} \end{cases}$$

(2) Let n be an integer >0, then the isogeny $[n]$ is a non-constant endomorphism of E. The isogeny $[1] = \mathbb{1}_E$ generates a subring of $\text{End}_k(E)$ isomorphic to $\mathbb{Z}$ since we have

$$\begin{cases} [m] + [n] = [m+n] \\ [m] \circ [n] = [mn]. \end{cases}$$

(3) When (E, O) is defined over $\mathbb{Q}$ we show that **in general** $\text{End}_{\bar{\mathbb{Q}}}(E)$ is reduced to the ring generated by $[1]$, so that $\text{End}_{\bar{\mathbb{Q}}}(E) \cong \mathbb{Z}$.

However, there exist exactly 9 equivalence classes over $\bar{\mathbb{Q}}$ of elliptic curves (defined over $\mathbb{Q}$!) for which $\text{End}_{\bar{\mathbb{Q}}}(E) \underset{\neq}{\supset} \mathbb{Z}$. These elliptic curves were discovered by N.H. Abel. The elements of $\text{End}_{\bar{\mathbb{Q}}}(E) \backslash \mathbb{Z}$ are called the **complex multiplications** of E.

When E runs through the set of curves with complex multiplications defined over $\mathbb{Q}$, the fraction field of $\text{End}_{\bar{\mathbb{Q}}}(E)$ runs through the set of 9 principal quadratic imaginary fields (see Chapter 1, Problem 1) and $\text{End}_{\mathbb{Q}}(E)$ itself runs through the set of maximal orders (rings of integers) of these fields, i.e. $\mathbb{Z}[\omega]$ with

$$\omega = \sqrt{-d} \qquad \text{if } d = 1, 2$$

$$\omega = \frac{1 + \sqrt{-d}}{2} \qquad \text{if } d = 3, 7, 11, 19, 43, 67, 163.$$

Example 4.8.3 $E = y^2 - (x^3 + x)$.

Let $\varphi : (x, y) \mapsto (-x, iy)$.

Clearly, $\varphi \in \text{End}_{\mathbb{Q}(i)}(E) \subset \text{End}_{\bar{\mathbb{Q}}}(E)$ and $\varphi \neq \pm\mathbb{1}$.

If φ was in the image of $\mathbb{Z}$, we would have $\varphi = [n]$, with $n \in \mathbb{Z}$, hence

$$\#\text{Ker}\,\varphi = n^2.$$

But clearly, $\#\text{Ker}\,\varphi = 1$, so $n = \pm 1$. We already noted that this is impossible, so φ has complex multiplications. As $\varphi^2 = -\mathbb{1}$, we have $\varphi = \pm i$ and $\text{End}_{\bar{Q}}(E) \cong \mathbb{Z}[i]$. □

(4) When (E, O) is defined over a finite field $\mathbb{F}_q$, $\mathrm{End}_{\mathbb{F}_q}(E)$ is **never isomorphic to** $\mathbb{Z}$ (i.e. E always has complex multiplications, see [Sil] p. 137).

One then proves that $\mathrm{End}_{\bar{\mathbb{F}}_q}(E)$ is either an order of a quadratic imaginary field (of characteristic zero!) or an order of a quaternion algebra (we then say that E is **supersingular**, even though it is actually not even singular!).

4.9 ELLIPTIC CURVES OVER A FINITE FIELD

Let $\mathbb{F}_q$ be a finite field of cardinal q and characteristic p. We know that q is of the form p^h. Let E be an elliptic curve defined over $\mathbb{F}_q$ by a long Weierstrass equation

$$y^2 + a_1xy + a_3y = x^3 + a_2x^2 + a_4x + a_6.$$

Our goal is to explain that $\#E(\mathbb{F}_q)$ is not too far from $q+1$ (which is $\#\mathbb{P}_1(\mathbb{F}_q)$).

Why $q+1$? because this is what we find if E has a double point with distinct tangents when we add twice the double point to the simple points. It is also what we can hope to find, on average, when $a_1 = a_3 = 0$; there are as many quadratic residues in $\mathbb{F}_q^*$ as there are non-residues ...

The precise result was conjectured in 1924 by E. Artin[53] in his thesis, and was proved in 1934 by Hasse. It can be stated as follows.

Hasse's Theorem 4.9.1 *Let E be an elliptic curve defined over $\mathbb{F}_q$. Then the number of points of E in $\mathbb{P}_2(\mathbb{F}_q)$ satisfies the inequality*

$$|\#E(\mathbb{F}_q) - (q+1)| \leq 2\sqrt{q}.$$

Proof. **(1)** Let φ be the Frobenius endomorphism

$$\begin{cases} (x, y) \longmapsto (x^q, y^q) \\ O \longmapsto O. \end{cases}$$

Since $\mathbb{F}_q = \{z \in \bar{\mathbb{F}}_q;\ z^q = z\}$, we see that

$$E(\mathbb{F}_q) = \mathrm{Ker}(\varphi - \mathbb{1});$$

consequently

$$\#E(\mathbb{F}_q) = \#\mathrm{Ker}(\varphi - \mathbb{1}).$$

(2) Now, set $\psi = \varphi - \mathbb{1} \in \mathrm{End}(E)$.

In Section 4.8, we noted that if ψ is separable (which we assume here), we have

$$\#\mathrm{Ker}\ \psi = \deg(\psi).$$

Thus, we have a way to compute this cardinal.

[53] E. Artin 1898–1962.

(3) Since the quadratic map

$$(m, n) \longmapsto \deg(m\varphi + n\psi)$$

is positive **semi-definite**, we have

$$\deg(m\varphi + n\mathbb{1}) = (\deg \varphi)m^2 + smn + (\deg \mathbb{1})n^2,$$

with $s^2 \leq 4 \deg \varphi . \deg \mathbb{1} = 4 \deg \varphi$.

Now, it is clear that if $p > 2$, we have $\deg \varphi = q$. Indeed,

$$\deg \varphi = \left[\bar{\mathbb{F}}_q(x, y) : \bar{\mathbb{F}}_q(x^q, y^q)\right] = \left[\bar{\mathbb{F}}_q(x, y) : \bar{\mathbb{F}}_q(x, y^q)\right]\left[\bar{\mathbb{F}}_q(x, y^q) : \bar{\mathbb{F}}_q(x^q, y^q)\right].$$

The first factor is equal to 1 since y satisfies a quadratic equation over $\mathbb{F}_q(x)$, so the degree of y over $\bar{\mathbb{F}}_q(x, y^q)$ divides both 2 and q. The second factor is naturally equal to q. Note that this result can be extended to the case where $p = 2$ ([Sil] p. 30).

Thus, we have

$$s^2 \leq 4q.$$

(4) Finally, $\#\text{Ker } \psi = \deg(\varphi - \mathbb{1}) = q - s + 1$, so

$$|\#E(\mathbb{F}_q) - (q + 1)| \leq 2\sqrt{q}.$$

□

Remark 4.9.1 The inequality above is optimal. Indeed, consider the supersingular curve

$$E: \quad y^2 + y = x^3$$

defined over $\mathbb{F}_2$, and let us compute $E(\mathbb{F}_4)$. We can write $\mathbb{F}_4 = \{0, 1, \omega, \omega^2\}$ with

$$\omega^2 + \omega + 1 = 0.$$

We find that

$$E(\mathbb{F}_4) = \left\{O, (0, 0), (0, 1), (1, \omega), (1, \omega^2), (\omega, \omega), (\omega, \omega^2), (\omega^2, \omega), (\omega^2, \omega^2)\right\},$$

so $\#E(\mathbb{F}_4) = 9$. It follows that $s = -2\sqrt{q} = -4$ and

$$\begin{aligned} \deg(m\varphi + n\,\mathbb{1}) &= 4m^2 - 4mn + n^2 \\ &= (2m - n)^2. \end{aligned}$$

Thus, we have

$$\deg(\varphi + 2\,\mathbb{1}) = 0,$$

even though $(1, 2) \neq (0, 0)$; thus the quadratic form

$$(m, n) \longmapsto \deg(m\varphi + n\,\mathbb{1})$$

is **not** definite. But since $\deg : End_{\bar{k}}(E) \to \mathbb{Z}$ is **positive definite**, we have $\varphi = -2\,\mathbb{1}$.

Theorem 4.9.2 *Consider two elliptic curves (E_1, O_1) and (E_2, O_2) defined over $\mathbb{F}_q$, and let $\theta : (E_1, O_1) \to (E_2, O_2)$ be a non-constant isogeny* **defined over** *$\mathbb{F}_q$. Then we have*

$$\#E_1(\mathbb{F}_q) = \#E_2(\mathbb{F}_q).$$

Remark 4.9.2 In other words, if E_1 and E_2 are “isogenous” **over** $\mathbb{F}_q$, they have the same number of points in $\mathbb{P}_2(\mathbb{F}_q)$, where two elliptic curves are said to be “isogenous” if there exists a **non-constant** isogeny from one to the other.

Proof. **(1)** Let φ_1 (resp. φ_2) be the Frobenius endomorphism of E_1 (resp. E_2). We saw earlier that

$$\begin{cases} \#E_1(\mathbb{F}_q) = \deg(\varphi_1 - \mathbb{1}_{E_1}) \\ \#E_2(\mathbb{F}_q) = \deg(\varphi_2 - \mathbb{1}_{E_2}). \end{cases}$$

(2) Now, we also see that the following diagram is commutative, since θ is **defined** over $\mathbb{F}_q$:

$$\begin{array}{ccc} E_1 & \xrightarrow{\varphi_1} & E_1 \\ \theta\downarrow & & \theta\downarrow \\ E_2 & \xrightarrow{\varphi_2} & E_2 \end{array}$$

A second commutative diagram follows naturally from this one:

$$\begin{array}{ccc} E_1 & \xrightarrow{\varphi_1 - \mathbb{1}_{E_1}} & E_1 \\ \theta\downarrow & & \theta\downarrow \\ E_2 & \xrightarrow{\varphi_2 - \mathbb{1}_{E_2}} & E_2. \end{array}$$

A little reflection shows that

$$\deg\theta \cdot \deg(\varphi_1 - \mathbb{1}_{E_1}) = \deg(\varphi_2 - \mathbb{1}_{E_2}) \cdot \deg\theta,$$

so

$$\deg(\varphi_1 - \mathbb{1}_{E_1}) = \deg(\varphi_2 - \mathbb{1}_{E_2})$$

since $\deg\theta > 0$. □

Example 4.9.1 **(1)** If a and b are in $\mathbb{F}_p$ and

$$2b(a^2 - 4b) \neq 0,$$

then the elliptic curves

$$\begin{aligned} E_1 : \quad & y^2 - x(x^2 + ax + b) \\ E_2 : \quad & y^2 - x(x^2 - 2ax + a^2 - 4b) \end{aligned}$$

have the same number of points in $\mathbb{P}_2(\mathbb{F}_p)$.

(2) Consider the elliptic curves over $\mathbb{F}_3$ given by

$$\begin{aligned} E_1 : \quad & y^2 - (x^3 - x + 1) \\ E_2 : \quad & y^2 - (x^3 - x - 1). \end{aligned}$$

These curves are isomorphic over $\mathbb{F}_9$, but $\#E_1(\mathbb{F}_3) = 7$ and $\#E_2(\mathbb{F}_3) = 1$, so they are not isomorphic over $\mathbb{F}_3$!

Remark 4.9.3 We cannot move on without at least a brief mention of the famous result due to F.K. Schmidt, which states that every smooth cubic defined over a finite field $\mathbb{F}_q$ has a rational point over this field, and consequently can be equipped with the structure of an **elliptic curve over** $\mathbb{F}_q$.

4.10 TORSION ON AN ELLIPTIC CURVE DEFINED OVER $\mathbb{Q}$

We know that every Abelian group A contains a certain important subgroup, namely its *torsion subgroup*, which we write $T(A)$. By definition,

$$T(A) = \{x \in A; \text{ there exists } n \in \mathbb{N}^* \text{ such that } nx = O\}.$$

Example 4.10.1 If k denotes a commutative field, the torsion subgroup of $(k^*, \times)$ is the group $\mu(k)$ formed by the roots of unity lying in k.

When the group A is the group $E(k)$ of points on an elliptic curve defined over a number field k, the torsion group of A can be determined more or less easily. But when $k = \mathbb{Q}$ and the elliptic curve E is **given** by its equation, this determination is even easier, since as we will see, the Weierstrass coordinates of a torsion point are integral when the Weierstrass equation has integral coefficients.

Now, we have the following very nice result at our disposal (note that here R denotes an integral domain which is integrally closed in its fraction field K, i.e. such that if $x \in K$ is a root of a monic polynomial with coefficients in R, then x is in R).

Theorem 4.10.1 *Let E be an elliptic curve defined by a short Weierstrass equation*

$$y^2 = x^3 + a_2x^2 + a_4x + a_6, \tag{1}$$

with coefficients in a **ring** *R, assumed to be of characteristic $\neq 2$, integrally closed in its field of fractions K.*

If $P \in E(K)$ is such that its coordinates and the coordinates of $[2]P$ lie in R, then the ordinate of P is either zero or a divisor (in R) of the discriminant

$$D = -4a_2^3a_6 + a_2^2a_4^2 + 18a_2a_4a_6 - 4a_4^3 - 27a_6^2.$$

Proof. The quantity D is precisely the discriminant of

$$F(x) := x^3 + a_2x^2 + a_4x + a_6 = (x - e_1)(x - e_2)(x - e_3),$$

with $e_i \in \bar{K}$ for $i \in \{1, 2, 3\}$, i.e. we have

$$D = (e_1 - e_2)^2(e_1 - e_3)^2(e_2 - e_3)^2.$$

Let us show that if the ordinate y of $P = (x, y)$ is not equal to 0, then y divides D in R. Indeed, we saw in Section 4.4 that if $2P = (\xi, \eta)$, then

$$2x + \xi = \lambda^2 - a_2 \quad \text{with } \lambda = \frac{F'(x)}{2y}.$$

But the theory of the discriminant gives us the relation

$$D = U(X)F(X) + V(X)F'(X) \quad \text{with } U(X) \text{ and } V(X) \in R[X].$$

Since $\lambda^2 \in R$ and $\lambda \in K$, we see that $\lambda \in R$ and it follows that y divides $F'(x)$.

As $y^2 = F(x)$, we see that y divides $F(x)$, so y also divides D. □

Corollary 4.10.1 *If the elliptic curve E is defined over $\mathbb{Q}$ by a short Weierstrass equation with* **integral coefficients**, *and if P and $[2]P$ are integral points of $E(\mathbb{Q})$, then the ordinate y of P is either zero or a divisor of D.*

Proof. Indeed, we know that $\mathbb{Z}$ is integrally closed in $\mathbb{Q}$. □

Special case 4.10.1 When $a_2 = 0$, we traditionally write (1) in the form

$$y^2 = x^3 + Ax + B, \quad A \text{ and } B \text{ in } R, \tag{1'}$$

and we have $D = -(4A^3 + 27B^2)$.

Then if P and $[2]P$ have integral coordinates and $y \neq 0$, y^2 divides D in R.

Indeed, we can improve the expression of D using F and F', by replacing F' by F'^2, and we find that

$$D = (3X^2 + 4A)F'(X)^2 + U(X)F(X)$$

with $U(X) \in R[X]$. □

We assume from now on that $R = \mathbb{Z}$ and (1) has integral coefficients.

Our goal is now to show that if $P = (x, y) \in \mathbb{Q}^2$ is a torsion point of $E(\mathbb{Q})$, then the coordinates of P are integers. To obtain this result, we will show that if p is an *arbitrary* prime number, then the p-adic valuations of x and y are positive or zero.

Recall the definition of the p-adic valuation of $\mathbb{Q}$ (see Chapter 3):

$$v_p : \quad \mathbb{Q} \longrightarrow \mathbb{Z} \cup \{+\infty\}.$$

Definition 4.10.1 *If $r = p^n(u/v) \in \mathbb{Q}^*$ with u and $v \in \mathbb{Z} \setminus p\mathbb{Z}$, we set*

$$v_p(r) = n.$$

If $r = 0$, we write $v_p(r) = +\infty$.

Recall also the following proposition:

Proposition 4.10.1

(1) *For any x and $y \in \mathbb{Q}$, we have*

$$v_p(xy) = v_p(x) + v_p(y)$$
$$v_p(x + y) \geq \inf(v_p(x), v_p(y)) \quad \text{with equality if } v_p(x) \neq v_p(y).$$

(2) *The set $\mathbb{Z}_{(p)} = \{x \in \mathbb{Q};\ |x|_p \leq 1\}$ is a subring of $\mathbb{Q}$ called the valuation ring of v_p. It is a local ring whose maximal ideal is $\{x \in \mathbb{Q};\ |x|_p < 1\}$.*

The absolute value $|\ |_p$ makes it possible to define a p-adic topology on $\mathbb{Q}$ by defining the p-adic distance of two elements x and y in $\mathbb{Q}$ by the formula

$$d_p(x, y) = |x - y|_p.$$

Thus, the smaller the absolute value $|x|_p$, the nearer the rational number x is to 0.

Since our goal is to study the points $P = (x, y) \in E(\mathbb{Q})$ such that $v_p(x) \leq 0$, we are led to consider the elements of $E(\mathbb{Q})$ which are p-adically close to the projective point $(0, 1, 0) = O$.

For this, we dehomogenise the equation

$$Y^2Z = X^3 + a_2X^2Z + a_4XZ^2 + a_6Z^3$$

by setting $Y = 1$, and we find that

$$z = x^3 + a_2x^2z + a_4xz^2 + a_6z^3, \quad \text{with } a_i \in \mathbb{Z} \text{ for } i = 2, 4, 6. \tag{2}$$

The origin of E is then the point

$$O = (x, z) = (o, o),$$

and the opposite of the point $P = (x, z)$ is $\ominus P = (-x, -z)$.

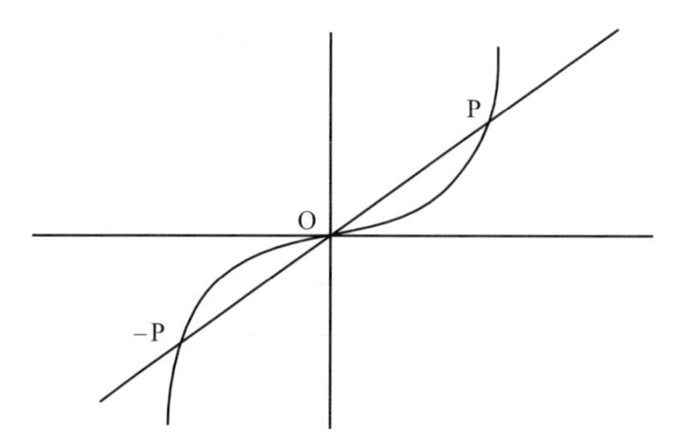

The "Abel-style" construction of the group law of $E(\mathbb{Q})$ which we gave in Section 4.4 is still valid: to construct $P_3 = P_1 \oplus P_2$, we first take the third point of intersection R of the line (P_1P_2) with E, and then take the point obtained from this one by a symmetry through the origin. Let $z = \alpha x + \beta$ be the equation of this line. By (2), we have

$$\begin{cases} \alpha = \dfrac{z_2 - z_1}{x_2 - x_1} = \dfrac{x_2^2 + x_1x_2 + x_1^2 + a_2(x_2 + x_1)z_2 + a_4z_2^2}{1 - a_2x_1^2 - a_4x_1(z_2 + z_1) - a_6\left(z_2^2 + z_1z_2 + z_1^2\right)} \\ \beta = z_1 - \alpha x_1. \end{cases} \tag{3}$$

The equation for the abscissae of the points of intersection of E and (P_1P_2) can be written

$$\begin{aligned} \alpha x + \beta &= x^3 + a_2x^2(\alpha x + \beta) + a_4x(\alpha x + \beta)^2 + a_6(\alpha x + \beta)^3 \\ &= (x - x_1)(x - x_2)(x - r), \end{aligned}$$

where r denotes the abscissa of the point R.

Thus, we have

$$x_1 + x_2 + r = -\frac{(a_2 + 2a_4\alpha + 3a_6\alpha^2)\beta}{1 + a_2\alpha + a_4\alpha^2 + a_6\alpha^3}. \tag{4}$$

Now, if $v_p(x) < 0$, we must necessarily have $v_p(y) < 0$ in equation (1), and in fact we find that

$$v_p(x) = -2\nu \quad v_p(y) = -3\nu, \qquad \text{with } \nu \in \mathbb{N}^*.$$

Thus, the points we wish to study are of the type (X, Y, Z) with $Y = 1$, $v_p(X) = \nu$, $v_p(Z) = 3\nu$.

Let E_{p^ν} denote the set of points (x, z) of (2) such that $|x|_p \le 1/p^\nu$ and $|z|_p \le 1/p^{3\nu}$: they form a p-adic neighbourhood of the point $(o, o) = O$, and we have the following result:

Lemma 4.10.1 *E_{p^ν} is a subgroup of $E(\mathbb{Q})$ when $\nu \in \mathbb{N}^*$.*

Proof. Assume that P_1 and P_2 lie in E_{p^ν}; then we have

$$|x_i|_p \le \frac{1}{p^\nu}, \quad |z_i|_p \le \frac{1}{p^{3\nu}} \qquad \text{for } i \in \{1, 2\}.$$

From (3) and the fact that $|a_i| \leq 1$ for $i \in \{2, 4, 6\}$, we deduce that

$$|\alpha|_p \leq \frac{1}{p^{2\nu}} \qquad |\beta|_p \leq \frac{1}{p^{3\nu}}.$$

Plugging this result into (4), we see that the absolute value of the second term is less than or equal to $|\beta|_p = 1/p^{3\nu}$.

As $|x_i|_p \leq 1/p^{\nu}$ for $i \in \{1, 2\}$ we deduce from the ultrametric inequality that $|r| \leq 1/p^{\nu}$.

Next, we note that the ordinate s of R is equal to $\alpha r + \beta$; we have

$$|s|_p = |\alpha r + \beta|_p \leq \frac{1}{p^{3\nu}}.$$

Thus, we see that R, and thus also $P_3 = P_1 \oplus P_2$ lie in E_{p^ν}. $\square$

Remark 4.10.1 We obtain a sequence of nested subgroups,

$$E_p \supset E_{p^2} \supset E_{p^3} \supset \cdots$$

whose intersection is reduced to $\{O\}$. We say that this sequence forms a filtration of $E(\mathbb{Q})$.

We will compare this filtration to the filtration of subgroups

$$p\mathbb{Z}_{(p)} \supset p^2\mathbb{Z}_{(p)} \supset p^3\mathbb{Z}_{(p)} \supset \cdots$$

of the valuation ring $\mathbb{Z}_{(p)}$, considered as an additive group.

Lemma 4.10.2 *The group $E_{p^\nu}/E_{p^{3\nu}}$ is isomorphic to a subgroup of $p^\nu\mathbb{Z}_{(p)}/p^{3\nu}\mathbb{Z}_{(p)}$.*

Proof. By the proof of lemma 4.10.1, we see that the map

$$\begin{cases} P \longmapsto \overline{x(P)} \in p^\nu\mathbb{Z}_{(p)}/p^{3\nu}\mathbb{Z}_{(p)} \\ E_{p^\nu} \longrightarrow p^\nu\mathbb{Z}_{(p)}/p^{3\nu}\mathbb{Z}_{(p)} \end{cases}$$

is such that

$$\overline{x(P_3)} = \overline{x(P_1 \oplus P_2)} = \overline{x(P_1)} + \overline{x(P_2)},$$

so it is a homomorphism of additive groups. Clearly, the kernel of this homomorphism is $E_{p^{3\nu}}$. Indeed, if $|x(P)|_p \leq 1/p^{3\nu}$ and $|z(P)|_p \leq 1/p^{3\nu}$ we deduce from (2) that $|z(P)|_p \leq 1/p^{9\nu}$!

Thus, the isomorphism theorem implies that $E_{p^\nu}/E_{p^{3\nu}}$ is isomorphic to the image of this homomorphism. $\square$

Lemma 4.10.3 *When $\nu \geq 1$, the only point of finite order in E_{p^ν} is O.*

Proof. Let $P \in E_{p^\nu}$ be a point of order $n > 1$. Since $P \neq O$, there exists $\mu \geq \nu$ such that

$$P \in E_{p^\mu} \quad \text{but } P \neq E_{p^{\mu+1}}.$$

(1) If n is not divisible by p, it follows that $[n]P = O$ that $nx(P) \in p^{3\mu}\mathbb{Z}_{(p)}$ hence $x(P) \in p^{3\mu}\mathbb{Z}_{(p)}$, which is absurd since $3\mu \geq \mu + 1$ if $\mu \geq 1$.

(2) If $n = pm$, then we see that $Q = mP$ is of order p. Then there exists $\lambda \geq \mu \geq 1$ such that $Q \in E_{p^\lambda}$ and $Q \notin E_{p^{\lambda+1}}$. It follows from $[p]Q = O$ that $px(Q) \in p^{3\lambda}\mathbb{Z}_{(p)}$, hence $x(Q) \in p^{3\lambda-1}\mathbb{Z}_{(p)}$.

But as $\lambda \geq \nu \geq 1$, we have $3\lambda - 1 \geq \lambda + 1$ so $x(Q) \in E_{p^{\lambda+1}}$, which is absurd. □

Nagell–Lutz Theorem 4.10.2 *Let E be an elliptic curve of equation*

$$y^2 = x^3 + a_2x^2 + a_4x + a_6,$$

with $a_i \in \mathbb{Z}$ *for* $i \in \{2, 4, 6\}$. *Set* $D = -4a_2^3a_6 + a_2^2a_4^2 + 18a_2a_4a_6 - 4a_4^3 - 27a_6^2$.

Then if (x, y) *is a torsion point of E lying in* $\mathbb{Q}^2$, *x and y are integers, and if y is non-zero, then y divides D (and in fact,* y^2 *divides D if* $a_2 = 0$*).*

Proof. We saw in lemma 4.10.3 that for every prime number p, E_p does not contain any torsion point apart from O.

It follows that the torsion points of $E(\mathbb{Q})$ which are different from the origin O have integral coordinates (x, y).

Since $\mathbb{Z}$ is integrally closed in $\mathbb{Q}$, it follows from the preceding theorem that if $y \neq 0$, then y divides D. □

Corollary 4.10.2 *An elliptic curve defined over* $\mathbb{Q}$ *admits at most a finite number of torsion points in* $\mathbb{P}_2(\mathbb{Q})$.

Proof. Indeed, we can choose for E a short Weierstrass equation with integral coefficients, and D admits only a finite number of divisors. □

An **infinitely more difficult** problem is to inquire whether we can **uniformly** bound the torsion of an elliptic curve defined over $\mathbb{Q}$ (or over an algebraic number field). The answer is yes (Mazur, Kamienny, Merel), but we restrict ourselves here to stating Mazur's result, which concerns the field of rational numbers.

Mazur's Theorem 4.10.3

(1) *If an elliptic curve E defined over* $\mathbb{Q}$ *contains a point of order n defined over* $\mathbb{Q}$, *then*

$$1 \leq n \leq 10 \quad or \quad n = 12.$$

(2) *More precisely, the torsion subgroup of* $E(\mathbb{Q})$ *is isomorphic to one of the following groups:*

(i) $\mathbb{Z}/m\mathbb{Z}$ *with* $1 \leq m \leq 10$ or $m = 12$.

(ii) $\mathbb{Z}/2\mathbb{Z} \times \mathbb{Z}/2m\mathbb{Z}$ *with* $1 \leq m \leq 4$.

Remark 4.10.2 There exists no elliptic curve E defined over $\mathbb{Q}$ such that $E(\mathbb{Q})$ contains a point of order 11.

4.11 MORDELL–WEIL THEOREM

Conjectured by H. Poincaré at the beginning of the 20th century, the most important theorem concerning the group $E(\mathbb{Q})$ of the points of an elliptic curve defined over $\mathbb{Q}$ was proved by L.J. Mordell[54] in 1922. He showed that for every curve E, this group is generated by a finite number of points. Some years later, in 1930, A. Weil extended this result to the group $E(K)$ of the points of an elliptic curve E defined over an algebraic number field K.

Mordell–Weil Theorem 4.11.1 *Let K be an algebraic number field, and E an elliptic curve defined over K. Then the group $E(K)$ is generated by a finite number of points.*

Corollary 4.11.1 *Under the same conditions, the torsion subgroup of $E(K)$ is finite.*

Proof of the corollary. We say that an Abelian group with a finite number of generators is of *finite type*, and a classical theorem shows that a subgroup of an Abelian group of finite type is also of finite type [Bou 2]. But a torsion group of finite type is a finite group. □

Remark 4.11.1 We will see in the exercises that if, contrarily, the cubic E is singular and has a simple rational point O, then $E_0(\mathbb{Q})$ is not an Abelian group of finite type.

4.12 BACK TO THE DEFINITION OF ELLIPTIC CURVES

(a) Introduction

In Section 2.4 of Chapter 2, concerning Liouville's theorem (theorem 2.4.5), we saw that a non-zero elliptic function $f \in \mathbb{C}(\wp, \wp')$ has as many zeros as poles in a suitable period parallelogram Π, and that the number of these is called the *order* of the function f.

We also saw that a function of order 0 is a constant ($f \in \mathbb{C}^*$) and that there exists no function of order 1.

Definition 4.12.1 *Let $n \in \mathbb{N}$ and let $O \in \Pi$ denote the origin of the complex plane. Let $\mathcal{L}(n(O))$ denote the $\mathbb{C}$-vector space of elliptic functions f attached to the lattice Λ admitting (at most) one pole at O of order $\leq n$, and let $\ell(n(O))$ denote the dimension of this vector space.*

Remark 4.12.1 If some non-zero f belongs to $\mathcal{L}(n(O))$, then the order of f is $\leq n$ since f admits only the pole O in Π, and the order of this pole must be $\leq n$.

Proposition 4.12.1 *For every $n \in \mathbb{N}^*$, we have*

$$\ell(n(O)) = n.$$

Proof. Let us use induction on n.

[54] L.J. Mordell (1888–1972).

(1) If $n = 1$, we saw that $\ell((O)) = 1$.

(2) Assume the property holds for $n-1$, and let α and $\beta \in \mathbb{N}$ be such that $2\alpha + 3\beta = n$. It is clear that $\mathfrak{p}^{\alpha}(\mathfrak{p}')^{\beta}$ is of order n and has a pole of order n at O.

It is also clear that

$$\mathcal{L}(n(O)) \supset \mathcal{L}((n-1)(O)) + \mathbb{C}\mathfrak{p}^{\alpha}(\mathfrak{p}')^{\beta},$$

and the sum of the vector spaces on the right is a direct sum.

Thus, it suffices to prove the inclusion in the other direction.

Let $f \in \mathcal{L}(n(O))$.

If the order of f is $\leq n-1$, then $f \in \mathcal{L}((n-1)(O))$. Otherwise, there exists $\lambda \in \mathbb{C}^*$ such that $f(z)$ and $\lambda\mathfrak{p}(z)^{\alpha}(\mathfrak{p}'(z))^{\beta}$ are infinitely large quantities which are equivalent when z tends to 0. Thus $f - \lambda\mathfrak{p}^{\alpha}(\mathfrak{p}')^{\beta} \in \mathcal{L}((n-1)(O))$, which was our goal.

Thus, we see that

$$\mathcal{L}(n(O)) = \mathcal{L}((n-1)(O)) \oplus \mathbb{C}\mathfrak{p}^{\alpha}(\mathfrak{p}')^{\beta},$$

and as $\ell((n-1)(O)) = n-1$ by the induction hypothesis, we see that

$$\ell(n(O)) = n.$$

□

(b) Generalisation

We now propose to generalise the above considerations to the case of a curve Γ defined over k, irreducible over $\bar{k}$ and contained in projective space of dimension d. We assume that $\Gamma(k)$ contains a **simple** point O. Let $k(\Gamma)$ denote the function field of the curve, i.e. the extension of k obtained by adjoining the coordinates of a generic point of Γ (i.e. a point with transcendental coordinates over k).

Definition 4.12.2 *Let $n \in \mathbb{N}$.*
Let $\mathcal{L}(n(O))$ denote the k-vector space of functions on C which admit (at most) a pole at O of order $\leq n$, and let $\ell(n(O))$ denote the dimension of this vector space.

Definition 4.12.3 *We say that the pair (Γ, O) is an elliptic curve defined over k if, for every $n \in \mathbb{N}^*$, we have $\ell(n(O)) = n$.*

It is not at all obvious that a smooth cubic defined over k and equipped with a rational point is an elliptic curve in the sense of definition 4.12.2. We need to admit the following result.

Theorem 4.12.1

(1) *Every smooth cubic defined over k and equipped with a rational point is an elliptic curve in the sense of definition 4.12.2.*

(2) *Every curve which is birationally equivalent to an elliptic curve over k and equipped with a rational point is an elliptic curve in the sense of definition 4.12.2.*

Example 4.12.1 **(1)** Let $k \in \mathbb{Q}^*$ and let B be the biquadratic

$$\begin{cases} kX^2 + Y^2 + 2TY = 0 \\ kX^2 - Z^2 - 2TZ = 0. \end{cases}$$

We see that $\Omega := (0, 0, 0, 1) \in B(\mathbb{Q})$.

Eliminating T, we obtain the projection of B in $\mathbb{P}_2(\mathbb{Q})$. It is given by

$$\Gamma : \qquad Z(kX^2 + Y^2) + Y(kX^2 - Z^2) = 0.$$

The curve Γ is a smooth cubic, and $O := (1, 0, 0) \in \Gamma(\mathbb{Q})$. Thus (Γ, O) is an elliptic curve, and (B, Ω), which is birationally equivalent to it, is also one.

(2) Assume that Γ is $Y^2 = X^4 + k^2$, with $k \in \mathbb{Q}^*$, and set $O := (0, k) \in \Gamma(\mathbb{Q})$.

We can write the equation of Γ in the form

$$(Y + X^2)(Y - X^2) = k^2.$$

Setting $Y + X^2 = S$ and $Y - X^2 = k^2/S$, we obtain

$$2X^2 = S - \frac{k^2}{S}.$$

Hence, after multiplication by S^2, we have

$$2(XS)^2 = S^3 - k^2 S.$$

If we set $XS = V/4$ and $2S = U$, we obtain the **Weierstrass equation**

$$W : \quad V^2 = U^3 - 4k^2 U.$$

The rational map is

$$(X, Y) \longmapsto (U, V) = (2(Y + X^2), 4X(Y + X^2)),$$

and the inverse map is

$$(U, V) \longmapsto (X, Y) = \left(\frac{V}{2U}, \frac{U^2 + 4k^2}{2U}\right).$$

Thus, we see that (Γ, O) is an elliptic curve over $\mathbb{Q}$, since $k \neq 0$ by hypothesis.

Theorem 4.12.2 *Let (Γ, O) be an elliptic curve defined over k, with function field $C = k(\Gamma)$. Then*

(1) *There exist two functions x and $y \in C$, such that the map*

$$\varphi \begin{cases} \Gamma \longrightarrow \mathbb{P}_2(k) \\ M \longmapsto (x(M), y(M), 1) \end{cases}$$

is a birational map from (Γ, O) to a Weierstrass curve W of equation

$$Y^2 Z + a_1 XYZ + a_3 YZ^2 = X^3 + a_2 X^2 Z + a_4 XZ^2 + a_6 Z^3 \tag{1}$$

defined over k, and such that $\varphi(O) = (0, 1, 0)$.

(2) *Every other pair (x', y') satisfying the same conditions can be written*

$$\begin{cases} x' = u^2 x + r \\ y' = u^3 y + sx + t \end{cases} \tag{2}$$

with $u, r, s, t \in k$ and $u \neq 0$.

Proof. **(1)** Since $\ell(2(O)) = 2$ and $\ell(3(O)) = 3$, there exist x and $y \in C$ such that

$$\mathcal{L}(2(O)) = k \oplus kx, \quad \mathcal{L}(3(O)) = k \oplus kx \oplus ky.$$

Thus, x must have a pole of order 2 at O (and no other poles), and y must have a pole of order 3 at O (and no others).

Consider the seven functions

$$1, x, y, x^2, xy, y^2, x^3.$$

We easily see that these functions belong to $\mathcal{L}(6(O))$.

Since $\ell(6(O)) = 6$ by hypothesis, these functions are linearly dependent over k, and we have

$$\lambda_0 + \lambda_1 x + \lambda_2 y + \lambda_3 x^2 + \lambda_4 xy + \lambda_5 y^2 + \lambda_6 x^3 = 0,$$

with $(\lambda_0, \ldots, \lambda_6)$ non-zero.

We cannot have $(\lambda_5, \lambda_6) = (0, 0)$, because the functions $1, x, y, x^2, xy$ have different orders at O (namely $0, -2, -3, -4, -5$), and all the λ_i would be zero.

For the same reason, we also cannot have $\lambda_5 \lambda_6 = 0$. Replacing x by $-\lambda_5 \lambda_6 x$ and y by $\lambda_5 \lambda_6^2 y$, we obtain an equation W of the desired form.

(2) To prove the birationality of the map $\Gamma \to W$ defined by $P \mapsto (x(P), y(P))$, it suffices to see that $C = k(x, y)$, since the field of the functions of a curve determines the curve up to birational equivalence.

Thus, we are led to prove that $C = k(x, y)$. For this, we note that since x admits only a double pole at O, we have

$$[C : k(x)] = 2.$$

As y admits only a triple pole at O, we have

$$[C : k(y)] = 3.$$

We deduce that $[C : k(x, y)]$ divides 2 and 3, so $C = k(x, y)$.

(3) For another choice of x and y, we necessarily have

$$\begin{cases} x' = u_1 x + r \\ y' = u_2 y + sx + t, \end{cases}$$

with $u_1 u_2 \neq 0$.

In order for y'^2 and x'^3 to have the same coefficient in the new Weierstrass equation, it is necessary and sufficient that $u_1^3 = u_2^2$.

Setting $u = u_2/u_1$, we have

$$\begin{cases} x' = u^2 x + r \\ y' = u^3 y + sx + t, \end{cases}$$

which gives the result.

(3) Finally if W had a multiple point, C would be a field of rational functions, and we would have $\ell(n(O)) = n + 1$, which is false. □

Remark 4.12.2

(1) The change of variables (2) is called an **admissible** change of variables for the Weierstrass form.

(2) The entire "philosophy" of this paragraph concerns the famous **Riemann–Roch theorem**, which is beyond the scope of this book. This theorem associates to every curve Γ defined over k and irreducible over $\bar{k}$, an integer g called the **genus** of Γ.
The genus can be defined in various ways, but what counts here is the following definition.

Definition 4.12.4 *An elliptic curve is a curve of genus one, defined over k, which has a rational point over k.*

Thus, saying that a Weierstrass smooth cubic is an elliptic curve comes down to showing that it is of genus one: for this, we can either check that the differential $\omega = dx/(2y + a_1 + a_3)$ has no zeros or poles, or note that a plane cubic which is not of genus one must be parametrisable in $\bar{k}(t)$, and use Exercise 4.2 when the characteristic of k does not divide 6.

4.13 FORMULAE

If we start from a long Weierstrass form

$$y^2 + a_1 xy + a_3 y = x^3 + a_2 x^2 + a_4 x + a_6 \tag{1}$$

defined over k, and if we desire to put this equation in "short" form by completing the square, then the cube, a certain number of new coefficients (b_2, b_4, b_6, b_8 and c_4, c_6) appear.

The coefficients b_i are given by

$$\begin{cases} b_2 = a_1^2 + 4a_2 \\ b_4 = a_1 a_3 + 2a_4 \\ b_6 = a_3^2 + 4a_6 \\ b_8 = a_1^2 a_6 - a_1 a_3 a_4 + 4a_2 a_6 + a_2 a_3^2 - a_4^2. \end{cases}$$

If the characteristic of k is different from 2, then with adequate coordinates, we have

$$y^2 = x^3 + \frac{b_2}{4}x^2 + \frac{b_4}{2}x + \frac{b_6}{4}. \tag{2}$$

Remark 4.13.1

(1) The coefficient b_8 does not appear in (2), and we can compute it in terms of b_2, b_4, b_6 by the formula

$$4b_8 = b_2 b_6 - b_4^2.$$

(2) The nice thing about this coefficient is that it greatly simplifies the computation of the **discriminant** Δ of the cubic (1), the very discriminant whose non-vanishing expresses the condition that the cubic is of **genus one**. Indeed, we have

$$\Delta = -b_2^2 b_8 - 8b_4^3 - 27b_6^2 + 9b_2 b_4 b_6.$$

The coefficients c_i are

$$\begin{cases} c_4 = b_2^2 - 24b_4 \\ c_6 = -b_2^3 + 36b_2 b_4 - 216b_6. \end{cases}$$

Then, with adequate coordinates, and if the characteristic of k does not divide 6,

$$y^2 = x^3 - \frac{c_4}{48}x - \frac{c_6}{864}, \tag{3}$$

Remark 4.13.2

(1) If we make an admissible change of coordinates by formulae (2) of the preceding paragraph (see top of page 214), we find that

$$u^4 c_4' = c_4, \qquad u^6 c_6' = c_6.$$

(2) The discriminant Δ is given in terms of the c_i by

$$12^3 \Delta = c_4^3 - c_6^2.$$

After an admissible change of coordinates, we thus have

$$u^{12}\Delta' = \Delta.$$

Definition 4.13.1 *The quantity*

$$j = \frac{c_4^3}{\Delta} = 12^3 \frac{c_4^3}{c_4^3 - c_6^2}$$

is **invariant** *under an admissible change of coordinates: j is called the* **modular invariant** *of the curve E and is written $j(E)$.*

Remark 4.13.3

(1) Every number j in k is the modular invariant of an elliptic curve defined over k.

(i) If $j = 0$, we can take

$$E: \quad y^2 = x^3 + 1.$$

(ii) If $j = 1728$, we can take

$$E: \quad y^2 = x^3 + x.$$

(iii) If $j \notin \{0, 1728\}$, we can take

$$E: \quad y^2 + xy = x^3 - \frac{36}{j - 1728}x - \frac{1}{j - 1728}.$$

(2) E is singular if and only if $\Delta = 0$.
When $\Delta = 0$ and $c_4 \neq 0$ E admits a double point in $\bar{k}$.
When $\Delta = 0$ and $c_4 = 0$, E admits a cusp.

Theorem 4.13.1 *A necessary and sufficient condition for two elliptic curves E and E' defined over k to be birationally equivalent over $\bar{k}$ is that they have the same modular invariant j.*

Proof. **(1)** If E and E' are birationally equivalent over $\bar{k}$, then they admit the same Weierstrass forms, so $j(E) = j(E')$.

(2) We prove the converse only for a field k of characteristic $\neq 2$ and 3 (it is left as an exercise for characteristics 2 and 3). We have

$$\begin{cases} E: & y^2 = x^3 + Ax + B \\ E': & y'^2 = x'^3 + A'x' + B' \end{cases}$$

and $j(E) = j(E')$ can be written

$$1728 \frac{4A^3}{4A^3 + 27B^2} = 1728 \frac{4A'^3}{4A'^3 + 27B'^2}.$$

Eliminating denominators, we obtain

$$A^3 B'^2 = A'^3 B^2.$$

(1) If $j = 0$, then $A = 0$ and $B \neq 0$. Since $j = j'$, we then have $A' = 0$ and $B' \neq 0$; we take $u = (B/B')^{1/6}$ for an admissible change of coordinates.
(2) If $j = 1728$, then $B = 0$ and $A \neq 0$. Since $j = j'$, we then have $B' = 0$ and $A' \neq 0$; we take $u = (A/A')^{1/4}$ for an admissible change of coordinates.
(3) If $j \notin \{0, 1728\}$ then $AB \neq 0$ and $A'B' \neq 0$ and we take $u = (A/A')^{1/4} = (B/B')^{1/6}$ for an admissible change of coordinates. □

Remark 4.13.4

(1) A closely related question is to compute the order of the group $Aut(E)$ of invertible elements of $End(E)$. If p is the characteristic of k, we find that

$$\begin{cases} \#Aut(E) = 2 & \text{if } j \notin \{0, 1728\} \\ \#Aut(E) = 4 & \text{if } j(E) = 1728 \text{ and } p \notin \{2, 3\} \\ \#Aut(E) = 6 & \text{if } j(E) = 0 \text{ and } p \notin \{2, 3\} \\ \#Aut(E) = 12 & \text{if } j(E) = 0 \text{ and } p = 3 \\ \#Aut(E) = 24 & \text{if } j(E) = 0 \text{ and } p = 2. \end{cases}$$

On this topic, see [Sil] p. 103.

(2) When $k = \mathbb{Q}$, we find that there are exactly 13 isomorphism classes (over $\bar{\mathbb{Q}}$) of elliptic curves defined over $\mathbb{Q}$ which admit complex multiplication (i.e. $End(E) \not\cong \mathbb{Z}$).

4.14 MINIMAL WEIERSTRASS EQUATIONS (OVER $\mathbb{Z}$)

Suppose we are given the elliptic curve $E : y^2 = x^3 + 1$ defined over $\mathbb{Q}$, and an arbitrary prime number p. We saw that the elliptic curve $E' : y^2 = x^3 + p^6$ is isomorphic to E over $\mathbb{Q}$.

However if $p > 3$, the first one has good reduction modulo p, while the second has (bad) reduction of additive type (the curve has a cusp at the origin).

By what we saw above, it is clear that every elliptic curve E defined over $\mathbb{Q}$ is isomorphic to a Weierstrass curve with integral coefficients of the type

$$y^2 + a_1xy + a_3y = x^3 + a_2x^2 + a_4x + a_6, \tag{1}$$

We then saw that the discriminant Δ of E is a non-zero integer (see Section 4.13). Now let p an arbitrary prime, and recall that the ring $\mathbb{Z}_{(p)}$ of p-integers is defined by the equality

$$\mathbb{Z}_{(p)} = \{x \in \mathbb{Q}; |x|_p \le 1\}.$$

We say that equation (1) is p-integral if its coefficients a_i are p-integers.

Definition 4.14.1 *Equation (1) is said to be* **minimal** *at p if*

(1) *This equation is p-integral.*

(2) *$v_p(\Delta)$ cannot decrease because of an admissible change of coordinates with coefficients in $\mathbb{Q}$ leading to a new p-integral equation.*

Example 4.14.1 If $v_p(\Delta) = 0$, then equation (1) is clearly p-minimal whenever the a_i are p-integers. □

From now on, we will write an **admissible change of coordinates** in the form

$$\begin{cases} x = u^2x' + r \\ y = u^3y' + su^2x' + t \end{cases} \tag{2}$$

with u, r, s, t in $\mathbb{Q}$. We then have

$$\begin{cases} ua_1' = a_1 + 2s \\ u^2a_2' = a_3 - sa_1 + 3r - s^2 \\ u^3a_3' = a_3 + ra_1 + 2t \\ u^4a_4' = a_4 - sa_3 + 2ra_2 - (t + rs)a_1 + 3r^2 - 2st \\ u^6a_6' = a_6 + ra_4 + r^2a_2 + r^3 - ta_3 - t^2 - rta_1 \\ u^3b_2' = b_2 + 12r \\ u^4b_4' = b_4 + rb_2 + 6r^2 \\ u^6b_6' = b_6 + 2rb_4 + r^2b_2 + 4r^3 \\ u^8b_8' = b_8 + 3rb_6 + 3r^2b_4 + r^3b_2 + 3r^4 \\ u^4c_4' = c_4 \\ u^6c_6' = c_6 \\ u^{12}\Delta' = \Delta. \end{cases} \tag{3}$$

Lemma 4.14.1 *Assume that the coefficients of equation (1) are p-integers. Then:*

(i) *If* $|\Delta|_p > p^{-12}$ *or* $|c_4|_p > p^{-4}$ *or* $|c_6|_p > p^{-6}$*, the equation is minimal,*
(ii) *If* $p > 3$ *and if* $|\Delta_p| \leq p^{-12}$ *and* $|c_4|_p \leq p^{-4}$*, the equation is not minimal.*

Proof. **(i)** We will show that if $|\Delta|_p > p^{-12}$, then equation (1) is p-minimal. The reasoning is analogous for the other two conditions: $|c_4|_p > p^{-4}$ and $|c_6|_p > p^{-6}$.
Indeed, suppose that (1) is not minimal. Then there exists an admissible change of coordinates of type (2), with coefficients in $\mathbb{Q}$, such that the new equation $(1)'$ has coefficients a_i' in $\mathbb{Z}_{(p)}$ and such that

$$|\Delta'|_p > |\Delta|_p.$$

Since $|\Delta|_p = |\Delta'|_p|u|_p^{12}$, it follows that $|u|_p^{12} < 1$, so $u_p \in \mathbb{Z}_{(p)}$. Now, we know that $|\Delta|_p > p^{-12}$ and $\Delta' \in \mathbb{Z}_{(p)}$ (since the a_i' are p-integers), so $|\Delta'|_p \leq 1$. It follows that

$$|u|_p^{12} = \frac{|\Delta|_p}{|\Delta'|_p} > p^{-12},$$

which implies that $|u|_p = 1$. But this is a contradiction!

(ii) Suppose now that $p > 3$ and

$$|\Delta|_p \leq p^{-12}, \qquad |c_4|_p \leq p^{-4}.$$

Since $1728\Delta = c_4^3 - c_6^2$, it follows that $|c_6|_p \leq p^{-6}$ (and conversely, $|c_6|_p \leq p^{-6}$ and $|c_4|_p \leq p^{-4}$ imply that $|\Delta_p| \leq p^{-12}$, etc.). Now, recall from Section 4.13, equation (3), that after an admissible change of coordinates (if $p > 3$), we obtain

$$y^2 = x^3 - \frac{c_4}{48}x - \frac{c_6}{864}.$$

If we make the admissible change,

$$\begin{cases} x = p^2 x' \\ y = p^3 y', \end{cases}$$

we find

$$y^2 = x^3 - \frac{c_4}{48p^4}x - \frac{c_6}{864p^6},$$

whose coefficients are p-integers and whose discriminant $\Delta' = \Delta p^{-12}$ is such that

$$|\Delta'|_p > |\Delta|_p.$$

The existence of this curve shows that, under these conditions, (1) is not minimal. □

Lemma 4.14.2 *Consider an elliptic curve E defined over $\mathbb{Q}$ and a prime number p.*

- **(i)** *The equation of E can be made minimal at p by an admissible change of coordinates (with coefficients in $\mathbb{Q}$).*
- **(ii)** *If the coefficients of E are already p-integers, then the coefficients of this admissible change are also p-integers.*
- **(iii)** *Two minimal equations at p coming from E are related by an admissible change of coordinates for which $|u|_p = 1$ and r, s, t are p-integers.*

Proof. We give the proof only for $p > 3$, and refer to [Kn] p. 292 for the other case.

- **(i)** Since we can assume that the equation of E has integral coefficients, we have $|\Delta|_p \leq 1$. Since $|\Delta|_p > 0$, there are only a finite number of possible $|\Delta'|_p$ between $|\Delta|_p$ and 1, which proves the existence of a minimal equation.
- **(ii)** If the new equation is minimal, the twelfth equation of (3) shows that $|u|_p \leq 1$.
  Since the a_i' are in $\mathbb{Z}_{(p)}$, the first, second and third equations of (3) show that s, r and t lie in $\mathbb{Z}_{(p)}$ (if $p = 2$ or 3, we need other equations).
- **(iii)** If the equation of E' is minimal, we saw in (ii) that $|u|_p \leq 1$, but the parameter u of the inverse transformation is u^{-1}, so if E was already minimal $|u^{-1}|_p \leq 1$, i.e. $|u|_p \geq 1$. Thus, it follows that $|u|_p = 1$. □

We will now define a **globally** minimal model.

Definition 4.14.2 *Equation (1) is said to be* **globally minimal** *if*
(1) *The equation has coefficients in $\mathbb{Z}$;*
(2) *At every prime p, the equation is minimal.*

Example 4.14.2 The equation $y^2 = x^3 + 1$ is globally minimal by lemma 4.14.1. □

To show that every elliptic curve admits a globally minimal equation, we will need the following form of the **Chinese remainder theorem** (see Problem 1 of Chapter 3).

Chinese Remainder Theorem 4.14.1 *Let $p_1, \ldots, p_n$ be a finite set of prime numbers and $\varepsilon_1, \ldots, \varepsilon_n$ be real numbers > 0. For every $i \in \{1, \ldots, n\}$, take p-integers $x_i \in \mathbb{Z}_{(p)}$.*

Then there exists $x \in \mathbb{Z}$ such that for every i we have

$$|x - x_i|_{p_i} \leq \varepsilon_i.$$

Néron's Theorem 4.14.2

(i) *Given an elliptic curve E defined over $\mathbb{Q}$ by a Weierstrass equation, there exists an admissible change of coordinates, with coefficients in $\mathbb{Q}$, such that the new equation is globally minimal.*

(ii) *Two globally minimal equations for the same curve E, are related by an admissible change of coordinates such that $u = \pm 1$ and r, s, t lie in $\mathbb{Z}$.*

Proof. Part (ii) follows immediately from lemma 4.14.2.

Thus, it remains to prove the **existence** of a globally minimal equation for E (part (i)).

Up to changing the equation of E, we can assume that the a_i lie in $\mathbb{Z}$, so $\Delta \in \mathbb{Z}$. Let p be a prime number dividing Δ. Then we can make an admissible change of coordinates, with coefficients $\{u_p, r_p, s_p, t_p\}$, such that the new equation (having coefficients $a_{i,p}$) is minimal at p. By lemma 4.14.2, the coefficients u_p, r_p, s_p, t_p are p-integers. Furthermore, we know that

$$|u_p|_p^{12} \, |\Delta_p|_p = |\Delta|_p,$$

where Δ_p denotes the discriminant of the new equation.

If we now set

$$u_p = p^{\lambda_p}\theta_p$$

with $\theta_p \in \mathbb{Z}_{(p)}$, $|\theta_p|_p = 1$ and $\lambda_p \geq 0$ (by lemma 4.14.2), and if we take $u = \prod_{p|\Delta} p^{\lambda_p} \in \mathbb{N}$, then if u is the first coefficient of an admissible change $\{u, r, s, t\}$, and if p divides Δ, we have

$$|\Delta'|_p = |u|_p^{-12} \, |\Delta|_p = |u_p|_p^{-12} \, |\Delta|_p = |\Delta_p|_p.$$

As $|\Delta'|_\ell = 1$ if ℓ is a prime number not dividing Δ, we see that the new equation is globally minimal insofar as its coefficients a_i' are *integers*. We realise this condition by suitably choosing r, s, t using the Chinese remainder theorem (theorem 4.14.1).

Indeed, we take r, s, t in $\mathbb{Z}$ such that

$$|r - r_p|_p \leq p^{-6\lambda_p}, \quad |s - s_p|_p \leq p^{-6\lambda_p}, \quad |t - t_p|_p \leq p^{-6\lambda_p};$$

then the formulae (3) show that the a_i' are integers.

To see this, we check that for every prime ℓ, we have

$$|a'_i|_\ell \leq 1.$$

If ℓ does not divide Δ, there is no problem because the a_i and r, s, t lie in $\mathbb{Z}$. If $\ell = p$ divides Δ, we need to look more closely, comparing the global transformation to the local transformation at p.

For example, let us show that $|a'_1|_p \leq 1$. By (3), we write

$$\begin{aligned} ua'_1 &= a_1 + 2s = a_1 + 2(s - s_p) + 2s_p \\ &= u_p a'_{1,p} + 2(s - s_p), \end{aligned}$$

hence

$$\begin{aligned} |ua'_1|_p &\leq \operatorname{Max}\{|u_p a'_{1,p}|_p, |2(s - s_p)|_p\} \\ &\leq \operatorname{Max}\{|u_p|_p, p^{-6\lambda_p}\} = |u_p|_p \end{aligned}$$

since $|u_p|_p = p^{-\lambda_p}$ by definition.

As $|u_p|_p = |u|_p$, by definition, we obtain

$$|a'_1|_p \leq 1.$$

We leave the cases $i = 2, 3, 4, 6$ as an exercise for the reader. □

Definition 4.14.3 *Let E be an elliptic curve defined over $\mathbb{Q}$, and birationally equivalent over $\mathbb{Q}$ to a Weierstrass cubic W whose equation is minimal. We say that the curve E is (globally)* **semi-stable** *if and only if W has good reduction or multiplicative reduction (see Section 4.5) at every prime number.*

Theorem 4.14.3 *In order for E to be (globally) semi-stable, it is necessary and sufficient that Δ and c_4 be relatively prime.*

Proof. Let us show that the condition is necessary.

Indeed, if the prime number p divides both Δ and c_4, we have $\bar{\Delta} = \bar{c}_4 = \bar{0}$ in $\mathbb{F}_p$. Thus, the curve $\widetilde{W}_p$ is of additive type, by a remark from Section 4.13.

Let us show that the condition is sufficient.

Indeed, if p does not divide Δ, $\widetilde{W}_p$ is of genus 1 by a remark from Section 4.13. Moreover, if p divides Δ, then p does not divide c_4.

We then have $\bar{\Delta} = \bar{0}$ but $\bar{c}_4 \neq \bar{0}$, and a remark from Section 4.13 shows that $\widetilde{W}_p$ is of multiplicative type. □

Remark 4.14.1

(1) It follows easily from (3) and Néron's theorem (theorem 4.14.2) that the choice of minimal equation W has no bearing on the question of whether E is semi-stable.

(2) We say that E is **semi-stable at** p if the reduction of W at p is good or of multiplicative type; this condition can be seen "locally", i.e. on an equation which is **minimal at** p.

4.15 HASSE–WEIL *L*-FUNCTIONS

The zeta function of an elliptic curve is constructed on the model of the Riemann zeta function (which was studied intensively by Euler in the 18th century). Let us rapidly recall some properties of this function.

4.15.1 Riemann Zeta Function

We know that the Riemann zeta function is defined over $\mathbb{C}$, for $\mathrm{Re}(s) > 1$, by the formula

$$\zeta(s) = \sum_{n \geq 1} \frac{1}{n^s}, \tag{1}$$

and Euler found that the zeta function can be decomposed into an "Euler product":

$$\zeta(s) = \prod_{p \text{ premier}} \frac{1}{1 - p^{-s}}. \tag{2}$$

This equation tells us that every natural integer $n \geq 1$ can be decomposed in a unique way as a product of primes.

Thus, the zeta function (often called the ζ function) expresses, in a synthetic and hidden form, important properties of $\mathbb{Z}$. However, it is important to note that in (1), the only integers n which appear on the right-hand side are ≥ 1. Thus, the ζ function can be considered to be attached to the set of "norms" of the non-zero ideals of $\mathbb{Z}$, i.e. the set of numbers

$$|n| = \mathrm{Card}(\mathbb{Z}/(n)).$$

Since we know some rings which arc, in some sense, "simpler" than $\mathbb{Z}$, namely the rings of p-adic integers $\mathbb{Z}_p$, all of whose non-zero ideals are of the form (p^ν), we can construct an analogue of the Riemann zeta function for these rings. Indeed, for $\mathrm{Re}(s) > 0$, we set

$$\zeta_p(s) = \sum_{\nu \geq 0} \frac{1}{p^{\nu s}} = \frac{1}{1 - p^{-s}}. \tag{3}$$

Then, equation (2) is quite simply

$$\zeta(s) = \prod_{p \text{ premier}} \zeta_p(s) \quad \text{for } \mathrm{Re}(s) > 1. \tag{4}$$

Moreover, if we set

$$\Lambda_p(s) = p^{-s/2} \zeta_p(s),$$

we obtain the obvious functional equation

$$\Lambda_p(s) = -\Lambda_p(-s). \tag{5}$$

We can then wonder whether the Riemann zeta function has analogous properties. In Chapter 5 (or see [Og 1]), we will see the proof of the following result.

Theorem (Riemann, 1859) 4.15.1 *The Riemann zeta function defined by equation (1) for $Re(s) > 1$ can be analytically continued to a meromorphic function of the complex plane.*

The only pole of this function is at $s = 1$, and its residue is 1.

Moreover, the function Λ defined by

$$\Lambda(s) := \pi^{-s/2}\Gamma\left(\frac{s}{2}\right)\zeta(s) \tag{6}$$

satisfies the functional equation

$$\Lambda(s) = \Lambda(1 - s). \tag{7}$$

Remark 4.15.1

(1) The function Γ is the Euler gamma function, defined by

$$\Gamma(s) := \int_0^\infty e^{-t}t^{s-1}\,dt$$

(it is the Mellin transform of $t \mapsto e^{-t}$, see Chapter 5).

It admits a meromorphic continuation to all of $\mathbb{C}$, and satisfies the functional equation

$$\Gamma(s + 1) = s\Gamma(s).$$

(2) The ζ function vanishes at every even integer < 0. The famous **Riemann conjecture** ("Riemann hypothesis") asserts that all the other zeros of ζ lie on the line $\mathrm{Re}(s) = \frac{1}{2}$ (i.e. on the symmetry axis of the functional equation (7)).

4.15.2 Artin Zeta Function

Before defining the Hasse–Weil zeta function of of the minimal model of an elliptic curve defined over $\mathbb{Q}$, let us begin by considering certain fields which are in some sense "simpler" than $\mathbb{Q}$, namely the finite fields $\mathbb{F}_p$. This is what Artin did in his thesis [Ar 2].

Artin took an elliptic curve over $\mathbb{F}_p$ in affine Weierstrass form

$$E: \quad Y^2 + a_1XY + a_3Y - \left(X^3 + a_2X^2 + a_4X + a_6\right),$$

and associated to it the Dedekind domain

$$A = \mathbb{F}_p[X, Y]/(E)$$

which is the quotient of the polynomial ring $\mathbb{F}_p[X, Y]$ by the ideal generated by E.

Naturally, A is a quadratic extension of $\mathbb{F}_p[X]$ and we know that $\mathbb{F}_p[X]$ is analogous to $\mathbb{Z}$. As for the quadratic extensions of $\mathbb{Z}$, we can define the norm of a non-zero ideal of A by

$$N(\mathfrak{a}) = \mathrm{Card}(A/\mathfrak{a});$$

thus we define the zeta function of A by

$$\zeta_A(s) = \sum_{\substack{\mathfrak{a} \text{ ideal of } A \\ \mathfrak{a} \neq 0}} \frac{1}{N(\mathfrak{a})^s} \tag{8}$$

when $\mathrm{Re}(s) > 1$ (a proof of the convergence of the right-hand side of (8) when $\mathrm{Re}(s) > 1$ can be found in [Ar 2]). The fact that A is a Dedekind domain proves the existence of an Euler product for ζ_A, i.e. of an equation of the form

$$\zeta_A(s) = \prod_{\substack{\wp \text{ prime ideal of} A \\ \wp \neq 0}} \frac{1}{1 - N(\wp)^{-s}}. \tag{9}$$

However, the zeta function of A is not really the zeta function of the curve E, since this curve is not really an elliptic curve: we have "forgotten" the point at infinity $(0, 1, 0)$ of the projective model of E! This corresponds to an ideal of degree 1, so to the factor

$$\zeta_\infty(s) = \frac{1}{1 - p^{-s}},$$

and we are led to the following definition.

Definition 4.15.1 *For $s \in \mathbb{C}$ such that $Re(s) > 1$, we set*

$$\zeta_E(s) = \frac{1}{1 - p^{-s}} \zeta_A(s).$$

Artin shows the following result.

Theorem (Artin) 4.15.2 *Set $T = p^{-s}$ and let N_E denote the cardinal of the group of points of E in $\mathbb{F}_p$. Then*

$$\zeta_E(s) = \frac{1 - a_E T + pT^2}{(1 - T)(1 - pT)},$$

where a_E is defined by the relation

$$a_E = p + 1 - N_E.$$

Corollary ("Riemann Hypothesis") 4.15.1 *The roots of the function ζ_E lie on the line $Re(s) = 1/2$.*

Proof. We need to show that the roots θ_1 and θ_2 of the polynomial $pT^2 - a_E T + 1$ are of module $p^{-1/2}$.

This is easily seen if we know that they are non-real or identical. Indeed, if $\theta_2 = \bar{\theta}_1$, we know that $|\theta_2| = |\theta_1|$, hence $|\theta_j| = |\theta_1 \theta_2|^{1/2} = (p^{-1})^{1/2}$ when $j \in \{1, 2\}$.

Now, theorem 4.9.1 shows that $|a_E| \leq 2p^{1/2}$, so θ_1 and θ_2 are non-real or identical. □

As the polynomial $1 - a_E T + pT^2$ will play an important role in what follows, we are led to the following definition (which is not the same as that given by Silverman [Sil] p. 360).

Definition 4.15.2 *The L-function of E is defined by the formula*

$$L_E(s) = (1 - a_E p^{-s} + p \cdot p^{-2s})^{-1}.$$

Corollary 4.15.2 *Changing s to* $1 - s$, *we have the functional equations:*

(i) $p^{-s} L_E(s) = p^{s-1} L_E(1 - s)$,
(ii) $\zeta_E(s) = \zeta_E(1 - s)$.

Proof. Clearly, (ii) follows from (i), and (i) is obtained by a simple computation. □

4.15.3 Hasse–Weil *L*-function

Now, let us take an elliptic curve defined over $\mathbb{Q}$ by a globally minimal equation $E = 0$ as in Section 4.14. We cannot proceed directly as in Section 4.15.2, since the norm of a non-zero prime ideal of $A = \mathbb{Q}[X, Y]/(E)$ is always infinite. Then, instead of (8), we can consider (4) and formally set

$$L_E(s) = \prod_p L_p(s), \tag{10}$$

where

$$L_p(s) = \begin{cases} L_{\widetilde{E}_p}(s) & \text{if } E \text{ has good reduction at } p, \\ (1 - p^{-s})^{-1} & \text{if } E \text{ has multiplicative reduction} \\ & \quad \text{with rational tangents at } p, \\ (1 + p^{-s})^{-1} & \text{if } E \text{ has multiplicative reduction} \\ & \quad \text{with irrational tangents at } p, \\ 1 & \text{if } E \text{ has additive reduction at } p. \end{cases} \tag{11}$$

In the formulae (11), the three last cases occur only for prime numbers p which divide the discriminant of E, i.e. only for a finite number of p. In the first case, $\widetilde{E}_p$ denotes the elliptic curve over $\mathbb{F}_p$ deduced from E by reduction modulo p.

By corollary 4.15.1 above, we have

$$L_{\widetilde{E}_p}(s) = \left(1 - \theta_p p^{-s}\right)^{-1} \left(1 - \bar{\theta}_p p^{-s}\right)^{-1},$$

where θ_p denotes an (arbitrary) root of $1 - a_{\widetilde{E}_p} T + pT^2$. Thus, we see that, up to a finite number of factors, $L_E(s)$ is the infinite product

$$\prod_{p \geq c} \left(1 - \theta_p p^{-s}\right)^{-1} \prod_{p \geq c} \left(1 - \bar{\theta}_p p^{-s}\right)^{-1}.$$

As $|\theta_p p^{-s}| = p^{-(1/2) - Re(s)}$, we see that this infinite product converges if $Re(s) > 3/2$.

Remark 4.15.2 Néron's theorem (theorem 4.14.2) shows that L_E does not depend on the choice of a globally minimal equation for the curve.

Hasse and Weil were the first to **conjecture** that, in analogy with the Riemann ζ function and the Artin L-functions, the functions L_E associated to elliptic curves E defined over $\mathbb{Q}$ possess meromorphic continuations to the whole complex plane.

This conjecture is now proved, as a consequence of the theorems of Wiles and his disciples.

Their second conjecture, which also follows from the results of Wiles and his disciples, is that the function L_E satisfies a functional equation, but in order to write it down, we first need to introduce the notion of the conductor of the curve E.

Definition 4.15.3 *Let E be a globally minimal equation of an elliptic curve defined over $\mathbb{Q}$. The* **conductor** *N_E is the product*

$$N_E = \prod_{p\, prime} p^{f_p}, \tag{12}$$

where

$$f_p = \begin{cases} 0 & \text{if } E \text{ has good reduction at } p, \\ 1 & \text{if } E \text{ has multiplicative reduction at } p, \\ 2 + \delta_p & \text{with } \delta_p \geq 0 \text{ if } E \text{ has additive reduction at } p. \end{cases}$$

Naturally, this definition is incomplete, since we have not yet defined δ_p precisely. We limit ourselves here to saying that $\delta_p = 0$ if $p > 3$, referring the reader to Ogg, [Og 2] p. 361, for the case where $p \in \{2, 3\}$.

To write the **functional equation of** L_E, we will modify this series by multiplying it by a normalisation factor containing the Euler gamma function and the conductor N_E. Set

$$\Lambda_E(s) := \left(\frac{\sqrt{N_E}}{2\pi}\right)^s \Gamma(s)\, L_E(s). \tag{13}$$

The desired equation functional is given by

$$\Lambda_E(s) = w\, \Lambda_E(2 - s)$$

with $w = \pm 1$.

Remark 4.15.3 The famous **Birch–Swinnerton–Dyer conjecture** asserts that the rank r of the Mordell–Weil group of an elliptic curve E defined over $\mathbb{Q}$ is equal to the order of the zero of $L_E(s)$ at the point $s = 1$ (note that this point is situated on the symmetry axis of the involution $s \leftrightarrow 2 - s$ which appears in the functional equation (14)).

We can give a more down to earth aspect to this assertion by writing it as follows:

$$\prod_{\substack{p\le R\\ p\mid\Delta}} \frac{\mathrm{Card}\big(\widetilde{E}_p(\mathbb{F}_p)\big)}{p} \sim C\,(\log R)^r$$

for a constant C.

COMMENTARY

The last ten years have witnessed an enormous explosion in the English-language literature devoted to the subject of elliptic curves. For the point of view which concerns us here, the most relevant books are those of Knapp [Kn], Silverman [Sil], Cassels [Ca] and Silverman and Tate [S-T]. But many others, for example that of D. Husemöller [Hus], can be extremely useful.

Before that, there may have been a shortage of books, but there was certainly no lack of excellent articles; we cite in particular those of Cassels [Cas 3] and Tate [Tat].

An introduction to the notions of algebraic geometry used in this chapter can be found in "Fulton's little book" [Fu]. But for a veritable initiation into modern algebraic geometry, the reader can consult the books by Hartshorne [Ha], as well as Silverman [Sil]. We also recommend the excellent book by D. Perrin [Per].

Curves of genus one are implicitly present in a large number of problems considered by Fermat, as A. Weil showed in [Wei]. But the systematic study of cubics (of genus zero or one) was apparently inaugurated by Newton. In the "divergent parabola" (which is the Weierstrass cubic if the genus is one), he was able to recognize the archetype of all plane cubics. Later, in his book *Geometrica organica*, his distant disciple C. MacLaurin explained the basic principles ("Bézout's theorem", the nine-point theorem) which we used to define a group law on a genus one cubic having at least one rational point. To conclude, let us mention that E. Artin's thesis [Ar 2] is a magnificent introduction to the study of elliptic curves over finite fields.

For deep results which are not proved within this book, we refer to [Sil] for Mordell–Weil, to [Se 3] for Serre's theorem and to [M] for Mazur's theorem.

Exercises and Problems for Chapter 4

4.1 Let k be a field of characteristic not equal to 2, in which every number is a square (for example $k = \mathbb{C}$), and take p and $q \in k[t]$ such that there exist four distinct projective points $(\lambda_i, \mu_i) \in \mathbb{P}_1(k)$ such that for every $i \in \{1, 2, 3, 4\}$, $\lambda_i p + \mu_i q$ is a square in $k[t]$. We propose to show that p and q are constants.

(α) Up to replacing p and q by a_1p+b_1q and a_2p+b_2q with $(a_1, b_1) \neq (a_2, b_2)$ in $\mathbb{P}_1(k)$, we may assume that $(\lambda_1, \mu_1) = (1, 0)$, $(\lambda_2, \mu_2) = (0, 1)$, $(\lambda_3, \mu_3) = (1, -1)$, $(\lambda_4, \mu_4) = (1, -\lambda)$.

(β) Assume that $\max\{\deg p, \deg q\}$ is minimal among all the possible relations such that $\max\{\deg p, \deg q\} > 0$. Show that setting $p = u^2$, $q = v^2$, we find a new quadruple of squares which contradicts the minimality of $\max\{\deg p, \deg q\}$.

(γ) Conclude.

4.2 Let k be a field of characteristic not equal to 2, in which every number is a square. Consider the cubic W_λ of the equation

$$y^2 = x(x-1)(x-\lambda)$$

for some $\lambda \in k$. We propose to show that if $\lambda \notin \{0, 1\}$, then this cubic is not parametrisable in $k(t)$.

(α) Show that in the contrary case, there exist four polynomials p, q, r, s in $k[t]$ such that $x = p/q$, $y = r/s$ (irreducible fractions) and

$$r^2 \sim p(p-q)(p-\lambda q).$$

(α) Deduce that $p, q, p-q, p-\lambda q$ are associated to squares in $k[t]$.

(β) Conclude using Exercise 4.1.

4.3 **Short Weierstrass form**

An elliptic curve is said to be in short Weierstrass form up to a birational transformation if its image under this transformation can be written

$$W: \quad Y^2Z = X^3 + a_4XZ^2 + a_6Z^6$$

with a_4 and $a_6 \in k$, the field of definition of the cubic. The characteristic of k is assumed to be different from 2.

(α) Assume that P is a rational point of the cubic C, and that it is not an inflection point. Let O be the (other) intersection point of the tangent to C at P with C.

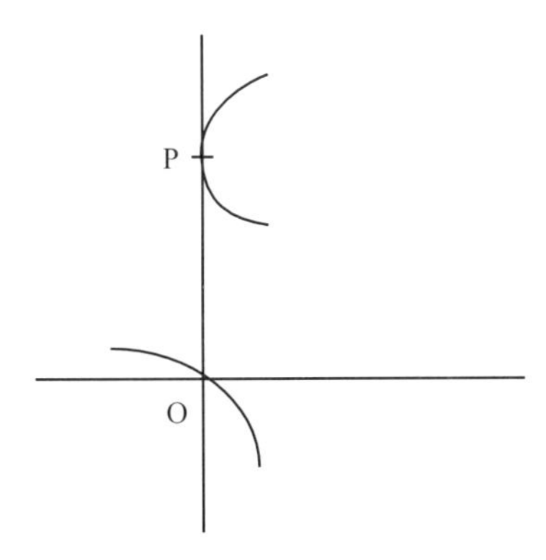

Show that if we take the origin at O and OP for the y-axis, the affine equation of C is

$$C(x, y) = C_1(x, y) + C_2(x, y) + C_3(x, y) = 0,$$

where $C_i(x, y)$ is homogeneous of degree i (for $i = 1, 2, 3$), and that

$$C_2(0, 1)^2 - 4C_1(0, 1)C_3(0, 1) = 0.$$

(β) Cut C by the line $y = tx$ and discard the fixed point O. Show that if we set

$$s := 2C_3(1, t)x + C_2(1, t),$$

the point (t, s) lies on the **cubic**

$$s^2 = C_2(1, t)^2 - 4C_1(1, t)C_3(1, t).$$

(γ) Put this last cubic in short Weierstrass form W when k is of characteristic not equal to 3.
(δ) Show that if C has a multiple point, the cubic W is "unicursal", i.e. parametrisable in $k(\theta)$.
(ε) Deduce that the polynomial

$$x^3 + a_4 x + a_6$$

then admits a double root in k (use Exercise 4.2).

4.4 Consider an affine quartic defined over a field k of characteristic zero, which admits the rational point (a, b) and has equation

$$y^2 = r_0 + r_0 x + r_1 x^2 + r_3 x^3 + r_4 x^4.$$

(α) Setting $u = 1/(x - a)$ and $v = y/(x - a)^2$ (this change of variables sends the point (a, b) to infinity), show that we have

$$v^2 = a_0 + a_1 u + a_2 u^2 + a_3 u^3 + a_4 u^4$$

and that a_4 is a square in k.
(β) Write the right-hand side of this equation in the form $G(u)^2 + H(u)$, with $G(u) = g_2 u^2 + g_1 u + g_0$, $H(u) = h_1 u + h_0$.
(γ) Set $t = v + G(u)$ and $tu = s$. Show that

$$2g_2 s^2 + 2g_1 ts + 2g_0 t^2 = t^3 - h_1 s - h_0 t.$$

(δ) Put this cubic in Weierstrass form W.
(ε) What can we say about W if Q is unicursal? (use Exercise 4.2).

4.5 Consider two quadrics Q_1 and Q_2 of $\mathbb{P}_3(k)$ passing through the point $(0, 0, 0, 1)$ which intersect along a non-degenerate biquadratic.

(α) Show that Q_1 and Q_2 can be written

$$\begin{cases} TL + R = 0 \\ TM + S = 0, \end{cases}$$

where L and M are in $k[X, Y, Z]_1$ and R and S in $k[X, Y, Z]_2$.
(β) Show that if L and M are linearly dependent, then $Q_1 \cap Q_2$ is unicursal over $\bar{k}$.
(γ) Show that if L and M are linearly independent, $Q_1 \cap Q_2$ projects (eliminating T) onto the cubic $LS - RM$, and that this cubic has a rational point.
(δ) Deduce that $Q_1 \cap Q_2$ is either unicursal or birationally equivalent to an elliptic curve.

4.6 In $\mathbb{R}^2$, draw the curves of equations

$$C_1 : (x^2 + y^2)^2 + 3x^2y - z^3 = 0$$
$$C_2 : (x^2 + y^2)^4 - 4x^2y^2 = 0;$$

it may be helpful to use polar coordinates.

4.7 Let C and C' be two plane cubics such that, in $\mathbb{P}_2(\bar{k})$, we have

$$C \cdot C' = \sum_{i=1}^{9} P_i.$$

(α) Assume that there exists a conic Γ such that

$$C \cdot \Gamma = \sum_{i=1}^{6} P_i.$$

Show that P_7, P_8 and P_9 are collinear.

(β)

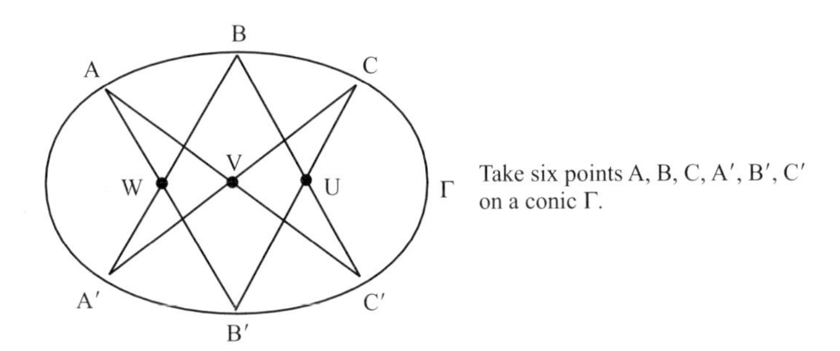

Take six points A, B, C, A′, B′, C′ on a conic Γ.

Let C and C' denote the decomposed cubics $(AB') \cup (CA') \cup (BC')$ and $(B'C) \cup (A'B) \cup (C'A)$. Show that the intersection points U, V, W of BC' and CB', CA' and AC', AB' and BA' are collinear (Pascal's theorem).

(γ) Prove the same result when the conic Γ decomposes into two lines (Pappus' theorem, 4th century A.D.).

(δ) Show the converse of Pascal's theorem, and deduce a way to construct each point of a conic passing through five given points, using only a ruler.

(ε) Describe what Pascal's theorem becomes when $A = B'$. Deduce a construction of the tangent to a conic at a point using only a ruler.

(φ) Same question when $A = B'$, $B = C'$ and $C = A'$.

4.8 Let C be an irreducible cubic and D a line. We assume that k is algebraically closed.

(α) Let $D \cdot C = P_1 + P_2 + P_3$, P_i be distinct.
Let T_i be the tangent to C at P_i, and set

$$T_i \cdot C = 2P_i + Q_i.$$

Show that Q_1, Q_2, Q_3 are collinear.

(β) Show that if a line passes through two inflection points of a cubic, then it passes through a third inflection point (Maclaurin's theorem).

4.9 Desboves' Formulae
Consider the cubic

$$F(X) = a_1X_1^3 + a_2X_2^3 + a_3X_3^3 + dX_1X_2X_3 = 0.$$

Show that $F(X)$ is non-singular if and only if

$$27\, a_1a_2a_3 + d^3 \neq 0.$$

Assume that $F(x) = 0$, and show that the third point t of intersection of the cubic and of the tangent at x is given by the formulae

$$t_i = x_i\big(a_{i+1}x_{i+1}^3 - a_{i+2}x_{i+2}^3\big), \quad i = 1, 2, 3.$$

If x and y are two distinct points of $F(X)$, show that the third point of intersection z of the line (xy) with the cubic is

$$z_i = x_i^2 y_{i+1}y_{i+2} - y_i^2 x_{i+1}x_{i+2}.$$

4.10 Starting from $(2, -1, -1)$ on $X^3 + Y^3 + 7Z^3 = 0$, find 10 distinct points.

4.11 Let k be a field of arbitrary characteristic, and C the singular cubic $y^2 - x^3$. Let $C_0(k)$ denote the set of non-singular points of C in $\mathbb{P}_2(k)$.

(α) Show that $C_0(k)$ can be rationally parametrised using the parameter $t = x/y$.
(β) Give a necessary and sufficient condition for the points of parameters t_1, t_2, t_3 to be collinear.
(γ) Let O be the point of parameter o and t_1 and t_2 the parameters of P_1 and $P_2 \in E_0(k)$.
Compute the value of the parameter of $P_1 \oplus P_2$.
What can we say about the group $(E_0(k), \oplus)$?

4.12 Let k be a field of characteristic not equal to 2, C the singular cubic $y^2 - x^2(x+1)$ and $C_0(k)$ the set of non-singular points of C in $\mathbb{P}_2(k)$.

(α) Show that $C_0(k)$ can be parametrised using $t = x/y$. What are the forbidden values of t?
(β) Set $t = (\theta - 1)/(\theta + 1)$, and give a parametrisation of $C_0(k)$ using θ. What are the forbidden values of θ?
(γ) Give a necessary and sufficient condition for the points of parameters θ_1, θ_2 and θ_3 to be collinear.
(δ) Let O be the point of parameter 1 and θ_1, θ_2 the parameters of P_1 and $P_2 \in C_0(k)$.
Compute the value of the parameter of $P_1 \oplus P_2$.
What can we say about the group $(C_0(k), \oplus)$?

4.13 Let k be a field of characteristic not equal to 2. Assume that k contains an element a which is not a square, and consider the singular cubic Γ of equation $y^2 - x^2(x+a) = 0$.

(α) Set $\alpha = \sqrt{a} \in \bar{k}$. Show that $k(\alpha)/k$ is a Galois extension of degree 2.
(β) Set $x = \alpha^2\xi$, $y = \alpha^3\eta$, and show that $(x, y) \in \Gamma$ is equivalent to $(\xi, \eta) \in C$, where C is a cubic whose equation we will determine.

(γ) Use the preceding exercise to give a parametrisation of $C_0(k(\alpha))$.
(δ) Set $\mathrm{Gal}(k(\alpha)/k) = \{\mathrm{id}, \sigma\}$.
Let $P \in C_0(k(\alpha))$ be of parameter θ.
Show that $P \in C_0(k)$ if and only if $N(\theta) = \theta \cdot \sigma(\theta) = 1$.
(ε) Show that the map N given by

$$\begin{cases} k(\alpha)^* \longrightarrow k(\alpha)^* \\ x \longmapsto x \cdot \sigma(x) \end{cases}$$

is a group homomorphism and that $(C_0(k), \oplus)$ is isomorphic to $\mathrm{Ker}N$.
(φ) Show that if $k = \mathbb{F}_7$, then $C_0(k)$ contains 8 points.
What is the structure of this group?

4.14 Let C_1 and C_2 be two curves in the projective plane, with no common components and of respective degrees d_1 and d_2.

Assume that C_1 and C_2 intersect at $d_1 d_2$ distinct points and that C_1 is smooth.

Let D be a third curve in the projective plane of degree $d_1 + d_2 - 3$, which passes through all the points of $C_1 \cap C_2$ except for at most one. Show that D then necessarily passes through the remaining point.

4.15 Let C be the affine quartic $y^4 - xy - x^3$.

(α) Determine the tangents to C at the point O of coordinates $(0, 0)$.
(β) Let D be the line $x = 0$, compute $\mu_O(C, D)$.
(γ) Let L be the line $y = 0$, compute $\mu_O(C, L)$.
(δ) Let Δ be a line passing through O not equal to D or L. Compute $\mu_O(C, \Delta)$.

4.16 Consider the elliptic curve $C = y^2 - (x^3 + x) \in \mathbb{Q}[x, y]$.

(α) Show that in order for $(x, y) \in C(\bar{\mathbb{Q}})$ to be of order 3, it is necessary and sufficient that

$$3x^4 + 6x^2 - 1 = 0.$$

(β) Compute the points of order 3 of $C(\bar{\mathbb{Q}})$.
(γ) Show that the field K_3 which is generated over $\mathbb{Q}$ by the coordinates of these points is $\mathbb{Q}(b, i)$, where

$$b = \sqrt{\frac{2a}{\sqrt{3}}} \quad \text{and} \quad a = \sqrt{\frac{2\sqrt{3} - 3}{3}}.$$

(δ) Show that $[K_3 : \mathbb{Q}] = 16$.

4.17 Again consider the elliptic curve $C = y^2 - (x^3 + x)$ defined over $\mathbb{Q}$.

(α) Show that in order for $(x, y) \in C(\bar{\mathbb{Q}})$ to be of order 4, it is necessary and sufficient that

$$x^6 + 5x^4 - 5x^2 - 1 = 0.$$

(β) Noting that ± 1 are roots of this polynomial, decompose this polynomial as a product of irreducible polynomials in $\mathbb{Q}[x]$.
(γ) Compute the points of order 4 of $C(\bar{\mathbb{Q}})$.

(δ) Show that the field K_4 which is generated over $\mathbb{Q}$ by the coordinates of these points is $\mathbb{Q}(i, \sqrt{2})$.

(ε) Determine the Galois group of $K_4/\mathbb{Q}$.

4.18 Determine the Galois representations ρ_2 of $G_{\mathbb{Q}}$ in $\mathrm{GL}_2(\mathbb{F}_2)$ attached to the 2-division points of the following elliptic curves:

(α) $E_1 = y^2 - x(x^2 - 1)$.

(β) $E_2 = y^2 - (x^3 + x)$.
In this case, show that $K_2 = \mathbb{Q}(i)$, and take as a basis of $E_2[2]$ the points

$$P_1 = (0, 0) \quad \text{and} \quad P_2 = (i, 0).$$

(γ) $E_3 = y^2 - (x^3 - 2)$.
In this case, show that $K_2 = \mathbb{Q}(\sqrt{-3}, \sqrt[3]{2})$ and that $\mathrm{Gal}(K_2/\mathbb{Q}) \cong \mathfrak{S}_3$. What can we say about the image of ρ_2?

Let j be a primitive cubic root of unity, and set

$$P_1 = \left(\sqrt[3]{2}, 0\right), \qquad P_2 = \left(j\sqrt[3]{2}, 0\right).$$

Take (P_2, P_2) as a basis of $E_3[2]$.

Compute $\rho_2(\sigma)$ when $\sigma \in \mathrm{Gal}(K_2/\mathbb{Q})$ is defined by $\sigma(\sqrt{-3}) = \sqrt{-3}$ and $\sigma(\sqrt[3]{2}) = j\sqrt[3]{2}$.

4.19 Consider a field k of characteristic not equal to 2, and the elliptic curves

$$E_1: \quad y^2 = x(x^2 + a_1 x + b_1),$$
$$E_2: \quad y^2 = x(x^2 + a_2 x + b_2),$$

with $a_2 = -2a_1$, $b_2 = a_1^2 - 4b_1$, $b_1 b_2 \neq 0$.

(α) Express the rational map $E_1 \to E_2$ given by

$$(x, y) \overset{\varphi}{\longmapsto} \left(\frac{y^2}{x^2}, \frac{y(x^2 - b_1)}{x^2}\right)$$

in homogeneous coordinates. What can we say about the image of the points $(0, 1, 0)$ and $(0, 0, 1)$?

(β) We know that φ must be defined everywhere over E_1 (since E_1 is smooth).

Find a homogeneous expression of φ which is defined at $(0, 1, 0)$. Use the fact that $(X, Y, Z) \in E_1$.

Deduce that φ is an isogeny.

(γ) Find a homogeneous expression of φ which is defined at $(0, 0, 1)$. What is the image of this point under φ?

4.20 Assume that the map

$$z \overset{\varphi}{\longmapsto} a_0 z^n + a_1 z^{n-1} + \ldots + a_n$$

is a homomorphism $(\mathbb{C}^*, \times) \to (\mathbb{C}^*, \times)$ and that $a_0 \neq 0$.

(α) Show that $a_0 + a_1 + \cdots + a_n = 1$.
(β) What can we say about the zeros of the polynomial $\varphi(z)$?
(γ) Deduce a simplified expression of $\varphi(z)$.
(δ) Same question, replacing $\varphi(z)$ by a rational function in z.
(ε) Same question, replacing $\varphi(z)$ by a meromorphic function in z.

4.21 Compute $\#E(\mathbb{F}_q)$ for $p = 3$ and 7 and $E = y^2 - (x^3 + x + 1)$.
Determine the structure of the Abelian group $E(\mathbb{F}_p)$ in the two preceding cases.

Problem I

(Congruent Numbers)

Consider the elliptic curves

$$\begin{cases} E_1: & y_1^2 = x_1\left(x_1^2 + a_1x_1 + b_1\right), \\ E_2: & y_2^2 = x_2\left(x_2^2 + a_2x_2 + b_2\right), \end{cases}$$

both defined over a field K of characteristic different from 2, and assume that

$$a_2 = -2a_1, \qquad b_2 = a_1^2 - 4b_1, \qquad 2b_1b_2 \neq 0.$$

Recall that E_1 and E_2 are isogenous via the following isogenies of degree two:

$$\begin{cases} (x_1, y_1) \xrightarrow{\varphi} (x_2, y_2) = \left(\left(\dfrac{y_1}{x_1}\right)^2, \dfrac{y_1(x_1^2 - b_1)}{x_1^2}\right) \\ (x_2, y_2) \xrightarrow{\psi} (x_1, y_1) = \left(\left(\dfrac{y_2}{2x_2}\right)^2, \dfrac{y_2(x_2^2 - b_2)}{8x_2^2}\right). \end{cases}$$

Recall that, if Ω denotes the point $(0, 0)$, the kernels of φ and ψ are just the group $\{O, \Omega\}$.

I. (1) Assume here that $a_1^2 - 4b_1$ is not a square, and show that

$$(x_2, y_2) \in \operatorname{Im}\varphi \Rightarrow x_2 \text{ is a square } \textbf{in } K^*,$$

which we denote by $x_2 \in {}^2K^*$.

(2) Show that if $a_1^2 - 4b_1 \in {}^2K^*$, and if $(x_2, y_2) \in \operatorname{Im}\varphi$, then $x_2 \in {}^2K^*$ or $x_2 = 0$. In the latter case, give $\varphi^{-1}(\Omega)$.

(3) Show that if $(x_2, y_2) = (\lambda^2, \mu) = \varphi(x_1, y_1)$ with $\lambda \in K^*$ and $\mu \in K$, then we must have

$$\begin{cases} x_1 - \dfrac{b_1}{x_1} = \pm\dfrac{\mu}{\lambda} \\ x_1 + \dfrac{b_1}{x_1} = \lambda^2 - a_1 \\ y_1 = \pm\lambda x_1. \end{cases}$$

(4) With notation as in question (3), compute $\varphi^{-1}(\lambda^2, \mu)$ when $(\lambda^2, \mu) \in E_2(K)$, and deduce that $(\lambda^2, \mu) \in \operatorname{Im}\varphi$.

II. Assume that (a, b, c) is a primitive Pythagorean triple, i.e. that

$$\begin{cases} a^2 + b^2 = c^2, & abc \neq 0, \; a \text{ even}, \\ (a, b, c) \in \mathbb{N}^3, & \text{relatively prime.} \end{cases}$$

Set

$$E_2 : \quad y_2^2 = x_2\left(x_2^2 - c^2 x_2 + a^2 b^2\right).$$

(1) Is E_2 an elliptic curve?
(2) Find a curve E_1 such that E_2 is related to E_1 by the formulae of part I.
(3) Compute the 2-division points of E_2.
(4) Set $A = (a^2, 0)$, $B = (b^2, 0)$ and $C = (c^2, abc)$. Using a question from part I, show that Ω, A, B and C belong to $\operatorname{Im} \varphi$.
(5) Show that $\varphi^{-1}(C)$ is contained in the image of ψ (use question 3) of I for inspiration. Prove the equations

$$\psi(x_2, y_2) = (\lambda^2, \mu) \quad \Longrightarrow \quad \begin{cases} x_2 - \dfrac{b_2}{x_2} = \pm 4 \dfrac{\mu}{\lambda} \\ x_2 + \dfrac{b_2}{x_2} = 4\lambda^2 - a_2 \\ y_2 = \pm 2\lambda x_2. \end{cases}$$

Compute the points C_1, C_2, C_3, C_4 of $E_2(\mathbb{Q})$ which are such that $[2]C_i = C$ for $i \in \{1, 2, 3, 4\}$ (use the relation $\varphi \circ \psi = [2]$, and find points with integral coordinates).
(6) Taking a simple Pythagorean triple, show that C is not (in general) a point of finite order of E_2 (apply the Nagell–Lutz theorem (theorem 4.10.2)). Deduce that (in general) the points $\pm C_i$, with $i \in \{1, 2, 3, 4\}$, are not of finite order.
(7) Compute $\Omega \oplus C$ and show that C is **never** a point of finite order of $E_2(\mathbb{Q})$.
(8) Let p be a prime number not dividing $ab(a^2 - b^2)$. Show that if we reduce E_2 modulo p, then the order of $\bar{E}_2(\mathbb{F}_p)$ is divisible by 4.

III. Take the situation of II, but now consider the curve $E_{a^2, b^2, -c^2}$ given by

$$\Gamma_2 : \quad y_2^2 = x_2\left(x_2 - b^2\right)\left(x_2 - c^2\right).$$

(1) Is Γ_2 an elliptic curve?
(2) Find a curve Γ_1 which is related to Γ_2 by the formulae of part I.
(3) Compute the points of 2-torsion of Γ_2.
(4) Set $B = (b^2, 0)$, $C = (c^2, 0)$.
Using a question from I, show that Ω, B and C lie in $\operatorname{Im} \varphi$.
(5) Show (as in part II) that $\varphi^{-1}(C)$ is contained in the image of $\psi(E_1(\mathbb{Q}))$.
Compute the points C_1, C_2, C_3, C_4 of $\Gamma_2(\mathbb{Q})$ which are such that $[2]C_i = C$ for $i \in \{1, 2, 3, 4\}$. Do they lie in the image of φ?
(6) Is $\varphi^{-1}(B) \subset \psi(\Gamma_2(\mathbb{Q}))$?
Noting that $\varphi^{-1}(B) \subset \psi(\Gamma_2(\mathbb{Q}(i)))$, compute the points B_1, B_2, B_3, B_4 of $\Gamma_2(\mathbb{Q}(i))$ such that $[2]B_i = B$ for $i \in \{1, 2, 3, 4\}$.
(7) Compute all the points of $\Gamma_2[4](\mathbb{C})$, and determine the smallest field K_4 which contains them all.

(8) Show that if $(x, y) \in \Gamma_2[4]$, then if y is not zero, it divides the discriminant of the right-hand side of the equation of Γ_2 in $\mathbb{Z}[i]$.

(9) Assume that p is a prime number congruent to 1 modulo 4 and which does not divide abc.

Show that $\#\bar{\Gamma}_2(\mathbb{F}_p)$ is divisible by 16.

Show that if $p \in \{31, 47\}$, then

$$\#\bar{\Gamma}_2(\mathbb{F}_p) = p + 1.$$

IV. (1) Define a map

$$N: \quad E_2(K) \longrightarrow K^*/{}^2K^*$$

by the formulae

$$\begin{cases} N(x_2, y_2) = x_2({}^2K^*) & \text{if } x_2 \neq o \\ N(\Omega) = (a_1^2 - 4b_1)({}^2K^*) \\ N(O) = {}^2K^*. \end{cases}$$

Show that N is a group homomorphism (one can show that if x_1, x_2, x_3 are the abscissae of three collinear points of $E_2(K)$, then $N(x_1)N(x_2)N(x_3) = 1$).

What is the kernel of N?

(2) Now assume that $K = \mathbb{Q}$ and that the equation of E_2 has integral coefficients, and let r denote a square-free integer. Show that in order to have $r({}^2\mathbb{Q}^*) \in \operatorname{Im} N$, it is necessary and sufficient that r divide b_2. To do this, one may remark that

$$x_2 = rt^2, \qquad x_2^2 + a_2x_2 + b_2 = rs^2$$

with t and s in $\mathbb{Q}^*$, and show that if the prime number p divides r then it divides b_2.

(3) Deduce from question (2) that $E_2(\mathbb{Q})/\operatorname{Im}\varphi$ is a finite group, then that $E_1(\mathbb{Q})/[2]E_1(\mathbb{Q})$ is finite (recall that $\psi \circ \varphi = [2]$).

Problem 2

I. (1) In an affine Euclidean plane, consider a point S and two lines Δ and Δ'.

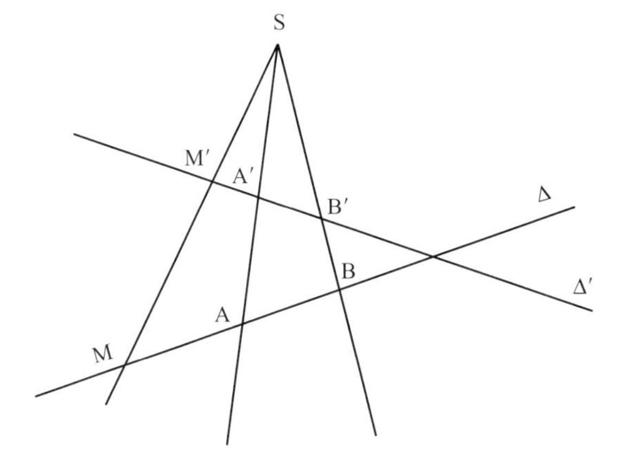

Prove the relation

$$\frac{\overline{M'A'}}{\overline{M'B'}} : \frac{\overline{MA}}{\overline{MB}} = \frac{\overline{SA'}}{\overline{SB'}} : \frac{\overline{SA}}{\overline{SB}}$$

when M, A, B lie on Δ and M', A', B' are the points on Δ' deduced from them by projection from the perspective of the point S.

(2) Deduce from the preceding question that every product $\bar{\omega}$ of quotients $\overline{MA}/\overline{MB}$, $\overline{NB}/\overline{NC}$, etc, in which M, A, B ; N, B, C; etc. are (respectively) collinear and in which A, B, C, etc. occur an equal number of times in the denominator and in the numerator, is an invariant of perspective.

(3) Check that if A, B, C, D are collinear, the **cross-ratio**

$$[A, B, C, D] = \frac{\overline{CA}}{\overline{CB}} : \frac{\overline{DA}}{\overline{DB}}$$

is an invariant of perspective.

(4) Let k be an algebraically closed field and $F(x, y)$ an affine curve of degree n in the plane k^2, given by

$$F(x, y) = F_n(x, y) + \cdots + F_0(x, y),$$

where $F_i(x, y)$ is homogeneous of degree i.

Show that if $A = (a, b)$ and $U = (u, v)$ are two distinct points of k^2, and if $z \in k\backslash\{-1\}$, then the point M on the line AU such that $\overrightarrow{MA} + z\overrightarrow{MU} = \vec{0}$ has coordinates

$$x = \frac{a + zu}{1 + z}, \qquad y = \frac{b + zv}{1 + z}.$$

Assume that A and U do not lie on the curve F, and that AU meets F in n distinct points $M_1, \ldots, M_n$ of parameters $z_1, \ldots, z_n$.

Show that

$$z_1 \cdots z_n = (-1)^n \frac{F(a, b)}{F(u, v)}, \quad \text{a quotient to be noted carefully } (-1)^n \frac{F(A)}{F(U)}.$$

Deduce that

$$\bar{\omega}_{AU} := \frac{\overline{M_1A}}{\overline{M_1U}} \cdot \frac{\overline{M_2A}}{\overline{M_2U}} \cdot \cdots \cdot \frac{\overline{M_nA}}{\overline{M_nU}} = \frac{F(A)}{F(U)}.$$

(5) Let $A_1, A_2, \ldots, A_r$, r be distinct points of k^2 not lying on the curve F. Set

$$\bar{\omega}_1 = \bar{\omega}_{A_1A_2}, \quad \bar{\omega}_2 = \bar{\omega}_{A_2A_3}, \quad \ldots, \quad \bar{\omega}_r = \bar{\omega}_{A_rA_1}.$$

Show that $\bar{\omega} = \bar{\omega}_1 \cdots \bar{\omega}_r$ is an invariant of perspective and check that $\bar{\omega} = 1$ (result dating back to 1806).

(6) Give details in the special case where $n = 1$ (Menelaüs, + 100), $n = 2$ and $n = 3$. Give a generalisation of this result in k^m with $m \geq 2$.

(7) Assume that the cubic C intersects the transversals Δ_1 and Δ_2 as follows (distinct points):

$$\begin{cases} \Delta_1 \cap C = \{M_{1,1}, \ M_{1,2}, \ M_{1,3}\} \\ \Delta_2 \cap C = \{M_{2,1}, \ M_{2,2}, \ M_{2,3}\}, \end{cases}$$

and for $i = 1, 2, 3$, let $M_{3,i}$ denote the third point of intersection of $M_{1,i}\, M_{2,i}$ with C.

Deduce from the results of (5) and (6) that the points $M_{3,1}$, $M_{3,2}$ and $M_{3,3}$ are collinear.

(8) Using (5) and (6), show that if six of the nine intersection points of a (plane) cubic with three transversals lie on a conic, then the three others are collinear.

(9) Suppose that a transversal Δ meets a cubic Γ in three distinct points F, G, H, and that the tangents to Γ at F and G cut Γ at A and B (Fig. 4.1). Let C denote the point where the line AB meets Γ. Show that the tangent to Γ at H passes C through C when none of the points occurring here are singular on Γ.

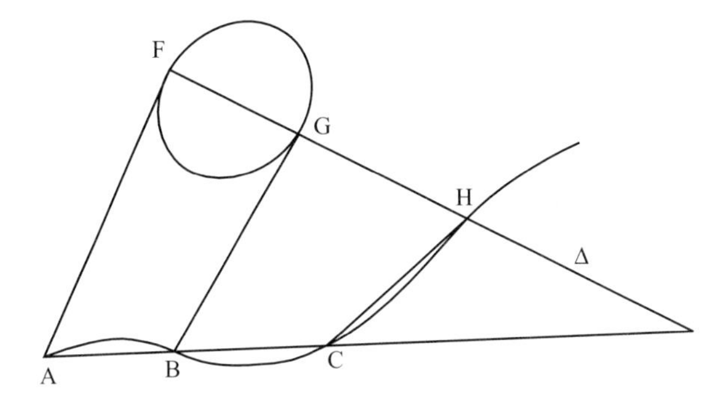

Figure 4.1.

(10) Let A be a given point on Γ (Fig. 4.2), and let AF, AG be tangents to Γ at F, G. If we join F to G, this line meets Γ at H and the tangent at H meets Γ at C. Show that the tangent to Γ at A passes through C when Γ is smooth (pass to the limit in (9)).

(11) Let A be a given point on Γ (Fig. 4.2), and let AF, AG, Af be tangents to Γ at F, G, f.

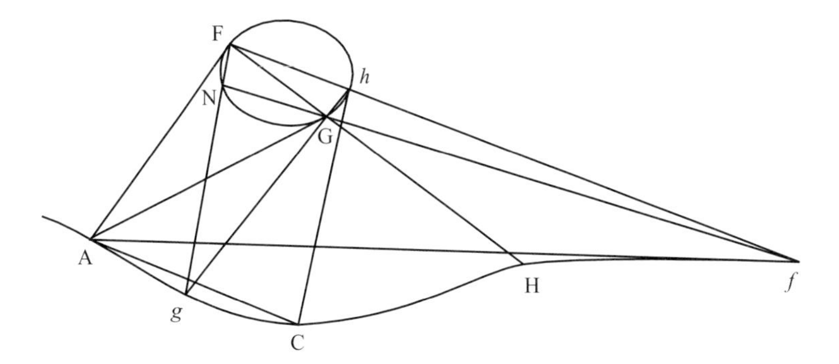

Figure 4.2.

If we join G to f, this line meets Γ at N, and NF meets Γ at g. Show (using (9) and (10)) that the tangent to Γ at N passes through C and that the tangent at g passes through A.

(12) From (11), deduce a construction of a fourth tangent coming out of A, given three of them. Show that it is the last one when Γ is smooth.

(13) Show that if A is an inflection point of Γ, the point C is equal to A. Show that the points F, G, f are collinear. In the case where Γ is smooth, take A as the origin of the group of points of $\Gamma(k)$.

II. **(1)** Consider the statement of theorem II of Article 13 of the text by Colin Maclaurin reproduced below. What is the geometric framework of this statement, projective or affine geometry? and over what field? What is meant by a "geometric line"? Is it possible to give a similar statement in characteristic p?

(2) In theorem I of Article 8 of the same text, what is meant by the "dimensions" of a line? Can we give a similar statement in characteristic p?

(3) Explain the Lemma of Article 8.

(4) When the line is defined over a field k of positive characteristic, how can we adapt the computation of the fluxions in Article 8? (One could consider studying Γ in the space $k[\varepsilon] \times k[\varepsilon]$ where $k[\varepsilon]$ denotes the ring $k[X]/(X^2)$).

(5) Check the assertion of line 6 of Article 13: "the point M will be in a right line".

(6) Recall (I, (13)) that if A is an inflexion point of a plane cubic Γ, the points of contact F, G, H of the tangents to Γ coming out of A are collinear, i.e. lie on a line L (Fig. 4.3). Let Δ be a transversal passing through A, which meets Γ at B and C, and L at P.

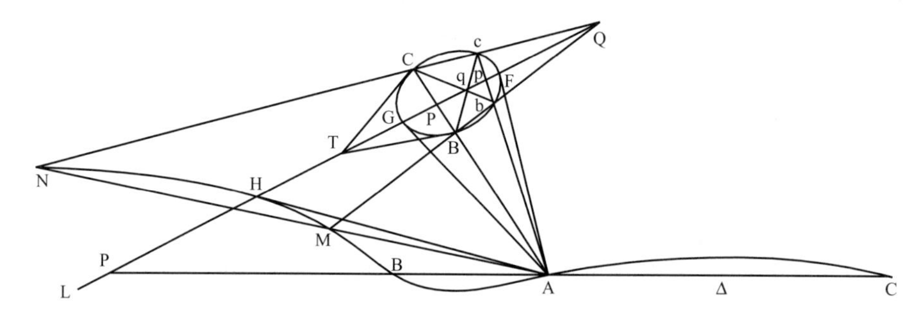

Figure 4.3.

Show that the division (A, P, B, C) is harmonic (one can show Descartes' relation $(1/\overline{AB}) + (1/\overline{AC}) = (2/\overline{AP})$).

CONCERNING THE GENERAL PROPERTIES OF GEOMETRICAL LINES (BY COLIN MACLAURIN [ML])

Concerning the lines of the second order, or the conic sections, the ancient and modern geometers have written very fully, concerning the figures which are referred to the superior orders of lines, little has been delivered before Newton. That most illustrious man, in his tract concerning the *Enumeration of Lines of the Third Order*, has revived this subject, which had long lain neglected, and has shown it to be worthy of the geometer's notice. For the general properties of these lines, which he has laid down, are so consonant to the known properties of the conic sections, that they seem to be conformable to the same law, and from his example many others have since been induced to make this subject their study, and have clearly comprehended and explained the analogy which there is between figures of such very different kinds. The pains which they have been at in the illustration and further investigation of these matters, have deservedly met with applause, since there is nothing in pure mathematics which can be called more beautiful, or that is more apt to delight a mind desirous of investigating truth than the agreement and harmony of different things, and the admirable connection of the succeeding with the preceding, where the more simple always open the way to those which are more difficult.

Most of the general properties of lines of the third order, delivered by Newton, relate to segments of parallels and asymptotes. Some other of their properties, of a different kind, I have briefly pointed out in my Treatise of Fluxions, lately published, Art. 324, and 401. The famous *Cotes* formerly discovered a most beautiful property of geometrical lines, hitherto unpublished, which has been communicated to me by the Rev. Dr. Robert Smith, Master of Trinity College, Cambridge, a gentleman not less remarkable for his learning and works, than for his fidelity and regard for his friends. While I had these under consideration, some other general theorems offered themselves; which, as they seem to conduce to the augmentation and illustration of this difficult part of geometry, I have thought fit to throw together, and briefly expound and demonstrate in order.

SECTION I

Of Geometrical Lines in General

(1) Lines of the second order are defined by the section of a geometrical solid, viz. a cone, where their properties are best derived by common geometry. But the nature of the figures which are referred to the superior order of lines, is different. To define and draw out their properties, general equations must be applied, expressing the relation of the coordinates (Fig. 4.4). Let x represent the abscissa AP, y the ordinate PM of the figure PMH, and let a, b, c, d, e, etc. denote any invariable coefficients; and having the angle APM given, if the relation of the coordinates x and y are defined by an equation which, besides the coordinates themselves, involves only invariable coefficients, the line FMH is called a geometrical one which indeed by some authors is called an algebraical

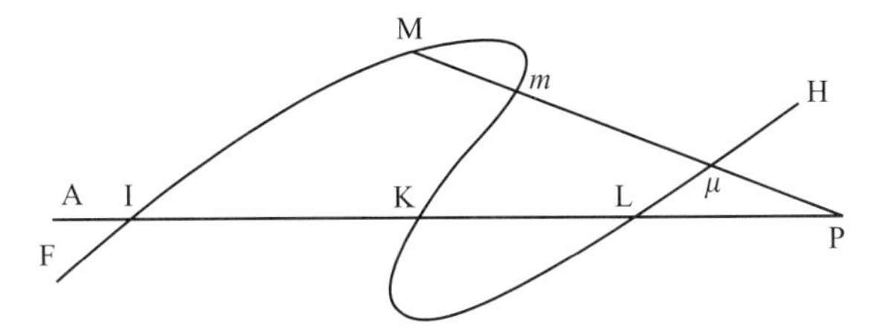

Figure 4.4.

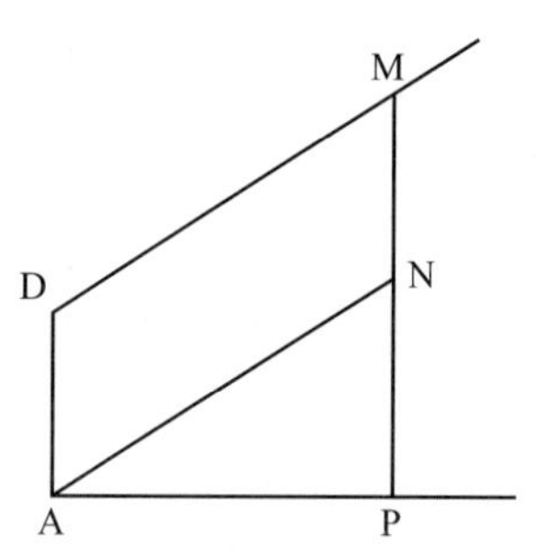

Figure 4.5.

line, by others a rational line. But the order of the line depends upon the highest index of x or y in the terms of the equation freed from fractions and surds, or upon the sum of the indices of both in a term where that sum is the greatest. For the terms x^2, xy, y^2 are equally referred to the second order; the terms x^3, x^2y, xy^2, y^3 to the third. Therefore the equation $y = ax + b$, or $y - ax - b = 0$, is of the first order, and denotes a line or the locus of the first order, which indeed is always a right line (Fig. 4.5). For let there be taken in the ordinate PM the right line PN, so that PN is to AP as $+a$ is to unity; let AD, parallel to PM, be made equal to $+b$, and DM, drawn parallel to AN, will be the locus to which the proposed equation will answer. For $PM = PN + NM = (a \cdot AP + AD) = ax + b$. But if the equation be of the form $y = ax - b$, or $y = -ax + b$, the right line AD, or PN, is to be taken on the other side of the abscissa AP for the contrary situation of right lines answers to the contrary signs of the coefficients. If the affirmative values of x denote right lines drawn from A, the beginning of the abscissa, to the right hand, the negative values will denote right lines drawn from the same beginning to the left; and in like manner if the affirmative values of y represent the ordinates constituted above the abscissa, the negative ones will denote the ordinates below the abscissa, drawn the opposite way.

The general equation for a line of the second order is of this form,

$$\left.\begin{array}{r} yy - axy + cx^2 \\ -by - dx \\ +e \end{array}\right\} = 0$$

and the general equation for lines of the third order is

$$y^3 - (ax + b)y^2 + (cx^2 - dx + e)y - fx^3 + gx^2 - bx + k = 0.$$

And by similar equations geometrical lines of superior orders are defined.

(2) A geometrical line may meet a right line in as many points as there are units in the number which denotes the order of the equation or line, and never in more. The number of times that any curve will meet its abscissa AP is determined by putting $y = 0$, in which case there remains only the last term of the equation into which y does not enter. For example, a line of the third order meets the abscissa AP in three points, when the equation $fx^3 - gx^2 + hx - k = 0$, has three real roots. In like manner in the general equation of any order, the highest index of the abscissa x is equal to the number which denotes the order of the line, but never greater, and of course expresses the number of times that the curve will meet the abscissa or any other right line. But since one root of a cubic equation is always real, and the same is true of an equation of the fifth or any odd order (because every imaginary root has necessarily its fellow) it follows that a line of the third or any other odd order cuts any right line, not parallel to the asymptote drawn in the same plane, in one point at least. But if the right line is parallel to the asymptote, in this case it is commonly said to meet the curve at an infinite distance. A line therefore of any odd order has necessarily two branches which may be produced *in infinitum*. But of a quadratic, or any other equation of an even number of roots, all the roots may be sometimes imaginary, therefore it may be that a right line drawn in the plane of a curve of an even order may never meet it.

(3) An equation of the second, or of any higher order is sometimes compounded of so many simple ones, freed from surds and fractions, multiplied into each other as often as the proposed dimensions of that equation express, in which case the figure FMH is not curvilinear, but is made up of so many right lines as described by the simple equations this determined as in (1). In like manner if a cubic equation be compounded of two equations multiplied into each other, one of which is a quadratic and the other a simple one, the locus will not be a line of the third order, properly so called, but a conic section joined with a right line. Now the properties which are generally demonstrated of geometrical lines of higher orders are to be affirmed also of geometrical lines of inferior orders, if the numbers denoting their orders, taken together, make up the number which denotes the order of the said superior line. Those which, for example, are generally demonstrated by lines of the third order, are also to be affirmed of three right lines drawn in the same plane, or of a conic section together with one right line described in the same plane. On the other hand, there can scarce be any property of a line of an inferior order be assigned sufficiently general to which some property of lines of superior orders do not correspond. But to derive these from those, not everyone can take the trouble to derive the latter from the former. This doctrine in a great measure depends upon the properties of general equations; which it is here only proper to mention.

(4) In every equation the coefficient of the second term is equal to the excess of the sum of the affirmative roots above the sum of the negative ones; and if that term is wanting, it is an indication that the sums of the affirmative and negative roots, or the sums of the ordinates constituted on different sides of the abscissa, are equal. Let the general equation be for a line of the order n,

$$y^n - (ax + b) \cdot y^{n-1} + (cxx - dx + e) \cdot y^{n-2} - \text{etc.} = 0,$$

suppose $u = y - (ax + b)/n$, for y let us substitute its value $u + (ax + b)/n$; and in the transformed equation the second term u^{n-1} will be wanting; as appears from the calculation, or from the doctrine of equations, everywhere delivered: and from hence it also appears, that by hypothesis every value of u is less than the corresponding value of y by $(ax + b)/n$; whence it follows that the sum of the values of u (whose number is n) falls short of the sum of the values of y (whose sum is $ax + b$) by the difference $[(ax + b)/n]n = ax + b$, so that the first sum vanishes, and the second term is wanting in the equation by which u is determined, or that the affirmative and negative values of u make equal sums. If therefore PQ be taken $= (ax + b)/n$, so that QM may $= u$, right lines on both sides of the point Q, terminated at the curve, will make the same sum (Fig. 4.6). Now the locus of

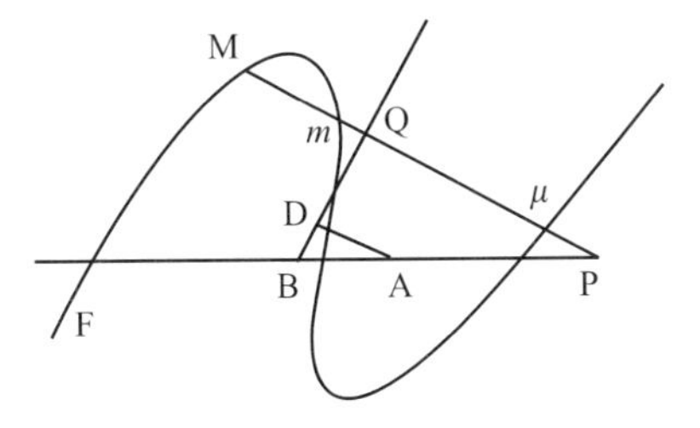

Figure 4.6.

the point Q is the right line BD which cuts the abscissa, produced beyond its beginning A, in B, so that $AB = b/a$, and the ordinate AD, parallel to PM, in D, so that $AD = 1/n \cdot b$: for if this right line meets the ordinate PM in the point Q, PQ will be to PB (or $b/a + x$) as AD to AB, or a to n; so that $PQ = (ax + b)/n$ as it ought to do. And hence it appears, that a right line may always be drawn which shall cut any number of parallels, meeting a geometrical line in as many points as the dimensions of the figure express, that the sum of the segments of every parallel, terminated at the curve on one side of the cutting line, may always be equal to the sum of the segments of the same on the other side of the cutting line. Now it is manifest that a right line which cuts any two parallels in this manner is necessarily that which will cut all other parallels in the same manner. And hence appears the truth of the *Newtonian* theorem, in which is contained the general property of geometrical lines, analogous to that well-known property of the conic sections. For in these a right line which bisects any two parallels, terminated at the section, is a diameter, and bisects all others parallel to these, and terminated at the section. And in like manner a right line, which cuts any two parallels, meeting a geometrical line in as many points as it has dimensions, so that the sum of the parts standing on one side of the cutting line and terminated at the curve may be equal to the sum of the parts of the same parallel standing on the other side of the cutting line terminated at the curve, will in the same manner cut all other right lines parallel to these.

(5) In every equation the last term, or that into which the root y does not enter, is equal to the product of all the roots multiplied into each other; from whence we are led to another property of geometrical lines, not less general than that above. Let the right line PM meet a line of the third order in M, m and μ, and it will be $PM \cdot Pm \cdot P\mu = fx^3 - gx^2 + hx - k$. Let the abscissa AP cut the curve in the three points I, K, L; and AI, AK, AL will be the values of the abscissa x, the ordinate being put $= 0$, in which case the general equation gives $fx^3 - gx^2 + hx - k = 0$ for determining these values. Therefore of the equation

$$x^3 - \frac{gx^2}{f} + \frac{hx}{f} - \frac{k}{f} = 0$$

the three roots are AI, AK, AL; and so this equation is compounded of the three $x - AI$, $x - AK$, $x - AL$ multiplied into each other; and

$$\begin{aligned} x^3 - \frac{gx^2}{f} + \frac{hx}{f} - \frac{k}{f} &= (x - AI) \cdot (x - AK) \cdot (x - AL) \\ &= (AP - AI) \cdot (AP - AK) \cdot (AP - Al) \\ &= IP \cdot KP \cdot LP = \frac{1}{f} \cdot PM \cdot Pm \cdot P\mu. \end{aligned}$$

Therefore the product of the ordinates PM, Pm, $P\mu$, terminated by the point P and the curve, is to the product of the segments IP, KP, IP, of the right line AP, terminated by the same point and the curve, in the invariable ratio of the coefficient f to unity. In like manner it is demonstrated, that having given the angle APM, if the right lines AP, PM, cut a geometrical line of any order in as many points as it has dimensions, the product of the segments of the first, terminated by P and the curve, will always be to the product of the segments of the latter, terminated by the same point and the curve, in an invariable ratio.

(6) In the preceding article we have supposed, with *Newton*, that the right line AP cuts a line of the third order in three points I, K, L; but that this famous theorem may be rendered more general, let us suppose that the abscissa AP cuts the curve in only one point, and let that be A (Fig. 4.7).

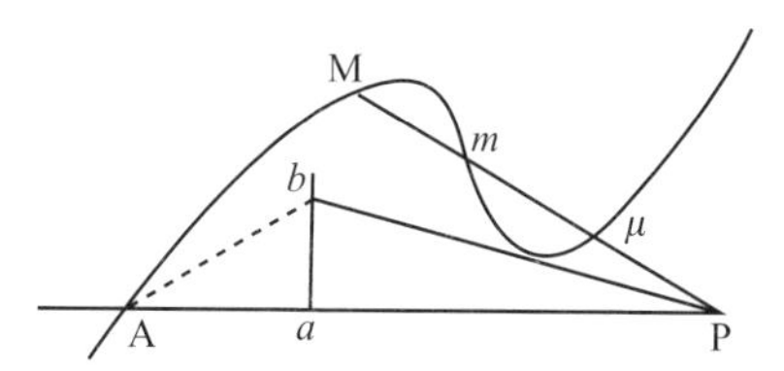

Figure 4.7.

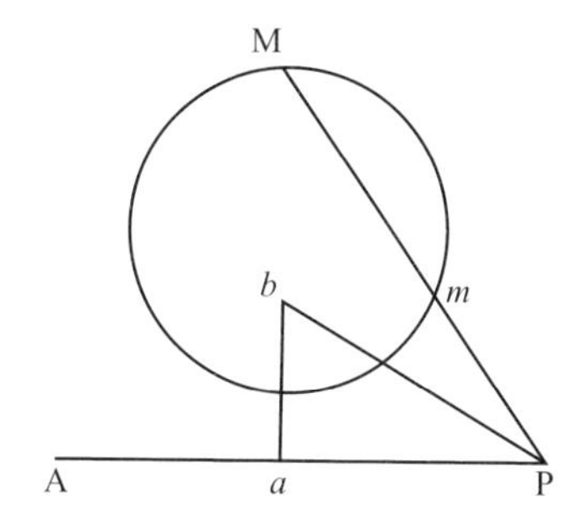

Figure 4.8.

Therefore because y vanishes, let x vanish also, the last term of the equation, in this case, will be

$$fx^3 - gx^2 + hx = fx\left(xx - \frac{gx}{f} + \frac{h}{f}\right) = fx\left[\left(x - \frac{g}{2f}\right)^2 + \frac{h}{f} - \frac{gg}{4ff}\right]$$

(if Aa be taken towards P equal $g/2f$, and at the point a be erected a perpendicular $ab = \sqrt{4fh - gg}/2f) = f \cdot AP \cdot (aP^2 + ab^2) = f \cdot AP \cdot bP^2$; when $PM \cdot Pm \cdot P\mu$ is equal to the last term $fx^3 - gx^2 + hx$, as in the preceding article, $PM \cdot Pm \cdot P\mu$ will be to $AP \cdot bP^2$ in the constant ratio of the coefficient f to unity. Now the value of the right line perpendicular to ab is always real, as often as the right line AP cuts the curve in one point only, for in this case the roots of the quadratic equation $fx^2 - gx + h$ are necessarily imaginary, so that $4fh$ is greater than gg, and the quantity $\sqrt{4fh - gg}$ real. When therefore any right line cuts a line of the third order in one point A only, the solid under the ordinates PM, Pm, $P\mu$ will be to the solid under the abscissa AP and the square of the distance of the point P from a given point b in a constant ratio. Ab, being joined is to Aa, as radius to the consine of the angle bAP, as $\sqrt{4fh}$ to g, and $Ab = \sqrt{h/f}$. But the same point b always agrees to the same right line AP, whatever be the angle which is contained by the abscissa and ordinate.

(7) Let the figure be a conic section, whose general equation is $yy-(ax - b)\cdot y+cxx-dx+e = 0$ as above; and if the roots of the equation $cxx - dx + e = 0$ be imaginary, the right line AP will not meet the section (Fig. 4.8). Now, in this case the quantity $4ec$ always exceeds dd; when $cxx - dx + e = c(x - d/2c)^2 + e - dd/4c$ (if Aa be taken $= d/2c$, and ab be erected perpendicular to the abscissa at a, so that $ab = \sqrt{4ec - dd}/2c) = c \cdot (aP^2 + ab^2) = c \cdot bP^2$, and $PM \cdot Pm = cxx - dx + e$, then $PM \cdot Pm$ is to bP^2 as c to unity. Therefore in any conic section, if the right line AP does not meet the section, the angle APM being given, the rectangle contained under right lines standing at the point b and terminated at the curve, is to the square of the distance of the point P from the given point b, in a constant ratio, which in a circle is that of equality. Now it is manifest that the same method may be applied to a line of the fourth order which the abscissa cuts in two points only, or to a line of any order which the abscissa cuts in points less by two than the number which denotes the order of the figure.

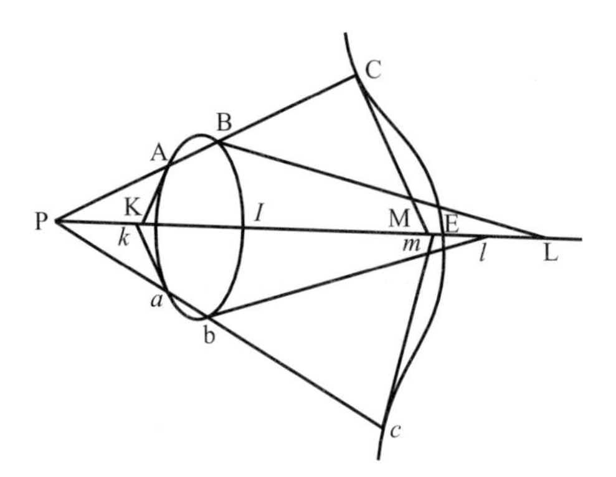

Figure 4.9.

(8) This being premised, I proceed to explain the less obvious properties of geometrical lines almost in the same order in which they occurred to me. Now I used the following lemma, derived from the doctrine of fluxions, and which I have demonstrated in my treatise on that subject; yet I have since observed that some of them may be demonstrated by common algebra.

Lemma If, of the quantities $x, y, z, u, \ldots,$ flowing together, and also the quantities $X, Y, Z, V, \ldots,$ the product of the former be to the product of the latter in any constant ratio, then

$$\frac{\dot{x}}{x}+\frac{\dot{y}}{y}+\frac{\dot{z}}{z}+\frac{\dot{u}}{u}+\cdots=\frac{\dot{X}}{X}+\frac{\dot{Y}}{Y}+\frac{\dot{Z}}{Z}+\frac{\dot{V}}{V}+\cdots.$$

Moreover, for brevity's sake, I call those quantities mutually *reciprocal*, when being multiplied into each other, the product is unity, so $1/x$ I call the *reciprocal* of x, and $1/y$ of y.

Theorem I *Let any right line, drawn through a given point, meet a geometrical line of any order in as many points as it has dimensions; and let right lines, touching the figure in these points, cut off from another right line, given in position and drawn through the same given point, as many segments terminated by this point; the reciprocals of these segments will always make the same sum, if the segments lying on the contrary side of the given point be affected with the contrary signs.*

Let P be the given point, PA and Pa any two right lines drawn from P, of which both meet the curve in as many points $A, B, C,$ and $a, b, c, \ldots,$ as it has dimensions (Fig. 4.9). Let the tangents $AK, BL, CM, \ldots,$ and $ak, bl, cm, \ldots,$ cut off from the right line EP, drawn through the point P, the segments $PK, PL, PM, \ldots$ and $Pk, Pl, Pm, \ldots$; I say that

$$\frac{1}{PK}+\frac{1}{PL}+\frac{1}{PM}+\cdots=\frac{1}{Pk}+\frac{1}{Pl}+\frac{1}{Pm}+\cdots,$$

and that this sum always remains the same, the point P remaining, and the right line PE being given in position.

For let us suppose the right lines ABC, abc to be carried by motions parallel to themselves, so that their concourse P proceeds in the right line PE given in position; since $AP \cdot PB \cdot CP \ldots,$ is always to $aP \cdot bP \cdot cp \ldots$ in a constant ratio, (5), let $\dot{AP}$ represent the fluxion of AP, $\dot{BP}$ the fluxion of BP, and $\dot{C}P$, $\dot{E}P$, ..., the fluxions of the right lines CP, EP, ..., respectively, that a useless multiplication of symbols may be avoided, then (8)

$$\frac{\dot{A}P}{AP} + \frac{\dot{B}P}{BP} + \frac{\dot{C}P}{CP} + \cdots = \frac{a\dot{P}}{aP} + \frac{b\dot{P}}{bP} + \frac{c\dot{P}}{cP} + \cdots.$$

But when the right line AP is carried by a motion parallel to itself, it is well-known that $\dot{A}P$, the fluxion of the right line AP, is to $\dot{E}P$, the fluxion of the right line EP, as AP to the subtangent PK, and so

$$\frac{\dot{A}P}{AP} = \frac{\dot{E}P}{PK}.$$

In like manner

$$\frac{\dot{B}P}{BP} = \frac{\dot{E}P}{PL}, \qquad \frac{\dot{C}P}{CP} = \frac{\dot{E}P}{PM}, \qquad \frac{a\dot{P}}{aP} = \frac{\dot{E}P}{Pk}, \qquad \frac{b\dot{P}}{bP} = \frac{\dot{E}P}{Pl} \quad \text{and} \quad \frac{c\dot{P}}{cP} = \frac{\dot{E}P}{Pm};$$

where

$$\frac{\dot{E}P}{PK} + \frac{\dot{E}P}{PL} + \frac{\dot{E}P}{PM} + \cdots = \frac{\dot{E}P}{Pk} + \frac{\dot{E}P}{Pl} + \frac{\dot{E}P}{Pm} + \cdots$$

and

$$\frac{1}{PK} + \frac{1}{PL} + \frac{1}{PM} + \cdots = \frac{1}{Pk} + \frac{1}{Pl} + \frac{1}{Pm} + \cdots,$$

Things are so whenever the points $K, L, M, \ldots$, and $k, l, m, \ldots$, are all on the same side of the point P, and so the fluxions of the right lines $AP, BP, CP, \ldots, aP, bP, cP, \ldots$, all have the same sign (Fig. 4.10). But if, other things remaining the same, some points M and m fall on the contrary side of P, then while the rest of the ordinates $AP, BP, \ldots$, increase, the ordinates CP and cP are necessarily diminished, and their fluxions are to be accounted subtractive, or negative; and so in this case

$$\frac{1}{PK} + \frac{1}{PL} - \frac{1}{PM} + \cdots = \frac{1}{Pk} + \frac{1}{Pl} - \frac{1}{Pm} + \cdots;$$

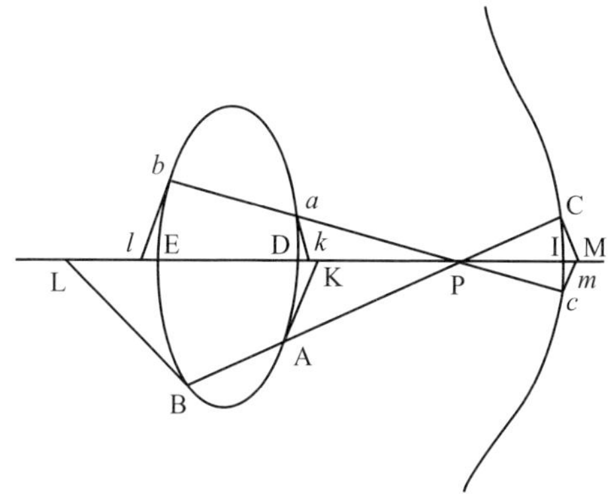

Figure 4.10.

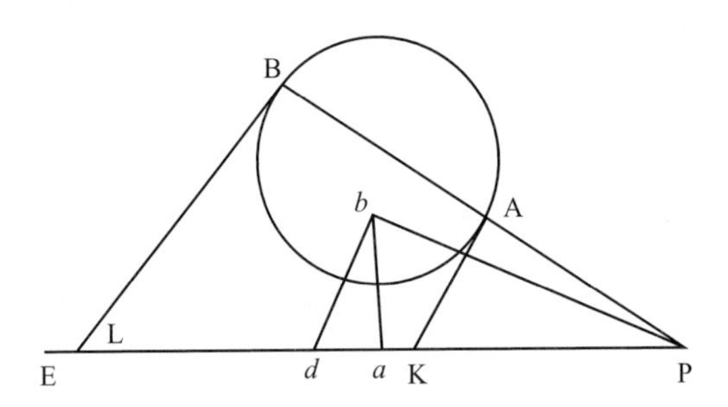

Figure 4.11.

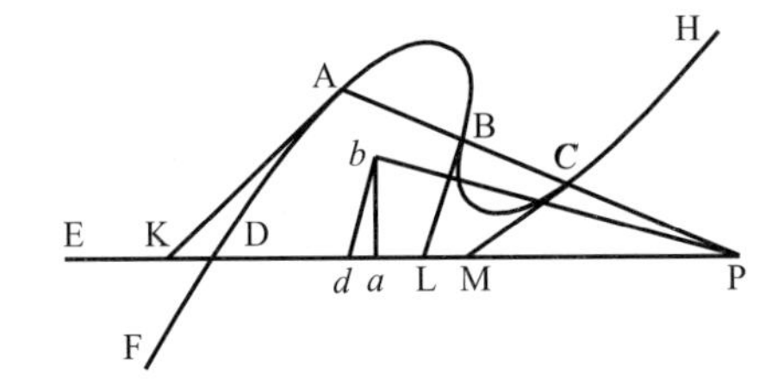

Figure 4.12.

and in general, in collecting these sums, the terms are to be affected with the same or contrary signs, as the segments fall on the same or contrary side of the given point P.

(9) If a right line PE meets a curve in as many points $D, E, I, \ldots$, as its dimensions express; the sum

$$\frac{1}{PK} + \frac{1}{PL} + \frac{1}{PM} + \cdots,$$

which we have shown to be constant or invariable, will be equal to the sum or aggregate

$$\frac{1}{PD} + \frac{1}{PE} + \frac{1}{PI} + \cdots,$$

i.e. to the sum of the reciprocals to the segments of the right line PE, given in position, and determined by the given point P and the curve; in which, if any segment is on the other side of the point P, its reciprocal is to be subtracted.

(10) If the figure (Fig. 4.11) be a conic section, which the right line PE nowhere meets, let the point b be found as in (7), and Pb joined, and at right angles to this let bd be drawn, cutting the right line PE in d; then will $1/PK + 1/PL = 2/Pd$. For $PA \cdot PB$ is to bP^2 in a constant ratio, and so (8)

$$\frac{\dot{A}P}{AP} + \frac{\dot{B}P}{BP} = \frac{2\dot{b}P}{bP},$$

whence (because $\dot{A}P$ is to $\dot{E}P$ as AP to PK, $\dot{B}P$ to $\dot{E}P$ as BP to PL, and $\dot{b}P$ to $\dot{E}P$ as bP to dP) $1/PK + 1/PL = 2/Pd$.

(11) In like manner if the right line EP meets a line of the third order in only one point D, let the point b be found as in (6), and let the right line bd, perpendicular to bP, meet the right line EP

in d, and because $AP \cdot BP \cdot CP$ is to $DP \cdot bP^2$ in a constant ratio (*ibid.*)

$$\frac{1}{PK} + \frac{1}{PL} + \frac{1}{PM} = \frac{1}{PD} + \frac{2}{Pd}.$$

But if Pb be perpendicular to the right line EP, $2/Pd$ will vanish (Fig. 4.12).

(12) The asymptotes of geometrical lines are determined from the given direction of their infinite branches, or legs, by this proposition; for they may be considered as tangents to the legs produced *in infinitum* (Fig. 4.13). Let the right line PA, parallel to the asymptote, meet the curve in the points $A, B, \ldots$ and the right line PE cut the curve in $D, E, I, \ldots$. Let PM be taken in this so that $1/PM$ may be equal to the excess by which the sum $1/PD + 1/PE + 1/PI + \cdots$ exceeds the sum $1/PK + 1/PL + \cdots$, and the asymptote will pass through M; but if these sums are equal, the curve will be a parabola, the asymptote going off *in infinitum*.

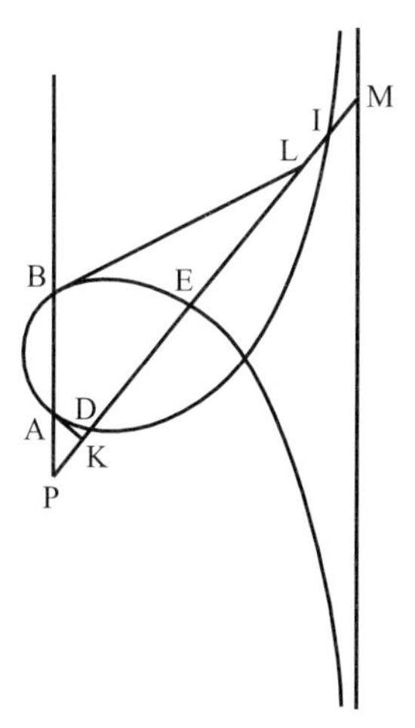

Figure 4.13.

(13) Theorem II *About the given point P let the right line PD revolve which meets a geometrical line of any order in as many points D, E, I, ..., as it has dimensions, and if in the same right line be always taken PM so that*

$$\frac{1}{PM} = \frac{1}{PD} \mp \frac{1}{PE} \mp \frac{1}{PI} \mp \cdots,$$

(where we suppose the signs of the terms to keep the rule repeatedly given) the locus of the point M will be a right line (Fig. 4.14).

For let there be drawn from the pole P any right line given in position PA, which let meet the curve in as many points $A, B, C, \ldots$, as it has dimensions. Let there be also drawn the right lines AK, BL, CN touching the curve in these points, which meet PD in as many points $K, L, N, \ldots$; and by (9)

$$\frac{1}{PD} \mp \frac{1}{PE} \mp \frac{1}{PI} \mp \cdots = \frac{1}{PK} \mp \frac{1}{PL} \mp \frac{1}{PN} \mp \cdots.$$

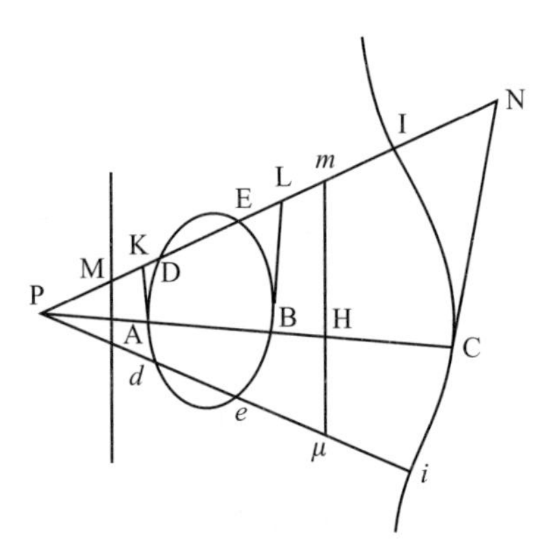

Figure 4.14.

Where $1/PM$ is equal to this sum, and when the line PA is given in position, and the right lines $AK, BL, CN, \ldots,$ remain fixed, while the right line PD revolves about the pole P, the point M will be in a right line, by the preceding Article; which may be determined by what has been shown above from the given tangents $AK, BL, \ldots.$

(14) As the right line Pm is a mean harmonical between the two lines PD and PE, when $2/Pm = 1/PD + 1/PE$; in like manner Pm may be called a *mean* harmonical between any right lines $PD, PE, PI, \ldots,$ whose number is n, when

$$\frac{n}{Pm} = \frac{1}{PD} \mp \frac{1}{PE} \mp \frac{1}{PI} \mp \cdots.$$

And if any right line drawn from a given point P cuts a geometrical line in as many points as it has dimensions, in which let Pm be always taken an harmonical mean between all the segments of the drawn line terminated by the point and the curve, the point m will be in a right line. For $1/PM$ will $=n/Pm$, and therefore Pm is to PM as n to unity; and since the point M is in a right line, by the preceding, the point m will also be in a right line. And this is *Cotes's* theorem, or nearly related to it.

(15) Let $a, b, c, d, \ldots,$ be the roots of an equation of the order n, v its last term into which the ordinate y does not enter, P the coefficient of the last term but one, M the harmonical mean between all the roots, or

$$\frac{n}{M} = \frac{1}{a} + \frac{1}{b} + \frac{1}{c} + \frac{1}{d} + \cdots.$$

Therefore since v is the product of all the roots $a, b, c, \ldots,$ multiplied into each other, and P is the sum of the products when all the roots, one expected, are multiplied into each other, P will $= v/a + v/b + v/c + v/d + \cdots = nv/M$, and therefore $M = nv/P$. So, if the equation be a quadratic, whose two roots are a and b, M will $= 2ab/(a+b)$ (having assumed the general equation for conic sections given in (1)) $=(2cxx - 2dx + 2e)/(ax - b)$. In a cubic equation, whose three roots are a, b, c, M will $= 3abc/(ab + ac + bc)$ (if there be assumed the general equation for lines of the third order there given)

$$= \frac{3fx^2 - 3gx^2 + 3hx - 3k}{cxx - dx + e}.$$

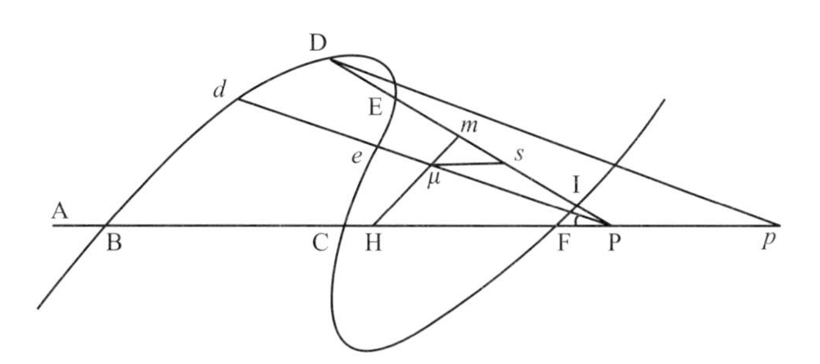

Figure 4.15.

(16) Let any two lines Pm and $P\mu$ (Fig. 4.15), drawn from the point P, meet a geometrical line in the points $D, E, I, \ldots$, and $d, e, i, \ldots$; and let Pm be an harmonical mean between the segments of the former terminated by the point P and the curve, and $P\mu$ an harmonical mean between the like segments of the latter line; let μm, being joined, meet the abscissa AP in H, then will $PH = nv\dot{x}/\dot{v}$ or PH is to Pm as P to $\dot{v}/\dot{x}$. For let the abscissa cut the curve in as many points $B, C, F, \ldots$, as it has dimensions; and since the last term of the equation (i.e. v) is to $BP \cdot CP \cdot FP \cdot \ldots$, in a constant ratio, as we have shown above (5) it will be (by 8)

$$\frac{\dot{v}}{v} = \frac{\dot{x}}{BP} \mp \frac{\dot{x}}{CP} \mp \frac{\dot{x}}{FP} \mp \cdots,$$

and therefore

$$\frac{n}{PH} = \frac{1}{BP} \mp \frac{1}{CP} \mp \frac{1}{FP} \mp \cdots = \frac{\dot{v}}{v\dot{x}},$$

and $PH = nv\dot{x}/\dot{v}$ (because the line $PM = nv/P) = PM \cdot (P\dot{x}/\dot{v})$. In conic sections it is PH to Pm as $ax - b$ to $2cx - d$; and in lines of the third order as $cxx - dx + e$ to $3fxx - 2gx + h$.

(17) If a demonstration of the preceding proposition be desired from principles purely algebraical, it may be had with the help of the following.

Lemma *Let the abscissa $AP = x$, the ordinate $PD = y$, V the last term of the equation defining the geometrical line $= Ax^n + Bx^{n-1} + Cx^{n-2} + \cdots$, P the coefficient of the last term but one $= ax^{n-1} + bx^{n-2} + cx^{n-3} + \cdots$, and let $Q = nAx^{n-1} + (n-1) \cdot Bx^{n-2} + (n-2) \cdot Cx^{n-3} + \cdots$ (which is the quantity we call $\dot{V}/\dot{x}$). Let there be drawn the ordinate Dp which makes any given angle ApD with the abscissa, and let the right lines PD, pD, and Pp be as the given ones l, r and k; let $pD = u$, $Ap = z$; and let the proposed equation be transformed into another expressing the relation between the ordinate u and abscissa z; and since $z = AP$, the last term V of the new equation will be equal V, but p the coefficient of the last term but one, will be equal to $(\mp Qk + Pl)/r$.*

For since PD $(= y)$ is to pD $(= u)$ as l to r, $y = lu/r$; but let Pp be to $pD (= u)$ as k to r, then $Pp = ku/r$, and $AP = x = Ap \pm Pp = z \pm ku/r$. Now these values being substituted for y and x in the proposed equation of the geometrical line, there will come out an equation determining the relation of the coordinates z and u. To determine the last term of this v and the last but one pu, it is sufficient to substitute these values in the last V, and in the last but one Py, of the proposed equation, and to collect the resulting terms in which the ordinate u is either not found, or of one dimension only; for the sum of these gives pu, and of those v. Let for x be substituted its value

$Z \pm ku/r$ in the quantity V or $Ax^n + Bx^{n-1} + Cx^{n-2} + \cdots$; and the resulting terms

$$Az^n \pm \frac{nAZ^{n-1}ku}{r} + Bz^{n-1} \pm (n-1) \cdot \frac{BZ^{n-2}ku}{r} + Cz^{n-2} \pm (n-2) \cdot \frac{CZ^{n-3}ku}{r} + \cdots,$$

will alone serve the purpose we are about. Then let be substituted for x the same value, and for y its value lu/r, in the quantity

$$Py = (ax^{n-1} + bx^{n-2} + cx^{n-3} + \cdots)y;$$

and the resulting terms alone

$$(az^{n-1} + bz^{n-2} + cz^{n-3} + \cdots)\frac{lu}{r}$$

are to be retained. Let it be supposed now that $z = x$, and the sum of the first be equal $V \pm Qku/r$, and of the latter the sum $= plu/r$. From where it is manifest that the last term of the new equation $v = V$, and the last but one $pu = [(Pl \pm Qk)/r] \cdot u$.

(18) Let Pm be an harmonical mean between the segments $PD, PE, PI, \ldots$, and $P\mu$ an harmonical mean between the segments $Pd, Pe, Pi, \ldots$, as in (16), let μm, being joined, cut the abscissa in H; and let us suppose $P\mu$ to be parallel to the ordinate pD. Let μs be drawn parallel to the abscissa, which meet the right line Pm in s; and Ps will be to $P\mu$ as PD to pD or as l to r, and μs to $P\mu$ as k to r. And since $P\mu = nN/p$ (by 17) $= nvr/Pl \pm Qk$, ms will equal

$$Pm - Ps = \frac{nv}{P} - \frac{nQl}{pr} = \frac{nv}{P} \pm \frac{vn\ell}{Pl \pm Qk} = \frac{nvQk}{P(Pl \pm Qk)}.$$

Now ms is to $s\mu$ as Pm to PH, i.e. $nvQk/[P(Pl \pm Qk)]$ to $nvk/(Pl \pm Qk)$ as Pm to PH; and so Q is to P as Pm to PH, or $PH = Pm \cdot P/Q$ or nv/Q. Since therefore the value of the right line PH does not depend upon the quantities l, k, and r; but these being changed, is always the same, the point μ will be at a right line given in position, as we have otherwise shown in Theorem II. Moreover also the value of the line PH is that which in (15) we have determined by another method; and the right line Hm cuts all right lines drawn through P harmonically, according to the definition of harmonical section given in general in (14).

Problem 3

Consider an integer $a > 2$ not divisible by the cube of an integer >1, and let (C_a) be the affine cubic

$$x^3 + y^3 = a. \qquad (C_a)$$

(1) Let $P = (\alpha, \beta)$ be a rational point of (C_a), and show that $\alpha\beta \neq 0$ and $\alpha \neq \beta$.

(2) Show that the abscissa γ of the third intersection point of (C_a) with its tangent at P is given by one of the equations

$$\begin{cases} 2\alpha + \gamma = \dfrac{3\alpha^4}{\alpha^3 - \beta^3} \\[2ex] \alpha^2\gamma = \dfrac{\beta^6 - a^2}{\beta^3 - \alpha^3}. \end{cases}$$

(3) Show that we cannot have $\alpha = \gamma$.

(4) **Without computing,** give the ordinate δ of the third point of intersection of (C_a) with its tangent at P.

(5) Assume that α and β are written as irreducible fractions

$$\alpha = \frac{x_1}{z_1} \qquad \beta = \frac{y_1}{z_2},$$

with $(x_1, z_1) = 1 = (y_1, z_2)$, $z_1 > 0$, $z_2 > 0$.

Show that $z_1 = z_2$.

(6) Show that

$$\begin{cases} \gamma = \dfrac{x_1}{z_1} \cdot \dfrac{x_1^3 + 2y_1^3}{x_1^3 - y_1^3} \\ \delta = \dfrac{y_1}{z_1} \cdot \dfrac{y_1^3 + 2x_1^3}{y_1^3 - x_1^3}. \end{cases}$$

(7) Let A be the positive greatest common divisor of the numbers $x_1(x_1^3 + 2y_1^3)$, $y_1(y_1^3 + 2x_1^3)$, $z_1(x_1^3 - y_1^3)$.

Show that $A \in \{1, 3\}$.

(8) Set

$$x_2 = \frac{x_1(x_1^3 + 2y_1^3)}{A}, \qquad y_2 = \frac{y_1(y_1^3 + 2x_1^3)}{A}, \qquad z_2 = \left| \frac{z_1(x_1^3 - y_1^3)}{A} \right|.$$

Show that $(x_2/z_2,\ y_2/z_2)$ is a rational point of (C_a), and that we have $z_2 > z_1$ (there are two cases to consider: $A = 1$ and $A = 3$).

(9) Deduce that the number of rational points of the cubic (C_a) is equal to zero or infinity.

(10) Give an example of a cubic (C_a) which has an infinity of rational points (i.e. give a).

(11) Show that (C_3) contains no rational points.

Problem 4

(Congruent Numbers)

We say that an integer $n \geq 1$ is a **congruent number** if there exists a right triangle whose sides have rational lengths and whose area is equal to n.

(1) Show that if the sides of the above triangle are a, b, c, then there exists $\lambda \in \mathbb{Q}_+^*$ such that

$$\frac{a}{c} = \frac{1 - \lambda^2}{1 + \lambda^2}, \qquad \frac{b}{c} = \frac{2\lambda}{1 + \lambda^2}.$$

Show that the point $(-n\lambda,\ n^2(1 + \lambda^2)/c)$ lies on the cubic

$$(C_n): \quad y^2 = x(x^2 - n^2).$$

(2) Conversely, let $(x, y) \in C_n(\mathbb{Q})$ be a rational point of the curve (C_n) such that $y \neq 0$. Show that $a = |(n^2 - x^2)/y|$, $b = |2nx/y|$ and $c = |(n^2 + x^2)/y|$ are the sides of the right angle and the hypotenuse of a right triangle.

(3) Given an arbitrary $n \geq 1$, we know that the curve (C_n) can be parametrised using Weierstrass functions $(\wp(u), 1/2\wp'(u))$. Consider the group law on $C_n(\mathbb{C})$, and determine all the points $(x, y) \in C_n(\mathbb{C})$ such that $(x, y) \oplus (x, y) = O$, where O denotes the identity element of the group (the point at infinity of $C_n(\mathbb{C})$).

Let $C_n[2]$ denote this set, and show that $C_n[2]$ is a group. What is its order? Is it cyclic? What can we say about y if $(x, y) \in C_n[2]$?

(4) We propose to determine the torsion subgroup T of the group $C_n(\mathbb{Q})$, i.e. the set of points of finite order of the group $C_n(\mathbb{Q})$. For this, we will use the following result of Nagell: if a point $P = (x, y) \neq 0$ is a point of finite order of $C_n(\mathbb{Q})$, then x and y are integers and y^2 is zero or divides $4n^6$.

Show that T is a finite group whose order m is divisible by 4.

(5) Let p be a prime number congruent to 3 modulo 4.

Show that $-\bar{1}$ is not a square in $\mathbb{F}_p$.

Set

$$P(X) = X(X^2 - \bar{n}^2),$$

where $\bar{n}$ denotes the class of n modulo p.

Show that for every $x \in \mathbb{F}_p$, either $P(x)$ or $P(-x)$ is a square of $\mathbb{F}_p$.

(6) Using (5) and not forgetting the point at infinity, show that the curve

$$Y^2 = X(X^2 - \bar{n}^2) \qquad (\bar{C}_n)$$

has $p + 1$ points in $\mathbb{P}_2(\mathbb{F}_p)$.

(7) Let $\mathbb{Z}_{(p)}$ denote the ring $\mathbb{Z}_{(p)} := \{c/d \in \mathbb{Q}\,;\, p \text{ does not divide } d\}$.

For $(x, y) \in C_n(\mathbb{Q}) \setminus \{0\}$, prove the relation

$$\{x \in \mathbb{Z}_{(p)}\} \iff \{y \in \mathbb{Z}_{(p)}\}.$$

(8) Let us define a map $\varphi : C_n(\mathbb{Q}) \to \bar{C}_n(\mathbb{F}_p)$ as follows. If $P = (x, y) \in (\mathbb{Z}_{(p)})^2$, set

$$\varphi(P) = (\bar{x}, \bar{y}) \in (\mathbb{F}_p)^2,$$

where $\bar{x}$ and $\bar{y}$ are the classes of x and y modulo p.

Otherwise, set $\varphi(P) = \bar{O}$ where $\bar{O}$ denotes the point at infinity of homogeneous coordinates $(\bar{0}, \bar{1}, \bar{0})$ of $\bar{C}_n(\mathbb{F}_p)$.

Assume the fact that φ is a group homomorphism when $\bar{C}_n(\mathbb{F}_p)$ is equipped with an "Abel-style" group structure, which is possible if p does not divide $2n$.

Show that if p is sufficiently large, the restriction of φ to T is injective, and show that $\#\bar{C}_n(\mathbb{F}_p)$ is divisible by 4.

(9) Assume that $m = 3^r \cdot m'$ with $r \geq 0$ and $(3, m') = 1$. Recall that there exists an infinity of prime numbers in the arithmetic progression $3 + m'\mathbb{Z}$. Show that if p is a prime belonging to this progression and not dividing $2n$, then

$$\begin{cases} p \equiv -1 \bmod m' \\ p \equiv 3 \bmod m', \end{cases}$$

and deduce that $m' = 4$.

Use questions (4), (6) and (8).

(10) Show that $r > 0$ is impossible, by considering the arithmetic progression $7 + m\mathbb{Z}$, which also contains an infinity of prime numbers. Conclude.

(11) Show that if $n \geq 1$ is a congruent number, then there exists an infinity of right triangles whose sides have rational lengths and whose area is equal to n.

(12) Do you know a congruent number?

5

MODULAR FORMS

Tyger ! Tyger ! burning bright
In the forests of the night,
What immortal hand or eye
Could frame thy fearful symmetry?
(William Blake)

The originator of the famous wisecrack about the **five** operations in arithmetic – addition, subtraction, multiplication, division and ... **modular functions** – would appear to have been the mathematician M. Eichler. As we will see, the fifth and last "operations" are actually functions, which have so incredibly many symmetries that one may really wonder how they can possibly exist!

5.1 BRIEF HISTORICAL OVERVIEW

The first publication of fragments of theta functions appeared in 1713, with the posthumous appearance of the *Ars Conjectandi* by Jakob Bernoulli, a book in which one encounters expressions such as

$$\sum_{n=0}^{\infty} m^{(1/2)n(n+3)}, \qquad \sum_{n=0}^{\infty} m^{(1/2)n(n+1)}, \qquad \sum_{n=0}^{\infty} m^{n^2}.$$

Thirty-five years later, theta functions reappeared in another form in the *Introductio in Analysin Infinitorum* by L. Euler (1748, Volume I, Section 304). In Chapter XVI of his book, Euler considered the problem of partitions of the integers, i.e. the problem of expressing every number $n \in \mathbb{N}$ using integers belonging to a certain set $A = \{\alpha, \beta, \gamma, \dots\}$.

Euler noted that when we take the infinite product

$$(1 + x^{\alpha} z)(1 + x^{\beta} z)(1 + x^{\gamma} z) \cdots = 1 + Pz + Qz^2 + \cdots, \tag{1}$$

P is the sum of the powers $x^{\alpha}, x^{\beta}, x^{\gamma}, \dots$ Q is the sum of the powers $x^{\alpha+\beta}, x^{\alpha+\gamma}, \dots, x^{\beta+\gamma}$, ... and so on. Thus, the coefficient of z^m is a polynomial in x in which the term Nx^n has a coefficient N equal to the number of times that we can write n as a sum of m **distinct** terms of A.

Rule 5.1.1 *If the term $Nx^n z^m$ appears in (1), this means that there are N different ways to represent n as sum of m* **distinct** *terms of A.*

Next, Euler noted that when we take the infinite product

$$\frac{1}{(1-x^{\alpha}z)(1-x^{\beta}z)(1-x^{\gamma}z)\cdots} = 1 + Pz + Qz^2 + \cdots, \tag{2}$$

P is the sum of the powers $x^{\alpha}, x^{\beta}, x^{\gamma}, \ldots$ Q is the sum of the powers $x^{\alpha+\alpha}, x^{\alpha+\beta}, \ldots$ and so on. Thus the coefficient of z^m is a polynomial in x in which the term Nx^n has a coefficient N equal to the number of times that n can be written as a sum of m **equal or distinct** terms of A.

Rule 5.1.2 *If the term $Nx^n z^m$ appears in (2), this means that there are N different ways of representing n as sum of m* **equal or distinct** *terms of A.*

Finally, Euler passed from theory to practice, by taking the sequence $\alpha, \beta, \gamma, \ldots$ to be the sequence $\mathbb{N}^*$ of the natural integers, and using a simple but efficient trick to establish the following result.

Euler's Theorem 5.1.1 *The following identities hold in the ring of formal series $\mathbb{Z}[[x, z]]$:*

$$\prod_{m=1}^{\infty}(1 + x^m z) = \sum_{n=0}^{\infty} \frac{x^{(1/2)n(n+1)}}{(1-x)\cdots(1-x^n)} z^n, \tag{3}$$

$$\prod_{m=1}^{\infty}(1 - x^m z)^{-1} = \sum_{n=0}^{\infty} \frac{x^n}{(1-x)\cdots(1-x^n)} z^n. \tag{4}$$

Proof. Euler's trick consists in obtaining a functional equation for the infinite products by substituting xz for z.

(1) Let $F(x, z)$ denote the first product. Then

$$F(x, xz) = \prod_{m=1}^{\infty}(1 + x^{m+1}z) = \frac{F(x, z)}{1 + xz},$$

so we obtain the functional equation

$$F(x, z) = (1 + xz)F(x, xz).$$

If we set

$$F(x, z) = \sum_{n=0}^{\infty} c_n(x) z^n \in (\mathbb{Z}[[x]])[[z]],$$

we obtain the induction relations

$$\begin{cases} c_0(x) = 1 \\ (1 - x^n)c_n(x) = x^n c_{n-1}(x) \quad \text{if } n \geq 1, \end{cases}$$

which give the result.

(2) Let $G(x, z)$ denote the second product. Then

$$G(x, xz) = \prod_{m=1}^{\infty}(1 - x^{m+1}z)^{-1} = (1 - xz)G(x, z),$$

which gives the functional equation

$$G(x, xz) = (1 - xz)G(x, z).$$

If we set

$$G(x, z) = \sum_{n=0}^{\infty} d_n(x)z^n \in (\mathbb{Z}[[x]])[[z]],$$

we obtain the induction relations

$$\begin{cases} d_0(x) = 1 \\ (1 - x^n)d_n(x) = xd_{n-1}(x), \end{cases}$$

which give the result.

(3) Convergence questions.

Until now, we used exactly the same reasoning as Euler. But in order for infinite products and the series to make sense, we need to define the topology of $\mathbb{Z}[[x, y]]$.

For this, we note that if $\mathfrak{m}$ denotes the ideal of $\mathbb{Z}[[x, y]]$ generated by x and y (which is a prime ideal), then

$$\bigcap_{\nu=1}^{\infty} \mathfrak{m}^{\nu} = \{0\}.$$

By taking the $\{\mathfrak{m}^{\nu}\}_{\nu\in\mathbb{N}}$ for a fundamental system of neighbourhoods of zero, we put the structure of a topological ring on $\mathbb{Z}[[x, y]]$. We say that a sequence $(s_n)_{n\in\mathbb{N}}$ of elements of $\mathbb{Z}[[x, y]]$ is a Cauchy sequence if for every $\nu \in \mathbb{N}$, there exists N such that m and $n \geq N$ implies $s_m - s_n \in \mathfrak{m}^{\nu}$. Then one of the essential features of this topology is that it makes the topological ring $\mathbb{Z}[[x, y]]$ complete, i.e. every Cauchy sequence of elements of $\mathbb{Z}[[x, y]]$ is convergent (see [Bou 3]).

Since in the identities of the statement of the Euler's theorem, the partial products of the infinite products (on the left) and the partial sums of the series (on the right) form Cauchy sequences, both sides are meaningful, both in (3) and in (4).

Moreover, points (1) and (2) of the proof show that these two sides are congruent modulo $\mathfrak{m}^{\nu}$ for every $\nu \in \mathbb{N}$. Thus, they are equal. □

Remark 5.1.1 The reader familiar with the topology of the complete local ring $\mathbb{Q}[[x, y]]$ will recognise that the topology of $\mathbb{Z}[[x, y]]$ is induced by that of $\mathbb{Q}[[x, y]]$.

Corollary (Euler) 5.1.1

(i) *The number of different ways of expressing $n \in \mathbb{N}$ as a sum of integers between 1 and m (inclusive) is equal to the number of different ways of expressing $n + (m(m+1))/2$ as a sum of m unequal numbers ≥ 1.*

(ii) *The number of different ways of expressing $n \in \mathbb{N}$ as a sum of m integers between 1 and m (inclusive) is equal to the number of different ways of expressing $n + m$ as a sum of m numbers ≥ 1.*

(iii) *The number of different ways of writing n as a sum of m numbers is equal to the number of different ways of writing $n + (m(m-1))/2$ as a sum of m unequal numbers ≥ 1.*

Proof.

(1) Assertion (iii) follows formally from assertions (i) and (ii).

(2) In formula (3), the coefficient N of the term $Nx^e z^m$ is equal to the number of different ways of expressing e as a sum of unequal numbers ≥ 1. But if $e = n + (m(m+1))/2$, we see that this term comes from the expansion of

$$\frac{x^{(1/2)m(m+1)}}{(1-x)\cdots(1-x^m)} z^m.$$

Consequently, N is also equal to the coefficient of x^n in the expansion of $1/((1-x)\cdots(1-x^m))$, so to the number of ways of expressing n as sum of numbers between 1 and m (inclusive).

(3) In formula (4), the coefficient N of the term $Nx^e z^m$ is equal to the number of different ways of expressing e as a sum of m numbers ≥ 1. But if $e = n + m$, we see that this term comes from the expansion of

$$\frac{x^m}{(1-x)\cdots(1-x^m)} z^m.$$

Consequently, N is also equal to the coefficient of x^n in the expansion of $1/((1-x)\cdots(1-x^m))$. □

Remark 5.1.2 When z is replaced by 1 in the second identity of the preceding theorem, we obtain

$$\prod_{m=1}^{\infty}(1-x^m)^{-1} = \sum_{n=0}^{\infty} \frac{x^n}{(1-x)\cdots(1-x^n)} = \sum_{n=0}^{\infty} p(n)x^n.$$

Thus $p(n)$ represents the number of ways of writing n as a sum of integers ≥ 1, with no restrictions on these integers.

Definition 5.1.1 *The function $n \mapsto p(n)$ is called the* **partition function**, *and its generating series $\sum_{n=0}^{\infty} p(n)x^n$ is written $P(x)$. Thus we have*

$$P(x) = \prod_{m=1}^{\infty} (1 - x^m)^{-1} \in \mathbb{Z}[[x]].$$

The partition function satisfies strange congruences called **Ramanujan congruences**[55]. For example, we have

$$\begin{cases} p(5n+4) \equiv 0 \bmod 5 \\ p(7n+5) \equiv 0 \bmod 7 \\ p(11n+5) \equiv 0 \bmod 11 \end{cases}$$

(see exercises).

Euler's true follower was Carl Gustav Jacobi [56], who in his *Fundamenta Nova Theoriae Functionum Ellipticarum* (1829), gave a modern theory of the theta functions. On April 24, 1828, Jacobi discovered the surprising formula

$$\left(\sum_{-\infty}^{\infty} z^{n^2} \right)^4 = 1 + 8 \sum_{m=1}^{\infty} \left(\sum_{\substack{d \mid m \\ 4 \nmid d}} d \right) z^m, \tag{5}$$

from which he deduced the equally surprising following corollary (which is equivalent to formula (5)).

Jacobi's Theorem 5.1.2 *The number of representations of an integer $n \geq 0$ as a sum of four squares is eight times the sum of its divisors if n is odd and twenty-four times the sum of its odd divisors if n is even.*

Proof. The left-hand side of (5) can be written

$$\sum_{n_1, n_2, n_3, n_4 \in \mathbb{Z}} z^{n_1^2 + n_2^2 + n_3^2 + n_4^2} = \sum_{n \in \mathbb{N}} A_4(n) z^n,$$

where $A_4(n)$ denotes the number of representations of n as a sum of four squares (taking the order and sign of the n_i into account).

If $n > 0$, then by (5), we have

$$A_4(n) = 8 \sum_{\substack{d \mid n \\ 4 \nmid d}} d.$$

[55] S. Ramanujan 1887–1920.

[56] C.G. Jacobi 1804–1851.

Thus, if n is odd, the theorem is proved. But if n is even, we have

$$\sum_{\substack{d|n \\ 4\nmid d}} d = \sum_{\substack{d|n \\ d \text{ odd}}} d + 2 \sum_{\substack{d|n \\ d \text{ odd}}} d = 3 \sum_{\substack{d|n \\ d \text{ odd}}} d$$

which concludes the proof. □

It remains only to prove formula (5), which is done below.

Modular forms also appear in the writings of Abel[57], Gauss[58], Hermite[59] and many others. However, it was above all F. Klein[60] who, beginning in 1877, originated the study of the field of modular functions, considered as a function field of a Riemann surface, providing fundamental domains for the different **levels** n (see Section 5.3). Then, in 1881–82, Poincaré[61] developed the general study of the discrete subgroups of $SL_2(\mathbb{R})$.

5.2 THE THETA FUNCTIONS

We saw in Chapter 2 that every **entire** function which is doubly periodic for a period lattice Λ is constant. Thus, if we want to obtain **non-constant, entire** functions, we must weaken the requirement of periodicity.

(a) Definition of the theta functions

We would like to define functions which are **entire** (i.e. holomorphic on all of $\mathbb{C}$) and semi-periodic for the period lattice

$$\wedge = \mathbb{Z} + \mathbb{Z}\tau,$$

where τ denotes a complex number in the **Poincaré upper half-plane** $\mathcal{H} = \{z \in \mathbb{C}; \Im z > 0\}$.

The properties of "semi-periodicity" we take here are

$$\begin{cases} \Theta(z+1) = \Theta(z) \\ \Theta(z+\tau) = F(z)\Theta(z), \end{cases} \tag{1}$$

where $F(z)$ is a factor to be determined.

Remark 5.2.1 If the function Θ is not identically zero, then clearly $F(z+1) = F(z)$. The usual choice of F is

$$F(z) = \frac{1}{ce^{2i\pi z}}, \quad c \in \mathbb{C}^*,$$

and this is the choice we adopt from now on. □

[57] N.H. Abel (1802–1829).
[58] K.F. Gauss (1777–1855).
[59] C. Hermite (1822–1901).
[60] F. Klein (1849–1925).
[61] J.H. Poincaré (1854–1912).

Since Θ is holomorphic of period 1, it can be developed as a Fourier series[62] and we have

$$\Theta(z) = \sum_{-\infty}^{\infty} a_n e^{2i\pi nz},$$

where the a_n are the **Fourier coefficients** of Θ.

The expansion of Θ as a Fourier series expresses the first condition of (1).

The second condition leads to

$$F(z) \sum_{-\infty}^{\infty} a_n e^{2i\pi nz} = \sum_{-\infty}^{\infty} a_n e^{2i\pi n(z+\tau)}.$$

With the **above choice** of $F(z)$, and by the uniqueness of the Fourier coefficients of Θ, we obtain the relation

$$a_{n+1} = ce^{2i\pi n\tau} a_n, \tag{2}$$

hence the expression

$$\Theta(z) = a_0 \sum_{-\infty}^{\infty} c^n e^{i\pi n(n-1)\tau + 2i\pi nz}. \tag{3}$$

It is traditional **here** to set

$$q = e^{i\pi\tau}, \tag{4}$$

and to note that since $\tau \in \mathcal{H}$, we have

$$0 < |q| < 1.$$

Thus, we see that

$$\Theta(z) = a_0 \sum_{-\infty}^{\infty} c^n q^{n(n-1)} e^{2i\pi nz} \tag{3'}$$

and in particular,

$$\Theta(0) = a_0 \sum_{-\infty}^{\infty} c^n q^{n(n-1)}. \tag{3''}$$

Conversely, if we take $a_0 = 1$ and $c = q$, we see that the right-hand side of (3′) converges uniformly on every compact subset $\mathbb{C}$, so it defines an entire function.

Definition 5.2.1 *Let $q \in \mathbb{C}$ be such that $0 < |q| < 1$. The function θ_3 is defined by*

$$\theta_3(z) = \sum_{-\infty}^{\infty} q^{n^2} e^{2i\pi nz} = 1 + 2\sum_{n\geq 1} q^{n^2} \cos(2\pi nz). \tag{5}$$

[62] J. Fourier (1768–1830).

Thus, we also have the following result.

Proposition 5.2.1 *θ_3 is an entire function satisfying the properties*

$$\begin{aligned}\theta_3(z+1) &= \theta_3(z)\\ \theta_3(z+\tau) &= q^{-1}e^{-2i\pi z}\theta_3(z).\end{aligned}$$

Another classical choice of a_0 and c is given by

$$a_0 = 1, \qquad c = -q,$$

which gives

$$\theta_4(z) = \sum_{-\infty}^{\infty}(-1)^n q^{n^2} e^{2i\pi nz} = 1 + 2\sum_{n\geq 1}(-1)^n q^{n^2}\cos(2\pi nz) = \theta_3\left(z+\frac{1}{2}\right). \tag{5'}$$

Jacobi also introduced the functions θ_1 and θ_2 defined by

$$\begin{cases}\theta_1(z) = -iq^{1/4}e^{\pi iz}\theta_3\left(z+\dfrac{1+\tau}{2}\right)\\ \theta_2(z) = q^{1/4}e^{\pi iz}\theta_3\left(z+\dfrac{\tau}{2}\right).\end{cases}$$

Thus, we have the expansions

$$\begin{cases}\theta_1(z) = -i\displaystyle\sum_{-\infty}^{\infty}(-1)^n q^{(n+1/2)^2}e^{(2n+1)i\pi z}\\ \qquad = 2\displaystyle\sum_{n\geq 0}(-1)^n q^{(n+1/2)^2}\sin(2n+1)\pi z\\ \theta_2(z) = \displaystyle\sum_{-\infty}^{\infty} q^{(n+1/2)^2}e^{(2n+1)i\pi z}\\ \qquad = 2\displaystyle\sum_{n\geq 0} q^{(n+1/2)^2}\cos(2n+1)\pi z.\end{cases} \tag{5''}$$

Naturally, the functions θ_1 and θ_2 do not have the same multipliers as θ_3 and θ_4. Let us summarise the elementary properties of these functions.

Relations 5.2.1 Set $\lambda = q^{-1/4}e^{-\pi i z}$, then we obtain the table:

	$z+\frac{1}{2}$	$z+\frac{\tau}{2}$	$z+\frac{1+\tau}{2}$
θ_1	$\theta_2(z)$	$i\lambda\theta_4(z)$	$\lambda\theta_3(z)$
θ_2	$-\theta_1(z)$	$\lambda\theta_3(z)$	$-i\lambda\theta_4(z)$
θ_3	$\theta_4(z)$	$\lambda\theta_2(z)$	$i\lambda\theta_1(z)$
θ_4	$\theta_3(z)$	$i\lambda\theta_1(z)$	$\lambda\theta_2(z)$

(6)

Multipliers 5.2.1 Set $\mu = q^{-1}e^{-2i\pi z}$, then

	$z+1$	$z+\tau$	$z+1+\tau$
θ_1	$-\theta_1(z)$	$-\mu\theta_1(z)$	$\mu\theta_1(z)$
θ_2	$-\theta_2(z)$	$\mu\theta_2(z)$	$-\mu\theta_2(z)$
θ_3	$\theta_3(z)$	$\mu\theta_3(z)$	$\mu\theta_3(z)$
θ_4	$\theta_4(z)$	$-\mu\theta_4(z)$	$-\mu\theta_4(z)$

(6′)

Remark 5.2.2

(1) θ_1 and θ_2 actually belong to the lattice $\mathbb{Z}2 + \mathbb{Z}\tau$, but the squares of the theta functions all belong to the lattice $\mathbb{Z} + \mathbb{Z}\tau$, and all have the same multipliers.

(2) It follows that the quotients of squares of theta functions are elliptic functions for the lattice $\mathbb{Z} + \mathbb{Z}\tau$. We will see below that these functions are of order 2. It follows that three functions of this type are linearly dependent (see Chapter 2).

(3) The notation adopted here goes back to Jacobi, except that he wrote θ instead of θ_4. This notation appears to have become universal now, although it is not optimal. Indeed, the symmetries of the set of four theta functions are those of the Klein group $\mathbb{Z}/2\mathbb{Z} \times \mathbb{Z}/2\mathbb{Z}$, and not those of the cyclic group $\mathbb{Z}/4\mathbb{Z}$. Hermite and Weber used the (preferable) notation:

$$\theta_{\mu,\nu}(z) = \sum_{-\infty}^{\infty} (-1)^{n\nu} q^{(1/2)(2n+\mu)^2} e^{i\pi(2n+\mu)z}$$

for $(\mu, \nu) \in \{0, 1\}^2$, which leads to the multipliers

$$\begin{cases} \theta_{\mu,\nu}(z+1) = (-1)^{\mu}\theta_{\mu,\nu}(z) \\ \theta_{\mu,\nu}(z+\tau) = (-1)^{\nu} q^{-1} e^{-2i\pi z}\theta_{\mu,\nu}(z). \end{cases}$$

(b) Relations between theta functions

We noted earlier that the squares of three theta functions must be linearly dependent.

There exists a marvellously simple method to prove these relations: it consists in noting that the series which define these functions are absolutely convergent and that the terms of a product can be rearranged according to the powers of q.

So we need to prove the relation

$$\theta_1(x, q)\theta_1(y, q) = \theta_3(x+y, q^2)\theta_2(x-y, q^2) - \theta_2(x+y, q^2)\theta_3(x-y, q^2), \qquad (7)$$

where x and y are two arbitrary elements of the field of complex numbers, and where we need to make the "mute variable" q appear.

Using (5″), we have

$$\theta_1(x,q)\theta_1(y,q) = -\sum_m\sum_n(-1)^{m+n}q^{(m+1/2)^2+(n+1/2)^2}e^{i(2m+1)x\pi+i(2n+1)y\pi}.$$

Set $m+n=r$ and $m-n=s$; then (r,s) runs through the pairs of integers of the same parity. We have

$$\left(m+\frac{1}{2}\right)^2+\left(n+\frac{1}{2}\right)^2=\frac{1}{2}(r+1)^2+\frac{1}{2}s^2$$

and

$$(2m+1)x+(2n+1)y=(r+1)(x+y)+s(x-y),$$

so

$$\theta_1(x,q)\theta_1(y,q) = -\sum{}'(-1)^r q^{(1/2)[(r+1)^2+s^2]}e^{i[(r+1)(x+y)+s(x-y)]\pi},$$

where $\sum'$ denotes the sum extended to the (r,s) of same parity. We deduce that

$$\begin{aligned}\theta_1(x,q)\theta_1(y,q) &= -\sum_r\sum_s q^{2(r+1/2)^2+2s^2}e^{i[(2r+1)(x+y)+2s(x-y)]\pi}\\ &\quad+\sum_r\sum_s q^{2r^2+2(s+1/2)^2}e^{i[2r(x+y)+(2s+1)(x-y)]\pi}\\ &= -\sum_r q^{2(r+1/2)^2}e^{i(2r+1)(x+y)\pi}\sum_s q^{2s^2}e^{i2s(x-y)\pi}\\ &\quad+\sum_r q^{2r^2}e^{i2r(x+y)\pi}\sum_s q^{2(s+1/2)^2}e^{i(2s+1)(x-y)\pi}\\ &= -\theta_2(x+y,q^2)\theta_3(x-y,q^2)+\theta_3(x+y,q^2)\theta_2(x-y,q^2).\end{aligned}$$

□

Making the translations by $1/2$, $\tau/2$ and $(1+\tau)/2$ on x and y, we obtain a complete system of relations deduced from (7):

$$\left\{\begin{aligned}\theta_1(x,q)\theta_1(y,q)&=\theta_3(x+y,q^2)\theta_2(x-y,q^2)-\theta_2(x+y,q^2)\theta_3(x-y,q^2)\\ \theta_1(x,q)\theta_2(y,q)&=\theta_1(x+y,q^2)\theta_4(x-y,q^2)+\theta_4(x+y,q^2)\theta_1(x-y,q^2)\\ \theta_2(x,q)\theta_2(y,q)&=\theta_2(x+y,q^2)\theta_3(x-y,q^2)+\theta_3(x+y,q^2)\theta_2(x-y,q^2)\\ \theta_3(x,q)\theta_3(y,q)&=\theta_3(x+y,q^2)\theta_3(x-y,q^2)+\theta_2(x+y,q^2)\theta_2(x-y,q^2)\\ \theta_3(x,q)\theta_4(y,q)&=\theta_4(x+y,q^2)\theta_4(x-y,q^2)-\theta_1(x+y,q^2)\theta_1(x-y,q^2)\\ \theta_4(x,q)\theta_4(y,q)&=\theta_3(x+y,q^2)\theta_3(x-y,q^2)-\theta_2(x+y,q^2)\theta_2(x-y,q^2).\end{aligned}\right. \tag{7'}$$

If we take $y=0$ in this last identity, we have

$$\theta_4(x,q)\theta_4(0,q)=\theta_3^2(x,q^2)-\theta_2^2(x,q^2). \tag{7''}$$

If we subtract the square of the first identity of (7′) from the square of the last one, we obtain

$$\begin{aligned}&\theta_4^2(x,q)\theta_4^2(y,q)-\theta_1^2(x,q)\theta_1^2(y,q)\\&\quad=\left[\theta_3^2(x+y,q^2)-\theta_2^2(x+y,q^2)\right]\left[\theta_3^2(x-y,q^2)-\theta_2^2(x-y,q^2)\right],\end{aligned}\tag{7'''}$$

and taking (7″) into account (omitting the argument q), we have

$$\theta_4(x+y)\theta_4(x-y)\theta_4^2(0)=\theta_4^2(x)\theta_4^2(y)-\theta_1^2(x)\theta_1^2(y),\tag{8}$$

which is the prototype of a series of relations which we obtain by making the canonical translations on x and y

$$\begin{cases}\theta_1(x+y)\theta_1(x-y)\theta_4^2(0)=\theta_3^2(x)\theta_2^2(y)-\theta_2^2(x)\theta_3^2(y)=\theta_1^2(x)\theta_4^2(y)-\theta_4^2(x)\theta_1^2(y)\\ \theta_2(x+y)\theta_2(x-y)\theta_4^2(0)=\theta_4^2(x)\theta_2^2(y)-\theta_1^2(x)\theta_3^2(y)=\theta_2^2(x)\theta_4^2(y)-\theta_3^2(x)\theta_1^2(y)\\ \theta_3(x+y)\theta_3(x-y)\theta_4^2(0)=\theta_4^2(x)\theta_3^2(y)-\theta_1^2(x)\theta_2^2(y)=\theta_3^2(x)\theta_4^2(y)-\theta_2^2(x)\theta_1^2(y)\\ \theta_4(x+y)\theta_4(x-y)\theta_4^2(0)=\theta_4^2(x)\theta_4^2(y)-\theta_1^2(x)\theta_1^2(y)=\theta_3^2(x)\theta_3^2(y)-\theta_2^2(x)\theta_2^2(y).\end{cases}\tag{8'}$$

Other relations analogous to (8′) can be deduced, by the same procedure, from another choice of the equations. We leave this as an exercise.

Replacing y by 0 in these relations, we obtain the predicted linear relations between the squares of the theta functions. For example, (8′) yields

$$\begin{cases}\theta_1^2(x)\theta_4^2(0)=\theta_3^2(x)\theta_2^2(0)-\theta_2^2(x)\theta_3^2(0)\\ \theta_2^2(x)\theta_4^2(0)=\theta_4^2(x)\theta_2^2(0)-\theta_1^2(x)\theta_3^2(0)\\ \theta_3^2(x)\theta_4^2(0)=\theta_4^2(x)\theta_3^2(0)-\theta_1^2(x)\theta_2^2(0)\\ \theta_4^2(x)\theta_4^2(0)=\theta_3^2(x)\theta_3^2(0)-\theta_2^2(x)\theta_2^2(0).\end{cases}\tag{8''}$$

The order three minors of this system of linear equations with respect to the $\theta_i^2(x)$ with $i=1,2,3,4$ must all be zero. In fact, if we replace x by 0 in the last relation of (8″), we find a fundamental relation between the constants (which implies that the determinant vanishes). This famous **Jacobi relation** is

$$\theta_3^4(0)=\theta_2^4(0)+\theta_4^4(0).\tag{8'''}$$

It follows from (8‴) that the system (8″) is of rank two; every relation is a linear combination of two others.

Remark 5.2.3 It follows from (8‴) that we have

$$\left(1+2\sum_{n\geq 1}q^{n^2}\right)^4=\left(2q^{1/4}\sum_{-\infty}^{\infty}q^{n^2+n}\right)^4+\left(1+2\sum_{n\geq 1}(-1)^nq^{n^2}\right)^4$$

for every $q \in \mathbb{C}^*$ such that $|q| < 1$. If we now set $q = z^4$, we see that the holomorphic functions in the interior of the unit disk defined by

$$\begin{cases} X(z) = 2z \sum_{-\infty}^{\infty} z^{4(n^2+n)} = \theta_2(0, z^4) \\ Y(z) = 1 + 2 \sum_{n \geq 1} (-1)^n z^{4n^2} = \theta_4(0, z^4) \\ Z(z) = 1 + 2 \sum_{n \geq 1} z^{4n^2} = \theta_3(0, z^4) \end{cases}$$

form a non-trivial solution of Fermat's equation

$$X^4 + Y^4 = Z^4.$$

Nevanlinna theory (see the problems of Chapter 6) shows that these three functions cannot be extended to $\mathbb{C}$, because if $n \geq 4$, there exist no solutions in non-trivial entire functions to the equation

$$X^n + Y^n = Z^n.$$

(c) The identity $\theta_1'(0) = \pi\theta_2(0)\theta_3(0)\theta_4(0)$.

Differentiating the second equation of $(7')$ **with respect to** x, and replacing x and y by zero, we have

$$\theta_1'(0, q)\theta_2(0, q) = 2\theta_1'(0, q^2)\theta_4(0, q^2). \tag{9}$$

But the relations $(7')$ also give

$$\begin{cases} \theta_2^2(0, q) = 2\theta_2(0, q^2)\theta_3(0, q^2) \\ \theta_3(0, q))\theta_4(0, q) = \theta_4^2(0, q^2). \end{cases} \tag{9'}$$

From (9) and $(9')$, we deduce the relation

$$\frac{\theta_1'(0, q)}{\theta_2(0, q)\theta_3(0, q)\theta_4(0, q)} = \frac{\theta_1'(0, q^2)}{\theta_2(0, q^2)\theta_3(0, q^2)\theta_4(0, q^2)},$$

and by induction on n, we obtain

$$\frac{\theta_1'(0, q)}{\theta_2(0, q)\theta_3(0, q)\theta_4(0, q)} = \frac{\theta_1'(0, q^{2^n})}{\theta_2(0, q^{2^n})\theta_3(0, q^{2^n})\theta_4(0, q^{2^n})}.$$

Since $|q| < 1$, the limit of q^{2^n} as n tends to infinity is 0, so we have

$$\frac{\theta_1'(0, q)}{\theta_2(0, q)\theta_3(0, q)\theta_4(0, q)} = \lim_{q \longrightarrow 0} \frac{\theta_1'(0, q)}{\theta_2(0, q)\theta_3(0, q)\theta_4(0, q)}. \tag{9''}$$

The series expansions of part (a) give

$$\begin{cases} \theta_1'(0,q) = \pi \sum_{-\infty}^{\infty} (-1)^n (2n+1) q^{(n+1/2)^2} \sim 2\pi q^{1/4} \\ \theta_2(0,q) = \sum_{-\infty}^{\infty} q^{(n+1/2)^2} \sim 2q^{1/4} \\ \theta_3(0,q) = \sum_{-\infty}^{\infty} q^{n^2} \sim 1 \\ \theta_4(0,q) = \sum_{-\infty}^{\infty} (-1)^n q^{n^2} \sim 1 \end{cases}$$

as q tends to zero.

The right-hand side of (9″) thus tends to π, and we deduce the relation

$$\theta_1'(0,q) = \pi \theta_2(0,q)\theta_3(0,q)\theta_4(0,q). \tag{10}$$

(d) Expression of the theta functions as infinite products

For $0 < |q| < 1$ and $\zeta \in \mathbb{C}^*$, set

$$\Phi(\zeta, q) = \prod_{n=1}^{\infty} (1 + q^{2n-1}\zeta)(1 + q^{2n-1}\zeta^{-1}). \tag{11}$$

We see easily that if $|q| \leq \rho < 1$ and if ζ belongs to a compact K of $\mathbb{C}^*$, then

$$(1 + q^{2n-1}\zeta)(1 + q^{2n-1}\zeta^{-1}) = 1 + u_n(\zeta),$$

with

$$|u_n(\zeta)| \leq \rho^{2n-1} M(\rho),$$

which shows that the infinite product (11) converges uniformly on K, to a holomorphic function in ζ (see Chapter 2, Section 2.4). Since K is arbitrary, we obtain a holomorphic function on $\mathbb{C}^*$.

The obvious "symmetries" of the function $z \mapsto \Phi(e^{2i\pi z}, q)$ are the following:

$$\begin{cases} \Phi(\zeta^{-1}, q) = \Phi(\zeta, q) \\ \Phi(q^2\zeta, q) = \dfrac{1}{q\zeta}\Phi(\zeta, q). \end{cases} \tag{12}$$

If we set $\zeta = e^{2i\pi z}$ and $q = e^{i\pi\tau}$, we see that $\Theta(z) = \Phi(e^{2i\pi z}, e^{i\pi\tau})$ satisfies equations (1), with

$$F(z) = \frac{1}{qe^{2i\pi z}}.$$

It follows from the computation of part (a) that $\Theta = a_0\theta_3$, hence

$$a_0\theta_3(z,q) = \prod_{n=1}^{\infty} (1 + q^{2n-1}\zeta)(1 + q^{2n-1}\zeta^{-1}),$$

and making the translations

$$z \longmapsto z + \frac{1}{2}, \qquad z \longmapsto z + \frac{\tau}{2}, \qquad z \longmapsto z + \frac{1+\tau}{2},$$

we obtain

$$\begin{cases} a_0\theta_1(z,q) = -i\nu\Phi(-qe^{2i\pi z}, q) \\ a_0\theta_2(z,q) = \nu\Phi(qe^{2i\pi z}, q) \\ a_0\theta_3(z,q) = \Phi(e^{2i\pi z}, q) \\ a_0\theta_4(z,q) = \Phi(-e^{2i\pi z}, q) \end{cases} \tag{13}$$

with $\nu = q^{1/4}e^{\pi i z}$.

To determine a_0, we will use the formula (10):

$$\theta_1'(0) = \pi\theta_2(0)\theta_3(0)\theta_4(0)$$

which was proved earlier.

Taking $z = 0$ in (13), we obtain

$$\begin{cases} a_0\theta_2(0) = 2q^{1/4}\prod_{n=1}^{\infty}(1+q^{2n})^2 \\ a_0\theta_3(0) = \prod_{n=1}^{\infty}(1+q^{2n-1})^2 \\ a_0\theta_4(0) = \prod_{n=1}^{\infty}(1-q^{2n-1})^2. \end{cases} \tag{13$'$}$$

Furthermore, the first equation of (13) can be written

$$a_0\theta_1(z) = 2q^{1/4}\sin\pi z\prod_{n=1}^{\infty}(1-2q^{2n}\cos 2\pi z + q^{4n}),$$

hence

$$a_0\frac{\theta_1(z)}{\sin\pi z} = 2q^{1/4}\prod_{n=1}^{\infty}(1-2q^{2n}\cos 2\pi z + q^{4n}).$$

Letting z tend to zero, we obtain

$$a_0\theta_1'(0) = 2\pi q^{1/4}\prod_{n=1}^{\infty}(1-q^{2n})^2. \tag{13$''$}$$

Plugging (13′) and (13″) into (10), we obtain

$$a_0^2 = \left(\frac{\prod_{n=1}^{\infty}(1+q^{2n})(1+q^{2n-1})(1-q^{2n-1})}{\prod_{n=1}^{\infty}(1-q^{2n})} \right)^2$$

$$= \left(\frac{\prod_{n=1}^{\infty}(1+q^{2n})(1+q^{2n-1})(1-q^{2n})(1-q^{2n-1})}{\prod_{n=1}^{\infty}(1-q^{2n})^2} \right)^2$$

$$= \left(\frac{\prod_{m=1}^{\infty}(1+q^{m})(1-q^{m})}{\prod_{n=1}^{\infty}(1-q^{2n})^2} \right)^2$$

$$= \left(\frac{1}{\prod_{n=1}^{\infty}(1-q^{2n})} \right)^2 .$$

Since a_0 is positive when q is real, we deduce from (13′) that

$$a_0 = \frac{1}{\prod_{n=1}^{\infty}(1-q^{2n})}, \tag{14}$$

and finally,

$$\begin{cases} \theta_1(z) = 2q^{1/4}\sin\pi z \prod_{n=1}^{\infty}(1-q^{2n})(1-2q^{2n}\cos 2\pi z + q^{4n}) \\ \theta_2(z) = 2q^{1/4} \prod_{n=1}^{\infty}(1-q^{2n})(1+2q^{2n}\cos 2\pi z + q^{4n}) \\ \theta_3(x) = \prod_{n=1}^{\infty}(1-q^{2n})(1+2q^{2n-1}\cos 2\pi z + q^{4n-2}) \\ \theta_4(z) = \prod_{n=1}^{\infty}(1-q^{2n})(1-2q^{2n-1}\cos 2\pi z + q^{4n-2}). \end{cases} \tag{15}$$

Corollary 5.2.1

(i) *All the zeros of the theta functions in $\mathbb{C}$ are simple.*

(ii) *The zeros of θ_1 (resp. $\theta_2, \theta_3, \theta_4$) are congruent to 0 (resp. $1/2, (1+\tau)/2, \tau/2$) modulo $\Lambda = \mathbb{Z} + \mathbb{Z}\tau$.*

Proof. It suffices to give the proof for θ_1.

By (13), we see that these zeros are those of $e^{2i\pi z} - q^{2n} = 0$, for $n \in \mathbb{Z}$.

These zeros are obviously simple and equal to zero modulo Λ, so we see that

$$z \equiv 0 \bmod \Lambda. \qquad \square$$

Remark 5.2.4 Naturally we can recover this result by the methods of Chapter 2, taking the integral of $\theta_i'(z)/\theta_i(z)\,dz$ along the boundary of a (translation of a) period parallelogram.

(e) The heat equation

One of the important roles that theta functions play in physics comes from the fact that they provide solutions of a partial differential equation called the **heat equation**. Indeed, we saw in (a) that our functions are of the form

$$\Theta(z, \tau) = \sum_{-\infty}^{\infty} (-1)^{\nu n} e^{(n+\mu/2)^2 \pi i \tau + 2i\pi(n+\mu/2)z},$$

with $(\mu, \nu) \in \{0, 1\}^2$.

This series is infinitely differentiable in z and τ on every compact subset, and we can differentiate it term by term. Thus, we see that the functions θ_1, θ_2, θ_3 and θ_4 satisfy the **heat equation**

$$\frac{\partial^2 \Theta}{\partial z^2} = 4\pi i \frac{\partial \Theta}{\partial \tau}. \tag{16}$$

Application 5.2.1 Set for $\alpha \in \{2, 3, 4\}$, $\theta_\alpha := \theta_\alpha(0)$, $\theta_1' = \theta_1'(0)$ and

$$f_\alpha(z, \tau) = \frac{\theta_1'}{\theta_\alpha} \frac{\theta_\alpha(z, \tau)}{\theta_1(z, \tau)}.$$

The function $z \mapsto f_\alpha(z, \tau)$ is a meromorphic function on $\mathbb{C}$ which admits a simple pole of residue equal to 1 at every point of the lattice $\Lambda = \mathbb{Z} + \mathbb{Z}\tau$. By Chapter 2, it cannot be doubly periodic, and in fact we have the table

z	$z+1$	$z+\tau$	$z+1+\tau$
f_2	f_2	$-f_2$	$-f_2$
f_3	$-f_3$	$-f_3$	f_3
f_4	$-f_4$	f_4	$-f_4$.

It follows, nonetheless, that the functions f_2^2, f_3^2, f_4^2 are elliptic of lattice Λ, and as their Laurent expansions at z starts with $1/z^2$, we have

$$f_\alpha^2(z, \tau) - f_\beta^2(z, \tau) = c_{\alpha\beta} \in \mathbb{C}. \tag{17}$$

The computation of the Laurent expansion of these functions begins with

$$f_\alpha^2(z, \tau) = \frac{1}{z^2} + \left(\frac{\theta_\alpha''}{\theta_\alpha} - \frac{1}{3}\frac{\theta_1'''}{\theta_1'} \right) + \cdots$$

where the values of the derivatives are taken at the origin, and thus we see that

$$c_{\alpha\beta} = \frac{\theta_\alpha''}{\theta_\alpha} - \frac{\theta_\beta''}{\theta_\beta}. \tag{18}$$

Now, the equations (8″) imply that

$$\begin{cases} \dfrac{\theta_3^2(x)}{\theta_1^2(x)}\theta_2^2(0) - \dfrac{\theta_2^2(x)}{\theta_1^2(x)}\theta_3^2(0) = \theta_4^2(0) \\[2ex] \dfrac{\theta_4^2(x)}{\theta_1^2(x)}\theta_2^2(0) - \dfrac{\theta_2^2(x)}{\theta_1^2(x)}\theta_4^2(0) = \theta_3^2(0) \\[2ex] \dfrac{\theta_4^2(x)}{\theta_1^2(x)}\theta_3^2(0) - \dfrac{\theta_3^2(x)}{\theta_1^2(x)}\theta_4^2(0) = \theta_2^2(0), \end{cases}$$

and taking (10) into account, we see that

$$c_{3,2} = \pi^2\theta_4^4(0), \qquad c_{4,2} = \pi^2\theta_3^4(0), \qquad c_{4,3} = \pi^2\theta_2^4(0),$$

i.e.

$$\begin{cases} \pi^2\theta_3^4(0) = \dfrac{\theta_4''}{\theta_4} - \dfrac{\theta_2''}{\theta_2} \\[2ex] \pi^2\theta_2^4(0) = \dfrac{\theta_4''}{\theta_4} - \dfrac{\theta_3''}{\theta_3} \\[2ex] \pi^2\theta_4^4(0) = \dfrac{\theta_3''}{\theta_3} - \dfrac{\theta_2''}{\theta_2}. \end{cases} \tag{19}$$

Remark 5.2.5 Adding the last two equations, we recover (8‴)!

Plugging the heat equation (16) into equations (19) gives

$$\begin{cases} \theta_3^4(0) = \dfrac{4i}{\pi}\dfrac{\partial}{\partial\tau}\log\dfrac{\theta_4}{\theta_2}(0) \\[2ex] \theta_2^4(0) = \dfrac{4i}{\pi}\dfrac{\partial}{\partial\tau}\log\dfrac{\theta_4}{\theta_3}(0) \\[2ex] \theta_4^4(0) = \dfrac{4i}{\pi}\dfrac{\partial}{\partial\tau}\log\dfrac{\theta_3}{\theta_2}(0). \end{cases} \tag{20}$$

Now, $\partial/\partial\tau = \partial/\partial q \cdot dq/d\tau = \pi i q(\partial/\partial q)$, so

$$\begin{cases} \theta_3^4(0) = -4q\dfrac{\partial}{\partial q}\log\dfrac{\theta_4}{\theta_2}(0) \\[2ex] \theta_2^4(0) = -4q\dfrac{\partial}{\partial q}\log\dfrac{\theta_4}{\theta_3}(0) \\[2ex] \theta_4^4(0) = -4q\dfrac{\partial}{\partial q}\log\dfrac{\theta_3}{\theta_2}(0). \end{cases} \tag{21}$$

Using the infinite products of the preceding part, we obtain from the first relation of (21) that

$$\theta_3^4 = \left(\sum_{-\infty}^{\infty} q^{n^2}\right)^4 = 4q\left(\frac{1}{4q} - 8\sum_{m=1}^{\infty}\frac{mq^{4m-1}}{1-q^{4m}} + 2\sum_{m=1}^{\infty}\frac{mq^{m-1}}{1-q^m}\right),$$

which gives the identity (5) of Section 5.1 (see Exercise 5.8).

(f) The Jacobi Transform

This transform establishes a fundamental link between $\theta_\alpha(z/\tau, -1/\tau)$ and $\theta_\alpha(z,\tau)$ for $\alpha = 1$ and 3. That such a link can exist is not too difficult to see, if we consider the quotient

$$H_1(z) = \frac{\theta_1(z/\tau, -1/\tau)}{\theta_1(z,\tau)}.$$

Indeed, this quotient has multipliers for the lattice Λ, as follows:

z	$z+1$	$z+\tau$
$H_1(z)$	$\nu H_1(z)$	$\mu H_1(z)$

with $\mu = qe^{2i\pi z}$, $\nu = e^{\pi i/\tau + 2i\pi z/\tau}$. Since the zeros of $\theta_1(z,\tau)$ and of $\theta_1(z/\tau, -1/\tau)$ belong to the lattice Λ and are simple, we see that H_1 has no poles (or zeros) in $\mathbb{C}$, and one can easily check that $H^*(z) := e^{-i\pi z^2/\tau}H_1(z)$ is an **entire** elliptic function of lattice Λ, so a constant C by Liouville's theorem 2.4.3 of Chapter 2.

Thus, we have

$$\begin{cases} \theta_1\left(\dfrac{z}{\tau}, -\dfrac{1}{\tau}\right) = Ce^{(i\pi z^2)/\tau}\theta_1(z,\tau) \\ \theta_2\left(\dfrac{z}{\tau}, -\dfrac{1}{\tau}\right) = iCe^{(i\pi z^2)/2}\theta_4(z,\tau) \\ \theta_3\left(\dfrac{z}{\tau}, -\dfrac{1}{\tau}\right) = iCe^{(i\pi z^2)/2}\theta_3(z,\tau) \\ \theta_4\left(\dfrac{z}{\tau}, -\dfrac{1}{\tau}\right) = iCe^{(i\pi z^2)/2}\theta_2(z,\tau). \end{cases} \tag{22}$$

We will use relation (10) for $\theta(z, -1/\tau)$:

$$\frac{\partial\theta_1}{\partial z}\left(\frac{z}{\tau}, -\frac{1}{\tau}\right)\Big|_{z=0} = \frac{1}{\tau}\theta_1'\left(0, -\frac{1}{\tau}\right),$$

hence

$$\frac{1}{\tau}\theta_1'\left(0,-\frac{1}{\tau}\right) = \frac{\pi}{\tau}\theta_2\left(0,-\frac{1}{\tau}\right)\theta_3\left(0,-\frac{1}{\tau}\right)\theta_4\left(0,-\frac{1}{\tau}\right),$$

and by (22),

$$C\theta_1'(0,\tau) = -\frac{i\pi}{\tau}C^3\theta_4(0,\tau)\theta_3(0,\tau)\theta_2(0,\tau).$$

Still using (10), but for $\theta(z,\tau)$, we have

$$C^2 = i\tau. \tag{23}$$

It remains only to give the determination of the square root of $i\tau$.

For this, we assume that $\tau \in i\mathbb{R}_+$ and we set $\tau = it$ with $t > 0$.

The third equation of (22) for $z = 0$ gives

$$iC(it) = \frac{\theta_3(0, i/t)}{\theta_3(0, it)} = \frac{\sum_{-\infty}^{\infty} e^{-\pi n^2/t}}{\sum_{-\infty}^{\infty} e^{-\pi n^2 t}} \in \mathbb{R}_+,$$

so we see that $C(it)$ is equal to $-i\sqrt{t}$.

Let $\sqrt{\tau/i}$ denote the determination of the square root which is positive on $i\mathbb{R}_+$. We have

$$\begin{cases} \theta_1\Big(\frac{z}{\tau}, -\frac{1}{\tau}\Big) = -i\sqrt{\frac{\tau}{i}}e^{(\pi i z^2)/t}\theta_1(z,\tau) \\ \theta_2\Big(\frac{z}{\tau}, -\frac{1}{\tau}\Big) = \sqrt{\frac{\tau}{i}}e^{(\pi i z^2)/\tau}\theta_4(z,\tau) \\ \theta_3\Big(\frac{z}{\tau}, -\frac{1}{\tau}\Big) = \sqrt{\frac{\tau}{i}}e^{(\pi i z^2)/\tau}\theta_3(z,\tau) \\ \theta_4\Big(\frac{z}{\tau}, -\frac{1}{\tau}\Big) = \sqrt{\frac{\tau}{i}}e^{(\pi i z^2)/\tau}\theta_2(z,\tau), \end{cases} \tag{24}$$

and in particular,

$$\begin{cases} \theta_2\Big(0, -\frac{1}{\tau}\Big) = \sqrt{\frac{\tau}{i}}\theta_4(0,\tau) \\ \theta_3\Big(0, -\frac{1}{\tau}\Big) = \sqrt{\frac{\tau}{i}}\theta_3(0,\tau) \\ \theta_4\Big(0, -\frac{1}{\tau}\Big) = \sqrt{\frac{\tau}{i}}\theta_2(0,\tau). \end{cases} \tag{25}$$

Remark 5.2.6 The Jacobi formulae (formula (24)) can also be obtained using the theory of Fourier series.

Theorem 5.2.1 *For $\tau \in \mathcal{H}$, we define the function $\varphi(\tau)$ by the relation*

$$\varphi(\tau) = \theta_2^8(0,\tau)\theta_3^8(0,\tau)\theta_4^8(0,\tau).$$

Then, we have

$$\varphi(\tau) = 2^8 q^2 \prod_{n=1}^{\infty}(1 - q^{2n})^{24} \tag{26}$$

and

$$\varphi(\tau + 1) = \varphi(\tau), \qquad \varphi\Big(-\frac{1}{\tau}\Big) = \tau^{12}\varphi(\tau).$$

Proof. By relation (10), we have

$$\varphi(\tau) = \left(\frac{\theta_1'(0)}{\pi}\right)^8.$$

By the first formula of (15), we have

$$\frac{\theta_1'(0)}{\pi} = 2q^{1/4}\prod_{n=1}^{\infty}(1-q^{2n})^3,$$

which gives equation (26).

It follows from (26) that $\varphi(\tau+1) = \varphi(\tau)$.

It follows from (25) that $\varphi(-1/\tau) = \tau^{12}\varphi(\tau)$. □

5.3 MODULAR FORMS FOR THE MODULAR GROUP $SL_2(\mathbb{Z})/\{I, -I\}$

From now on, we will make use of a new parameter q which will be the **square of that of the preceding section**:

$$q := e^{2i\pi\tau}.$$

In this way, we can write $\theta(\tau) := \theta_3(0, \tau)$ in the form

$$\theta(\tau) = \sum_{-n}^{n} q^{n^2/2}, \qquad q = e^{2i\pi\tau}. \tag{1}$$

Similarly, the function φ of theorem 5.2.1 can now be written

$$\varphi(\tau) = 2^8 q\Pi(1-q^n)^{24}. \tag{1'}$$

Thus, we have

$$\begin{cases} \varphi(\tau+1) = \varphi(\tau) \\ \varphi\left(-\dfrac{1}{\tau}\right) = \tau^{12}\varphi(\tau). \end{cases} \tag{2}$$

We now propose to establish properties analogous to (1′) and (2) for the Eisenstein series of the lattice $\Lambda = \mathbb{Z} + \mathbb{Z}\tau$.

5.3.1 Modular Properties of the Eisenstein Series

We saw that the elliptic functions of the lattice Λ are attached to the cubic

$$Y^2 = 4X^3 - g_4X - g_6, \tag{E}$$

where $g_4 = g_4(\Lambda) = 60G_4(\Lambda)$, $g_6 = g_6(\Lambda) = 140G_6(\Lambda)$ (see Chapter 2, Sections 2.6 and 2.7).

The formulae of Chapter 4, Section 4.13, give

$$\begin{cases} \Delta(\Lambda) = g_4(\Lambda)^3 - 27g_6(\Lambda)^2 \\ j(\Lambda) = 1728\dfrac{g_4(\Lambda)^3}{\Delta(\Lambda)}. \end{cases}$$

When we scale the lattice Λ by a factor of $\alpha \in \mathbb{C}^*$, we obtain the homogeneity relations

$$\begin{cases} G_{2k}(\alpha\Lambda) = \alpha^{-2k}G_{2k}(\Lambda) \\ \Delta(\alpha\Lambda) = \alpha^{-12}\Delta(\Lambda) \\ j(\alpha\Lambda) = j(\Lambda). \end{cases} \tag{3}$$

The relations (3) enable us to always use a lattice Λ of the type

$$\Lambda_\tau = \mathbb{Z} + \mathbb{Z}\tau = \mathbb{Z}\tau + \mathbb{Z},$$

with $\tau \in \mathcal{H} = \{z \in \mathbb{C}; \Im z > 0\}$, and the **true** Eisenstein series are the functions $\tau \mapsto G_{2k}(\Lambda_\tau)$.

Definition 5.3.1 *From now on, for an integer $k \geq 2$, set*

$$\begin{cases} G_{2k}(\tau) := G_{2k}(\Lambda_\tau) \\ \Delta(\tau) := \Delta(\Lambda_\tau) \\ j(\tau) := j(\Lambda_\tau). \end{cases}$$

Suppose we now make the change of basis

$$\begin{pmatrix} \omega_2 \\ \omega_1 \end{pmatrix} = \begin{pmatrix} a & b \\ c & d \end{pmatrix} \begin{pmatrix} \tau \\ 1 \end{pmatrix}$$

in Λ_τ, keeping the orientation of the basis. Then we have

$$\begin{cases} \tau' = \dfrac{\omega_2}{\omega_1} = \dfrac{a\tau + b}{c\tau + d} \\ ad - bc = 1, \end{cases}$$

and we see (by conservation of the orientation) that $\tau' \in \mathcal{H}$. Moreover, we have

$$\Lambda_\tau = \mathbb{Z}\omega_2 + \mathbb{Z}\omega_1 = \omega_1\Lambda_{\tau'},$$

and we deduce from the homogeneity formulae (3) that

$$\begin{cases} G_{2k}\left(\dfrac{a\tau + b}{c\tau + d}\right) = G_{2k}(\tau') = (c\tau + d)^{2k}G_{2k}(\tau) \\ \Delta\left(\dfrac{a\tau + b}{c\tau + d}\right) = \Delta(\tau') = (c\tau + d)^{12}\Delta(\tau) \\ j\left(\dfrac{a\tau + b}{c\tau + d}\right) = j(\tau') = j(\tau). \end{cases} \tag{4}$$

In particular, we have

$$\begin{cases} G_{2k}(\tau+1) = G_{2k}(\tau), & G_{2k}\Big(-\dfrac{1}{\tau}\Big) = \tau^{2k} G_{2k}(\tau) \\ \Delta(\tau+1) = \Delta(\tau), & \Delta\Big(-\dfrac{1}{\tau}\Big) = \tau^{12}\Delta(\tau) \\ j(\tau+1) = j(\tau), & j\Big(-\dfrac{1}{\tau}\Big) = j(\tau). \end{cases} \tag{4'}$$

Let us summarise the results above in a statement.

Proposition 5.3.1 *Let $\mathcal{H}$ denote the Poincaré upper half-plane and $SL_2(\mathbb{Z})$ the group $\{(\begin{smallmatrix} a & b \\ c & d \end{smallmatrix}) \in M_{2\times 2}(\mathbb{Z});\, ad - bc = 1\}$.*

(i) *$SL_2(\mathbb{Z})$ acts on $\mathcal{H}$ via the formula*

$$\begin{pmatrix} a & c \\ b & d \end{pmatrix} \tau = \frac{a\tau + b}{c\tau + d}.$$

(ii) *The functions G_{2k}, Δ, j are holomorphic on $\mathcal{H}$ of period 1.*

(iii) *These functions satisfy the functional equations*

$$\begin{cases} G_{2k}\Big(\dfrac{a\tau+b}{c\tau+d}\Big) = (c\tau+d)^{2k} G_{2k}(\tau) \\ \Delta\Big(\dfrac{a\tau+b}{c\tau+d}\Big) = (c\tau+d)^{12}\Delta(\tau) \\ j\Big(\dfrac{a\tau+b}{c\tau+d}\Big) = j(\tau). \end{cases}$$

Remark 5.3.1 For the proof of (ii), we refer to Chapter 2 (uniform convergence on every compact subset of $\mathcal{H}$).

To complete our study of the resemblance between the function φ and Eisenstein series, it remains to find the q-expansions of the latter.

Lemma 5.3.1 *Let f be a function defined in $\mathcal{H}$, holomorphic and periodic of period 1.*

Then there exists a unique holomorphic function g, defined in $C = \{z \in \mathbb{C};\, 0 < |z| < 1\}$, such that

$$g(e^{2i\pi\tau}) = f(\tau).$$

Proof. **(1)** Note that the map

$$\varphi \begin{cases} \mathcal{H} \longrightarrow \mathbb{C} \\ \tau \longmapsto e^{2i\pi\tau} \end{cases}$$

is holomorphic, locally invertible, and that its image $\varphi(\mathcal{H})$ is contained in C.

Thus, it suffices to complete the diagram

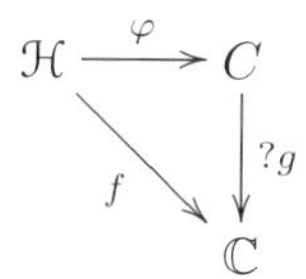

by a holomorphic function g.

(2) To construct g, we use two "charts" which cover C, and we construct local inverses of φ.

The charts are given by

$$\begin{cases} O_1 = \{z \in C; z \notin]-1, 0[\} \\ O_2 = \{z \in C; z \notin]0, 1[\}. \end{cases}$$

In O_1, we **choose** as the inverse of φ the function

$$z \overset{\psi_1}{\longmapsto} \frac{1}{2i\pi} \log_1(z),$$

with $\log_1(z) := \log|z| + i \arg_1(z)$, where we take $\arg_1(z) \in]-\pi, \pi[$.

In O_2 we **choose** as the inverse of φ the function

$$z \overset{\psi_2}{\longmapsto} \frac{1}{2i\pi} \log_2(z),$$

with $\log_2(z) := \log|z| + i \arg_2(z)$, where we take $\arg_2(z) \in [0, 2\pi[$.

Thus, we see that $\psi_1(z) = \psi_2(z)$ when $z \in \mathcal{H} \cap C$, but that $\psi_2(z) = \psi_1(z) + 1$ when $\Im z < 0$.

But since f is periodic of period 1, we have

$$f \circ \psi_1(z) = f \circ \psi_2(z)$$

for every $z \in C$, so $f \circ \psi_1 = f \circ \psi_2$ is the desired function g, holomorphic on D. □

Theorem 5.3.1

(1) *Let f be a complex-valued function defined over $\mathcal{H}$, holomorphic, periodic of period 1, and having a uniform limit a_0 when $\Im\tau$ tends to $+\infty$. Then f admits a q-expansion of the type*

$$f(\tau) = g(e^{2i\pi\tau}) = \sum_{n=0}^{\infty} a_n q^n, \qquad q = e^{2i\pi\tau}.$$

(2) *If* $k \geq 2$, *then*

$$G_{2k}(\tau) = 2\zeta(2k) + \frac{2(2i\pi)^{2k}}{(2k-1)!}\sum_{n=1}^{\infty} \sigma_{2k-1}(n)q^n,$$

where $\sigma_h(n) = \sum_{\substack{d>0 \\ d|n}} d^h$.

Proof. **(1)** We know by the lemma that

$$f(\tau) = g(e^{2i\pi\tau}),$$

where g is holomorphic in C.

Since $f(\tau)$ tends uniformly to a_0 as $\Im\tau$ tends to infinity, we see that $g(q)$ tends to a_0 as $|q|$ tends to zero. The function g is thus holomorphic in the open ball of centre 0 and radius 1.

The function g admits a Taylor expansion at the origin whose radius of convergence is greater than or equal to one. Thus,

$$f(\tau) = g(e^{2i\pi\tau}) = \sum_{n=0}^{\infty} a_n q^n.$$

(2) Starting from Euler's identity (see Chapter 1, formula (7) of Section 1.8), we have

$$\pi \operatorname{cotg} \pi\tau = \frac{1}{\tau} + \sum_{m=1}^{\infty}\left(\frac{1}{\tau+m} + \frac{1}{\tau-m}\right),$$

where the right-hand series converges uniformly on every compact subset of $\mathbb{C}$.

If $q = e^{2i\pi\tau}$ with $\tau \in \mathcal{H}$, then

$$\pi \operatorname{cotg} \pi\tau = i\pi\frac{q+1}{q-1} = i\pi - \frac{2i\pi}{1-q} = i\pi - 2i\pi\sum_{d=0}^{\infty} q^d,$$

so

$$\frac{1}{\tau} + \sum_{m=1}^{\infty}\left(\frac{1}{\tau+m} + \frac{1}{\tau-m}\right) = i\pi - 2i\pi\sum_{d=0}^{\infty} q^d.$$

Differentiating this equation $(2k-1)$ times, we have

$$\sum_{m=-\infty}^{\infty} \frac{1}{(\tau+m)^{2k}} = \frac{1}{(2k-1)!}(2i\pi)^{2k}\sum_{d=1}^{\infty} d^{2k-1}q^d. \tag{5}$$

Furthermore, we have

$$\begin{aligned} G_{2k}(\tau) &= \sum_{(\ell,m)\neq(0,0)} \frac{1}{(\ell\tau+m)^{2k}} \\ &= \sum_{m\neq 0} \frac{1}{m^{2k}} + \sum_{\ell\neq 0}\sum_{m=-\infty}^{\infty} \frac{1}{(\ell\tau+m)^{2k}} \\ &= 2\zeta(2k) + 2\sum_{\ell=1}^{\infty}\sum_{m=-\infty}^{\infty} \frac{1}{(\ell\tau+m)^{2k}}. \end{aligned}$$

Replacing τ by $\ell\tau$ in (5), we easily obtain

$$G_{2k}(\tau) = 2\zeta(2k) + \frac{2(2i\pi)^{2k}}{(2k-1)!}\sum_{d=1}^{\infty}\sum_{\ell=1}^{\infty} d^{2k-1}q^{d\ell},$$

and setting $d\ell = n$, we obtain

$$G_{2k}(\tau) = 2\zeta(2k) + \frac{2(2i\pi)^{2k}}{(2k-1)!}\sum_{n=1}^{\infty}\sigma_{2k-1}(n)q^n.$$

□

Remark 5.3.2 **(1)** If we define $E_{2k}(\tau)$ by

$$G_{2k}(\tau) = \frac{2(2i\pi)^{2k}}{(2k-1)!}E_{2k}(\tau), \tag{6}$$

then, using the expression of $\zeta(2k)$ in terms of π^{2k} and the Bernoulli numbers (cf. Chapter 1, formula (8) of Section 1.8.4), we easily see that $E_{2k}(\tau) \in \mathbb{Q}[[q]]$.

Here are the series expansions of the first Eisenstein forms.

$$E_4(\tau) = \frac{1}{240} + q + 9q^2 + 28q^3 + 73q^4 + 126q^5 + 252q^6 + \cdots$$

$$E_6(\tau) = -\frac{1}{504} + q + 33q^2 + 244q^3 + 1057q^4 + \cdots$$

$$E_8(\tau) = \frac{1}{480} + q + 129q^2 + 2188q^3 + \cdots$$

$$E_{10}(\tau) = -\frac{1}{264} + q + 513q^2 + \cdots$$

$$E_{12}(\tau) = \frac{691}{65520} + q + 2049q^2 + \cdots$$

$$E_{14}(\tau) = -\frac{1}{24} + q + 8193q^2 + \cdots$$

The constant term of E_k is equal to $-B_k/2k$.

(2) Recall that in Chapter 2, formula (27), we gave a proof of *Jacobi's formula*

$$\Delta(\tau) = (2\pi)^{12} q \prod (1 - q^n)^{24}. \tag{8}$$

Definition 5.3.2 *If we set*

$$\sum_{n=1}^{\infty} \tau(n) q^n := q \prod_{n=1}^{\infty} (1 - q^n)^{24},$$

the function $n \mapsto \tau(n)$ is called the **Ramanujan function**.

A computation shows that

n	1	2	3	4	5	6	7	8
$\tau(n)$	1	−24	252	−1472	4830	−6048	−16744	84880

We can show that $\tau(n) = O(n^6)$ and

$$\begin{cases} \tau(mn) = \tau(m)\tau(n) & \text{if } (m, n) = 1 \\ \tau(p^{k+1}) = \tau(p)\tau(p^k) - p^{11}\tau(p^{k-1}) & \text{for } p \text{ prime and } k \geq 1. \end{cases}$$

We express the property of $\tau(mn)$ by saying that the **function** τ is **multiplicative**.

The $\tau(n)$ satisfy remarkable congruences modulo 2^{12}, 3^6, 5^3, 7, 23 and 691. For example, we have

$$\begin{cases} \tau(n) \equiv n^2 \sigma_7(n) \bmod 3^3 \\ \tau(n) \equiv n\ \sigma_3(n) \bmod 7 \\ \tau(n) \equiv \quad \sigma_{11}(n) \bmod 691. \end{cases}$$

The Ramanujan–Petersson conjecture states that

$$|\tau(p)| < 2p^{11/2}.$$

This conjecture was proved by Deligne in 1974.

5.3.2 The Modular Group

We saw earlier that $SL_2(\mathbb{Z})$ acts on the Poincaré upper half-plane $\mathcal{H}$, but this action is not "faithful" since the matrix $\left(\begin{smallmatrix} -1 & 0 \\ 0 & -1 \end{smallmatrix}\right)$ acts trivially. One can check (exercise) that conversely, if $\left(\begin{smallmatrix} a & b \\ c & d \end{smallmatrix}\right)$ acts trivially on $\mathcal{H}$, then it belongs to the subgroup $\{I, -I\}$ of $SL_2(\mathbb{Z})$.

Definition 5.3.3 *The group $G = SL_2(\mathbb{Z})/\{I, -I\}$ is called the* **modular group**.

Remark 5.3.3

(1) The modular group G can be seen as the group of homographies $\tau \mapsto (a\tau + b)/(c\tau + d)$, with $\left(\begin{smallmatrix} a & b \\ c & d \end{smallmatrix}\right) \in SL_2(\mathbb{Z})$. The group $SL_2(\mathbb{Z})$ is sometimes called the **homogeneous modular group.** We use Greek letters to denote the homogeneous groups (e.g. $\Gamma = SL_2(\mathbb{Z})$), and Latin letters to denote the non-homogeneous groups (e.g. $G = SL_2(\mathbb{Z})/\{I, -I\}$).

(2) One can show ([J, S]) that the group $PSL_2(\mathbb{R})/\{I, -I\}$, which acts faithfully and transitively on $\mathcal{H}$, is the group of analytic automorphisms of $\mathcal{H}$. Thus the group G is a subgroup of $PSL_2(\mathbb{R})$.

Definition 5.3.4 *Let H be a subgroup of G, and let k be an integer.*
A **weakly modular form** *of weight k for the group Γ is a meromorphic function f defined over $\mathcal{H}$ which satisfies the functional equation:*

$$f\left(\frac{a\tau + b}{c\tau + d}\right) = (c\tau + d)^k f(\tau)$$

for every automorphism $\tau \mapsto (a\tau + b)/(c\tau + d)$ of $\mathcal{H}$ belonging to H. A **function** *is said to be* **weakly modular** *if it is a weakly modular form of weight zero.*

Example 5.3.1

(1) If k is odd, then $f = 0$.

(2) The formulae (4) show that G_{2k} is a weakly modular form of weight $2k$ for G. Similarly, Δ and j are weakly modular forms of respective weights 12 and 0 for G.

(3) The formulae (2) show that φ is a weakly modular form of weight 12 for G. Similarly, we see that θ^8 is a weakly modular form for the group H generated by $\tau \mapsto \tau + 2$ and $\tau \mapsto -1/\tau$ in G (see formulae (24) of Section 5.2).

Thus, we see that if f is weakly modular for G, it is periodic of period 1; thus it is given by its restriction to a vertical band of width 1. In fact, we can do much better.

Definition 5.3.5 *A* **fundamental domain** *of $\mathcal{H}$ for the group H is an open subset D of $\mathcal{H}$ which meets every orbit of H at exactly one point and whose closure $\bar{D}$ contains at least one point of each orbit.*

The first condition means that if τ_1 and τ_2 are two points contained in D which are **equivalent** under the action of H, then $\tau_1 = \tau_2$.

Lemma 5.3.2 *Let $\tau = x + iy \in \mathcal{H}$, with x and $y \in \mathbb{R}$ and $y > 0$. In the orbit of τ under the action of H, there are only a finite number of points "above" τ.*

Proof. Let $\tau' = x' + iy' = (a\tau + b)/(c\tau + d)$ be a point of the orbit of τ. An easy computation gives

$$y' = \frac{y}{|c\tau + d|^2}.$$

Thus, the condition $y' \geq y$ is equivalent to

$$|cz + d|^2 \leq 1,$$

so $c^2y^2 \leq |cz + d|^2 \leq 1$, which gives a finite number of $c \in \mathbb{Z}$ such that $|c| \leq 1/y$. Then the condition $(cx + d)^2 + cy^2 \leq 1$ gives a finite number of d. □

Theorem 5.3.2

(1) *The subset D of $\mathcal{H}$ defined by*

$$D = \{\tau \in \mathcal{H};\ |Re\,\tau| < \tfrac{1}{2},\ |\tau| > 1\}$$

is a fundamental domain for G.

(2) *Two distinct elements of $\bar{D}$ cannot be equivalent unless they both lie on the boundary of D.*

Here is a representation of D; it is in the interior of the "triangle" marked D.

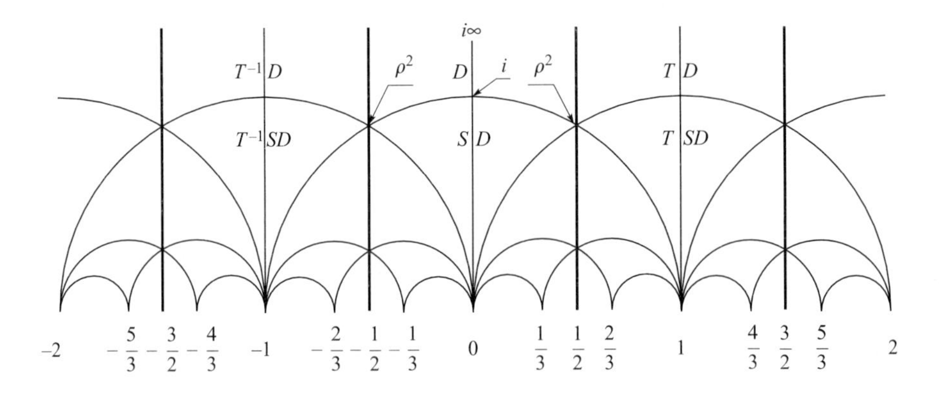

Proof. **(1)** We will begin by showing that D is the same set as

$$D_1 = \left\{\tau \in \mathcal{H};\ |Re\tau| < \tfrac{1}{2} \text{ and } |c\tau + d| > 1 \text{ if } (c, d) \neq (0, 0) \text{ and } (c, d) \neq (0, 1)\right\}.$$

Clearly, $D_1 \subset D$ (take $(c, d) = (1, 0)$).

Conversely, if $\tau \in D$, then

$$\begin{aligned}|c\tau + d|^2 &= (cx + d)^2 + c^2y^2\\ &= c^2(x^2 + y^2) + 2cdx + d^2 > c^2 - |cd| + d^2 \geq 1.\end{aligned}$$

(2) Set $S := \left(\begin{smallmatrix}0 & -1\\ 1 & 0\end{smallmatrix}\right)$. By lemma 5.3.2, there exists $\tau \in \mathcal{H}$ having ordinate maximal in its orbit; we can translate it into $\{\tau \in \mathcal{H},\ |Re(\tau)| \leq \frac{1}{2}\}$ and we must have $|\tau| \geq 1$, otherwise $\Im(S(\tau)) > \Im\tau$! Thus $\tau \in \bar{D}$.

(3) Let τ and $\tau' \in \bar{D}$ be such that $\tau' = (a\tau + b)/(c\tau + d)$. Since τ and $\tau' \in \bar{D}_1$, we must have $\Im\tau' \leq \Im\tau$ and $\Im\tau \leq \Im\tau'$, so $\Im\tau = \Im\tau'$. This gives

$$1 = (c\tau + d)^2 = (cx + d)^2 + c^2y^2 \geq c^2 - |cd| + d^2 \geq 1.$$

But $c^2 - |cd| + d^2 = 1$ implies that $(c, d) = (0, \pm 1)$ or $(c, d) = (\pm 1, 0)$, hence

$$\begin{pmatrix} a & b \\ c & d \end{pmatrix} = \pm \begin{pmatrix} 1 & 1 \\ 0 & 1 \end{pmatrix}^n \quad \text{or} \quad \begin{pmatrix} a & b \\ c & d \end{pmatrix} = \pm \begin{pmatrix} 0 & -1 \\ 1 & 0 \end{pmatrix}.$$

If, in the first case, $n \neq 0$, then τ and τ' must belong to the vertical boundaries of $\bar{D}$.

If $\tau' = S(\tau)$, then $|\tau'| \geq 1$, and $|\tau| \geq 1$ implies that $|\tau| = |\tau'| = 1$, so τ and τ' belong to the unit circle.

If τ or τ' is not on the boundary of $\bar{D}$, then we have $\tau' = T^{\circ}(\tau) = \tau$. □

Corollary 5.3.1 *The map $\bar{D} \to \mathcal{H}/G$ is surjective and its restriction to D is injective.*

Corollary 5.3.2 *When σ runs through G, the set of subsets $\sigma(\bar{D})$ forms a tiling of $\mathcal{H}$.*

Corollary 5.3.3 *Let H be a subgroup of finite index of G, and let $Hg_1, \ldots, Hg_n$ be a system of left cosets of G modulo H. Then $\Delta := g_1 D \cup g_2 D \cup \cdots \cup g_n D$ is a fundamental domain for H.*

Proof.

(1) Let $\tau \in \mathcal{H}$. Then there exists $\sigma \in G$ such that $\sigma(\tau) \in \bar{D}$. As $H\sigma$ is a left coset, there exists $i \in \{1, \ldots, n\}$ and $\gamma \in H$ such that $\sigma = \gamma g_i$. Thus, we have $\sigma(\tau) \in \gamma(g_i \bar{D}) \subset \gamma(\bar{\Delta})$.

(2) Now, if τ and τ' are two H-equivalent points of Δ, we can say that $\tau \in g_i D$ and $\tau' \in g_j D$, so $g_i^{-1}(\tau) = g_j^{-1}(\tau')$. Thus $i = j$ and $\tau = \tau'$. □

Theorem 5.3.3 *Let $n \in \mathbb{N}$ be such that $1 \leq n \leq 4$, and let Γ be the subgroup of $SL_2(\mathbb{R})$ generated by $\left(\begin{smallmatrix} 1 & \sqrt{n} \\ 0 & 1 \end{smallmatrix}\right)$ and $\left(\begin{smallmatrix} 0 & -1 \\ 1 & 0 \end{smallmatrix}\right)$. Then Γ consists of all the matrices of $SL_2(\mathbb{R})$ of the form*

$$\begin{pmatrix} a & b\sqrt{n} \\ c\sqrt{n} & d \end{pmatrix} \text{ or } \begin{pmatrix} a\sqrt{n} & b \\ c & d\sqrt{n} \end{pmatrix}.$$

To prove the theorem, we will need the following lemma (exercise).

Lemma 5.3.3

(1) *If $\alpha \in \mathbb{N}^*$ and if $1 \leq n \leq 3$, then we have $\mathbb{R} = \bigcup_{t \in \mathbb{Z}}]\alpha(t - 1/\sqrt{n}),\ \alpha(t + 1/\sqrt{n})[$.*

(2) *If α is odd and if $n = 4$, then $\mathbb{Z}$ is contained in this union of intervals.*

Note that this lemma is false when $n \geq 5$.

Proof. **(1)** Let Γ' be the set consisting of the matrices described in the statement. We easily see that Γ' is a subgroup of $SL_2(\mathbb{R})$ and that the generators of Γ belong to this subgroup. Thus, $\Gamma \subset \Gamma'$.

(2) We will now prove that $\Gamma' \subset \Gamma$. Since

$$\begin{pmatrix} a\sqrt{n} & b \\ c & d\sqrt{n} \end{pmatrix} = \begin{pmatrix} -b & a\sqrt{n} \\ -d\sqrt{n} & c \end{pmatrix} \begin{pmatrix} 0 & -1 \\ 1 & 0 \end{pmatrix},$$

it suffices to consider a matrix of the type $\begin{pmatrix} a & b\sqrt{n} \\ c\sqrt{n} & d \end{pmatrix}$.

We use the method of infinite descent on $|a|$. Assume that $|a|$ is minimal for the matrices of $\Gamma' \setminus \Gamma$.

We cannot have $a = 0$, otherwise $n = 1$ and $\pm\begin{pmatrix} 0 & -1 \\ 1 & d \end{pmatrix} = \pm\begin{pmatrix} 0 & -1 \\ 1 & 0 \end{pmatrix}\begin{pmatrix} 1 & d \\ 0 & 1 \end{pmatrix} \in \Gamma$.

Nor can we have $b = 0$, otherwise $a = d = \pm 1$ and

$$\pm\begin{pmatrix} 1 & 0 \\ c\sqrt{n} & 1 \end{pmatrix} = \mp\begin{pmatrix} 0 & -1 \\ 1 & 0 \end{pmatrix}\begin{pmatrix} 1 & -c\sqrt{n} \\ 0 & 1 \end{pmatrix}\begin{pmatrix} 0 & -1 \\ 1 & 0 \end{pmatrix} \in \Gamma.$$

Now, if $t \in \mathbb{Z}$, we have

$$\begin{pmatrix} a & b\sqrt{n} \\ c\sqrt{n} & d \end{pmatrix}\begin{pmatrix} 1 & t\sqrt{n} \\ 0 & 1 \end{pmatrix} = \begin{pmatrix} a & b'\sqrt{n} \\ c\sqrt{n} & d' \end{pmatrix}$$

with $b' = b + at$, $d' = d + ctn$.

We apply the lemma with $\alpha = |a|$, and we see that there exists $t \in \mathbb{Z}$ such that $|b'\sqrt{n}| < |a|$. Since $\begin{pmatrix} a & b'\sqrt{n} \\ c\sqrt{n} & d' \end{pmatrix} \notin \Gamma$, we know that $b' \neq 0$. Now, if $s \in \mathbb{Z}$, we have

$$\begin{pmatrix} 0 & -1 \\ 1 & 0 \end{pmatrix}\begin{pmatrix} a & b'\sqrt{n} \\ c\sqrt{n} & d' \end{pmatrix}\begin{pmatrix} 0 & -1 \\ 1 & 0 \end{pmatrix}\begin{pmatrix} 1 & s\sqrt{n} \\ 0 & 1 \end{pmatrix} = \begin{pmatrix} -d' & c'\sqrt{n} \\ b'\sqrt{n} & -a' \end{pmatrix}$$

with $c' = c - d's$, $a' = a - nb's$.

Taking $\alpha = |nb'|$, we see that there exists $s \in \mathbb{Z}$ such that $|a'| < |b'|\sqrt{n}$. Thus we obtain

$$\begin{pmatrix} a' & b'\sqrt{n} \\ c'\sqrt{n} & d' \end{pmatrix} = \begin{pmatrix} 0 & -1 \\ 1 & 0 \end{pmatrix}\begin{pmatrix} -d' & c'\sqrt{n} \\ b'\sqrt{n} & -a' \end{pmatrix}\begin{pmatrix} 0 & -1 \\ 1 & 0 \end{pmatrix} \in \Gamma' \setminus \Gamma,$$

and we have $|a'| < |b'|\sqrt{n} < |a|$, which contradicts the minimality of $|a|$. □

Corollary 5.3.4 *The modular group G is generated by $S := \begin{pmatrix} 0 & -1 \\ 1 & 0 \end{pmatrix}$ and $T := \begin{pmatrix} 1 & 1 \\ 0 & 1 \end{pmatrix}$.*

Proof. This is just the special case $n = 1$ of the theorem. □

We saw that the symmetries of the function θ^8 are generated by S and T^2. Let Γ_θ denote the group generated by S and T^2.

Corollary 5.3.5 *Let π_2 be the homomorphism*

$$\begin{pmatrix} a & b \\ c & d \end{pmatrix} \longmapsto \begin{pmatrix} \bar{a} & \bar{b} \\ \bar{c} & \bar{d} \end{pmatrix},$$

where $\bar{x}$ denotes the reduction of x modulo 2. Then π_2 passes to the quotient modulo $\{I, -I\}$ and defines a homomorphism

$$\varphi_2 : G \longrightarrow SL_2(\mathbb{F}_2).$$

Moreover, Γ_θ is the inverse image of the subgroup of $SL_2(\mathbb{F}_2)$ generated by $\pi_2(S) = \bar{S}$.

Proof. This is just the case $n = 4$ of theorem 5.3.3. □

The construction of corollary 5.3.5 can be generalised so as to construct subgroups, called "**congruence subgroups**" of G; these groups were considered by F. Klein starting in 1877.

Let N be an integer ≥ 1, which we call the *level* (German "Stufe") of the groups we will construct. As above, we associate to N a reduction homomorphism of the *homogeneous* modular group:

$$\begin{cases} SL_2(\mathbb{Z}) \xrightarrow{\pi_N} SL_2(\mathbb{Z}/N\mathbb{Z}) \\ \begin{pmatrix} a & b \\ c & d \end{pmatrix} \longmapsto \begin{pmatrix} \bar{a} & \bar{b} \\ \bar{c} & \bar{d} \end{pmatrix}. \end{cases}$$

Definition 5.3.6

(1) *The kernel of* π_N *is called the* **principal congruence subgroup** *of level* N*; it is written* $\Gamma(N)$*. Thus*

$$\Gamma(N) := \left\{ \begin{pmatrix} a & b \\ c & d \end{pmatrix} \in SL_2(\mathbb{Z});\ \begin{pmatrix} a & b \\ c & d \end{pmatrix} \equiv \begin{pmatrix} 1 & 0 \\ 0 & 1 \end{pmatrix} \bmod N \right\}.$$

(2) *The* **Hecke subgroups**[63] *of level* N *are given by*

$$\Gamma_0(N) := \left\{ \begin{pmatrix} a & b \\ c & d \end{pmatrix} \in SL_2(\mathbb{Z});\ c \equiv 0 \bmod N \right\},$$

$$\Gamma^0(N) := \left\{ \begin{pmatrix} a & b \\ c & d \end{pmatrix} \in SL_2(\mathbb{Z});\ b \equiv 0 \bmod N \right\}.$$

(3) *Generally speaking, a subgroup* Γ *of* $SL_2(\mathbb{Z})$ *is called a* **congruence subgroup** *if there exists* $N \geq 1$ *such that* $\Gamma \supset \Gamma(N)$.

Example 5.3.2 We easily check that if $T = \left(\begin{smallmatrix} 1 & 1 \\ 0 & 1 \end{smallmatrix}\right)$, then

$$T\Gamma_\theta T^{-1} = \Gamma^0(2),$$

which shows in particular that Γ_θ is not a normal subgroup of G.

Remark 5.3.4 Since all the Hecke groups contain $\{I, -I\}$, we allow ourselves to use the same notation for a homogeneous group and the non-homogeneous group which corresponds to it by passing to the quotient modulo $\{I, -I\}$. This is no longer possible for $\Gamma(N)$ when $N > 2$.

[63] E. Hecke (1887–1947).

Proposition 5.3.2 *Let $N \geq 1$ be an integer. Then $\Gamma(N)$ is a normal subgroup of $SL_2(\mathbb{Z})$, and the quotient group $SL_2(\mathbb{Z})/\Gamma(N)$ is isomorphic to $SL_2(\mathbb{Z}/N\mathbb{Z})$.*

Proof. It suffices to prove the surjectivity of π_N. Let $M = \left(\begin{smallmatrix} a & b \\ c & d \end{smallmatrix}\right) \in M_2(\mathbb{Z})$ be such that $\det M \equiv 1 \mod N$. We want to modify a, b, c, d in such a way that $\left(\begin{smallmatrix} a & b \\ c & d \end{smallmatrix}\right) \equiv \left(\begin{smallmatrix} a' & b' \\ c' & d' \end{smallmatrix}\right) \mod N$ and $a'd' - b'c' = 1$. We know that there exist U and $V \in SL_2(\mathbb{Z})$ such that UMV is diagonal, i.e.

$$UMV = \begin{pmatrix} a_1 & 0 \\ 0 & a_2 \end{pmatrix}.$$

Set

$$W = \begin{pmatrix} a_2 & 1 \\ a_2 - 1 & 1 \end{pmatrix}, \qquad X = \begin{pmatrix} 1 & -a_2 \\ 0 & 1 \end{pmatrix}, \qquad \tilde{M} = \begin{pmatrix} 1 & 0 \\ 1 - a_1 & 1 \end{pmatrix};$$

then we have $WUMVX \equiv \tilde{M} \pmod N$ since $a_1 a_2 \equiv 1 \mod N$. If we take $M' = (WU)^{-1}\tilde{M}(VX)^{-1}$, then $M' \equiv M \mod N$ and $M' \in SL_2(\mathbb{Z})$. □

Remark 5.3.5 We will see in Exercise 5.5 that the index of $\Gamma(N)$ in $SL_2(\mathbb{Z})$ is equal to

$$N^3 \prod_{p|N} \left(1 - \frac{1}{p^2}\right).$$

Theorem 5.3.4 *Let p be a prime.*

A system of representatives of the left cosets of $SL_2(\mathbb{Z})$ modulo $\Gamma^0(p)$ is given by $I, T, \ldots, T^{p-1}$ and S.

Proof.

(1) Clearly, the cosets $\Gamma^0(p)T^j$ are disjoint. Moreover, they are also disjoint from $\Gamma^0(p)S$ since $T^j S^{-1} = \left(\begin{smallmatrix} -j & 1 \\ -1 & 0 \end{smallmatrix}\right) \notin \Gamma^0(p)$.

(2) Let us show that every matrix $\left(\begin{smallmatrix} a & b \\ c & d \end{smallmatrix}\right) \in \Gamma(1)$ belongs to one of these classes.

(2.1) If p divides a, then $\left(\begin{smallmatrix} -b & a \\ -d & c \end{smallmatrix}\right) \in \Gamma^0(p)$ and we have

$$\begin{pmatrix} a & b \\ c & d \end{pmatrix} = \begin{pmatrix} -b & a \\ -d & c \end{pmatrix} \begin{pmatrix} 0 & -1 \\ 1 & 0 \end{pmatrix} \in \Gamma^0(p)S.$$

(2.2) If p does not divide a, we can apply Bachet–Bézout to (a, p); there exists j and $x \in \mathbb{N}$ such that $ja + xp = b$, and we can take $0 \leq j \leq p - 1$. We then have

$$\begin{pmatrix} a & b - aj \\ c & d - cj \end{pmatrix} \in \Gamma^0(p)$$

and

$$\begin{pmatrix} a & b \\ c & d \end{pmatrix} = \begin{pmatrix} a & b - aj \\ c & d - cj \end{pmatrix} \begin{pmatrix} 1 & j \\ 0 & 1 \end{pmatrix} \in \Gamma^0(p)T^j.$$

□

Corollary 5.3.6 *Let p be a prime. A system of representatives of the left cosets of $SL_2(\mathbb{Z})$ modulo $\Gamma_0(p)$ is given by*

$$S, ST, \ldots, ST^{p-1}, I = \begin{pmatrix} 1 & 0 \\ 0 & 1 \end{pmatrix}.$$

Proof. Since $\Gamma_0(n) = S\Gamma^0(n)S^{-1}$, we have

$$SL_2(\mathbb{Z}) = (S^{-1}\Gamma_0(p)S)A_1 \cup \cdots \cup (S^{-1}\Gamma_0(p)S)A_{p+1},$$

where the system $A_1, \ldots, A_{p+1}$ is the system of theorem 5.3.4. Multiplying by S on the left, we have

$$SL_2(\mathbb{Z}) = \Gamma_0(p)SA_1 \cup \cdots \cup \Gamma_0(p)SA_{p+1}.$$

□

Corollary 5.3.7 *A system of representatives of the left cosets $SL_2(\mathbb{Z})$ modulo Γ_θ is given by the three matrices $I, T^{-1}, T^{-1}S$.*

Proof. We have

$$SL_2(\mathbb{Z}) = \Gamma^0(2) \cup \Gamma^0(2)T \cup \Gamma^0(2)S,$$

and as $\Gamma^0(2) = T\Gamma_\theta T^{-1}$, we have

$$SL_2(\mathbb{Z}) = T\Gamma_\theta T^{-1} \cup T\Gamma_\theta I \cup T\Gamma_\theta(T^{-1}S).$$

Multiplying the two terms on the left by T^{-1}, we obtain the result. □

Example 5.3.3 We deduce from corollary 5.3.7 that a fundamental domain for Γ_θ is given by $D \cup T^{-1}(D) \cup T^{-1}S(D)$.

5.3.3 Definition of Modular Forms and Functions

We give the definition of modular functions only in the case where these functions are associated to **a congruence subgroup** of level N, i.e. functions which possess invariance properties relative to a subgroup H of $SL_2(\mathbb{Z})$ containing $\Gamma(N)$. In this definition, the *cusps* are the different orbits of $\mathbb{Q} \cup \{i\infty\}$ modulo H.

Definition 5.3.7 *A meromorphic function $f : \mathcal{H} \to \mathbb{C}$ is called a* **modular form** *which is* **meromorphic of weight** k **relative to** H *if the following conditions hold.*

(i) Modularity condition: *For every $\tau \in \mathcal{H}$ and every $\left(\begin{smallmatrix} a & b \\ c & d \end{smallmatrix}\right) \in H$, we have*

$$f\left(\frac{a\tau + b}{c\tau + d}\right) = (c\tau + d)^k f(\tau). \tag{8}$$

(ii) Condition of meromorphy at the cusps: *For every* $\left(\begin{smallmatrix} a & b \\ c & d \end{smallmatrix}\right) \in SL_2(\mathbb{Z})$, *the function*

$$(c\tau + d)^{-k} f\left(\frac{a\tau + b}{c\tau + d}\right)$$

admits an expansion in powers of $q^{1/N}$, *convergent in* $C = \{q \in \mathbb{C}; 0 < |q| < 1\}$, *in which there is only a finite number of terms with strictly negative exponents.*

Remark 5.3.6 We have proved the existence of an expansion in q only when $N = 1$. However, the proof of theorem 5.3.1 (and that of lemma 5.3.1) generalise easily to arbitrary N.

Definition 5.3.8

(i) *A* **modular form** *is a meromorphic modular form which is holomorphic everywhere, including at infinity.*

(ii) *A* **parabolic form**, *more usually called* a cusp form, *is a modular form which vanishes at the cusps.*

(iii) *A* **modular function** *is a meromorphic modular form of weight zero.*

Examples 5.3.4

(1) Theorem 5.3.1 shows that $G_{2k}(\tau)$ is a modular form of weight $2k$ for $SL_2(\mathbb{Z})$.

(2) Theorem 5.2.1 shows that $\varphi(\tau)$ is a cusp form of weight 12 for $SL_2(\mathbb{Z})$.

(3) Relations (1) of Section 5.3 and (24) of Section 5.2 show that $\theta^8(\tau)$ is a modular form of weight 4 for Γ_θ, a group which contains $\Gamma(2)$. □

Let H be a congruence subgroup of $SL_2(\mathbb{Z})$, and let $M_k(H)$ (resp. $S_k(H)$) denote the complex vector space of the forms (resp. the cusp forms) of weight k relative to H.

Notation 5.3.1 The following notation is often useful.

Let $f \in M_k(H)$ and $\gamma = \left(\begin{smallmatrix} a & b \\ c & d \end{smallmatrix}\right) \in GL_2(\mathbb{R})$, and set

$$f\,|_k\,\gamma(\tau) = \det \gamma^{k/2} (c\tau + d)^{-k} f(\gamma(\tau)). \tag{9}$$

The **modularity condition** on f can be written:

$$f\,|_k\,\gamma = f \quad \text{for every } \gamma \in H. \tag{8'}$$

The map $\gamma \mapsto f \mid_k \gamma$ has the following formal property:

$$f \mid_k \gamma_1\gamma_2 = (f \mid_k \gamma_1) \mid_k \gamma_2, \tag{10}$$

which can easily be checked by direct computation.

5.4 THE SPACE OF MODULAR FORMS OF WEIGHT k FOR $SL_2(\mathbb{Z})$

The quotient space $\widehat{\mathcal{H}/G}$, which is $\mathcal{H}/G$ compactified and which is equipped with an analytic structure, is a Riemann surface on which the functions are modular functions. However, we will not develop this aspect of the theory here, as it would lead us too far afield (see [Gu] and [Sh]); we restrict ourselves to giving a traditional presentation, using the theory of residues.

Let f be a meromorphic modular form on $\mathcal{H}$, not identically zero and of weight k. If τ_0 is a point of $\mathcal{H}$, the **order** of f at τ_0 is by definition the unique integer n such that $f/(\tau - \tau_0)^n$ is an invertible holomorphic function in the neighbourhood of τ_0; we denote this order by $v_{\tau_0}(f)$.

The modularity condition satisfied by f shows that if $\gamma \in G$, then

$$v_{\tau_0}(f) = v_{\gamma(\tau_0)}(f).$$

Denote this number by $v_P(f)$, where P denotes the projection of τ_0 (or $\gamma(\tau_0)$) to $\mathcal{H}/G$.

Finally, if $\tau_0 \in \mathbb{Q} \cup \{i\infty\}$, then we call $v_{\tau_0}(f)$ the degree in q of the expansion of the function g of theorem 5.2.1 Section 5.2, and theorem 5.3.1 Section 5.3.

Theorem 5.4.1 *Let f be a* **non-null meromorphic modular form of weight** k *relative to the group $SL_2(\mathbb{Z})$. Then*

$$v_\infty(f) + \frac{1}{2}v_i(f) + \frac{1}{3}v_\rho(f) + \sum_{P\in\mathcal{H}/G}{}^{*} v_P(f) = \frac{k}{12},$$

where the asterisk indicates that we limit this sum to the points of $\mathcal{H}/G$ different from i and ρ.

Remark 5.4.1

(i) In this formula, i (resp. ρ) denotes the projection of $e^{i\pi/2}$ (resp. $e^{2i\pi/3}$) to $\mathcal{H}/G$. The reason for which we divide the order of f at i (resp. ρ) by 2 (resp. 3) is that the order of the stabiliser of i (resp. ρ) in G is equal to 2 (resp. 3).

(ii) Note that $v_P(f)$ is almost always zero.

Indeed, the fundamental domain D of G is contained in $\{\tau \in \mathcal{H};\ \Im\tau \geq \sqrt{3}/2\}$. Thus, the image inverse of D in $\{q \in \mathbb{C};\ |q| < 1\}$ is contained in the compact set $K = \{q \in \mathbb{C};\ |q| \leq e^{-\sqrt{3}\pi}\}$. As $f(\tau) = g(e^{2i\pi\tau})$, where g is meromorphic in K, we see that g can have only a finite number of zeros or poles in K; a fortiori f has only a finite number of zeros or poles in D.

Proof. **(1)** Assume that all the zeros and poles of f in D lie strictly below of the line of ordinate α (we can always assume this), and let D_α denote the compact set

$$\{\tau \in D;\ \operatorname{Im}\tau \leq \alpha\}.$$

Assume furthermore (in the first part) that f has neither poles nor zeros on the boundary of D_α. We apply Cauchy's theorem, integrating df/f on this boundary:

$$\frac{1}{2i\pi}\int_{\partial D_\alpha}\frac{df}{f} = \sum_{P\in\mathcal{H}/G} v_P(f) = \sum_{P\in\mathcal{H}/G}{}^{*} v_P(f).$$

It remains only to evaluate the integral on the left.

Since f has the period 1, the integrals over AB and DE vanish.

The integral over EA can be computed simply by writing

$$f(\tau) = g(q).$$

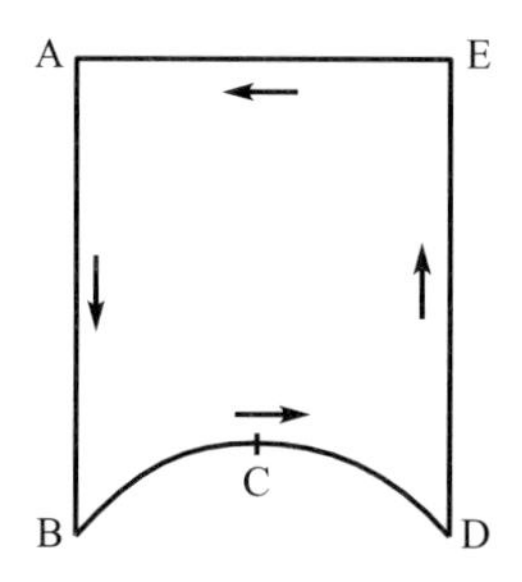

We obtain

$$\frac{1}{2i\pi}\int_E^A \frac{df}{f}(\tau) = \frac{1}{2i\pi}\int_{C^-}\frac{dg}{g}(q) = -v_\infty(f).$$

There still remains the segment BD, which we divide into BC and CD. We have

$$\int_{BD}\frac{df}{f} = \int_{BC}\frac{df}{f} + \int_{CD}\frac{df}{f}.$$

The functional equation of f gives

$$f\left(-\frac{1}{\tau}\right) = \tau^k f(\tau),$$

hence

$$\int_{CD}\frac{df}{f}(\tau) = \int_{CB}\frac{df}{f}\left(-\frac{1}{\tau}\right) = \int_{CB}\frac{d(\tau^k f(\tau))}{\tau^k f(\tau)}.$$

Finally, we have

$$\frac{1}{2i\pi}\int_{BD}\frac{df}{f} = \frac{1}{2i\pi}\int_{CB} k\frac{d\tau}{\tau} = \frac{k}{12}.$$

(2) If f has poles or zeros on AB (and DE), we deform D by making a little detour to avoid them, following a classical procedure in complex analysis.

The same holds if f has several poles or zeros on BC (and CD).

(3) If the pole or zero is at $e^{(\pi i)/2}$, $e^{(\pi i)/3}$ or $e^{(2\pi i)/3}$, we are led to the decomposition

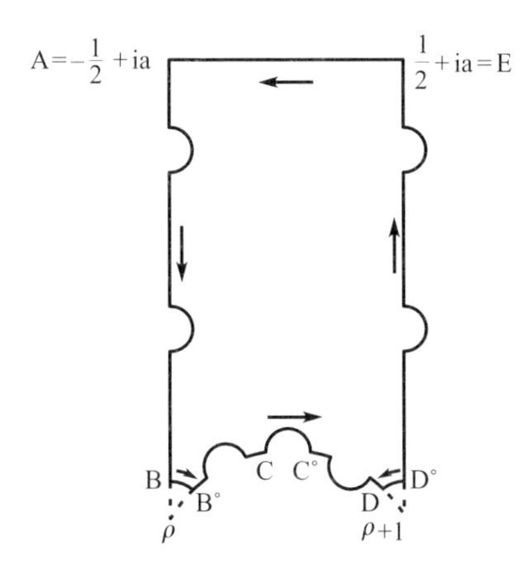

When $\overrightarrow{CC'}$ tends to the point i, we see that

$$\frac{1}{2i\pi}\int_C^{C'}\frac{df}{f}\longrightarrow\frac{1}{2i\pi}\int_C^{C'}v_i(f)\frac{d\tau}{\tau}=-\frac{1}{2}v_i(f).$$

When $\overrightarrow{BB'}$ (resp. $\overrightarrow{DD'}$) tends to the point ρ (resp. $-\overline{\rho}$), we see that

$$\frac{1}{2i\pi}\int_B^{B'}\frac{df}{f}\longrightarrow\frac{1}{2i\pi}\int_B^{B'}v_\rho(f)\frac{d\tau}{\tau}=-\frac{1}{6}v_\rho(f),$$

$$\frac{1}{2i\pi}\int_D^{D'}\frac{df}{f}\longrightarrow\frac{1}{2i\pi}\int_D^{D'}v_\rho(f)\frac{d\tau}{\tau}=-\frac{1}{6}v_\rho(f).$$

□

Corollary 5.4.1

(i) *The* **modular invariant**

$$j := 1728\frac{g_4^3}{\Delta} = 1728\frac{g_4^3}{g_4^3-27g_6^2}$$

defined in Section 5.3.1 is a modular function.

(ii) *It is holomorphic in $\mathcal{H}$ and admits a simple pole at $i\infty$, of residue equal to 1.*

(iii) *It induces a bijection $\mathcal{H}/G \twoheadrightarrow \mathbb{P}_1(\mathbb{C})=\mathbb{C}\cup\{\infty\}$.*

Proof. **(i)** This assertion comes from the fact that g_4^3 and g_6^2 are both modular forms of weight 12.

(ii) We saw in formula (8) that $\Delta(\tau)$ does not vanish when $\tau\in\mathcal{H}$.

Furthermore, we saw in Section 5.3.1 that

$$g_4 = 60G_4 = \frac{4}{3}\pi^4(1 + 240q + \cdots)$$

$$g_6 = 140G_6 = \frac{8}{27}\pi^6(1 - 504q + \cdots),$$

hence

$$\begin{aligned} g_4^3 - 27g_6^2 &= \frac{2^6}{3^3}\pi^{12}[(1 + 240q + \cdots)^3 - (1 - 504q + \cdots)^2] \\ &= \frac{2^6}{3^3}\pi^{12}(1728q + \cdots) \end{aligned}$$

which gives the result.

(iii) Let $\lambda \in \mathbb{C}$. The equation $j(\tau) = \lambda$ is equivalent to

$$f_\lambda(\tau) = 1728g_4^3(\tau) - \lambda\Delta(\tau) = (1728 - \lambda)g_4^3(\tau) + 27\lambda g_6^2(\tau) = 0.$$

Since f_λ is of weight 12, theorem 5.4.1 leads to solving the equation

$$n + \frac{n'}{2} + \frac{n''}{3} + \sum_{p \in \mathcal{H}/g} v_p(f_\lambda) = 1$$

with $n, n', n'' \geq 0$, which gives the solution

$$(n, n', n'') \in \{(1, 0, 0), (0, 2, 0), (0, 0, 3), (0, 0, 0)\}.$$

Thus f_λ of $\mathcal{H}/G$ vanishes at one and only one point of $\mathcal{H}/G$, in the sense of the analytic structure. □

Remark 5.4.2

(1) This result shows that if we want to construct a suitable analytic structure on $\widehat{\mathcal{H}/G}$ (i.e. such that j induces an isomorphism of analytic varieties), we are led to choose $(\tau - i)^2$ as a local parameter at i, and $(\tau - \rho)^3$ as a local parameter at ρ.

(2) The coefficient 1728 is chosen on purpose in order for the residue of j at infinity to be equal to 1.

Corollary 5.4.2 *$\widehat{\mathcal{H}/G}$ is a Riemann surface of genus zero.*

Proof. Indeed, it is isomorphic to $\mathbb{P}_1(\mathbb{C})$ by corollary 5.4.1. □

Now, we are ready to prove the important result which we have already used several times.

Theorem 5.4.2 *Every smooth Weierstrass cubic defined over the field of complex numbers is parametrisable by elliptic functions.*

Proof. Let $C = Y^2 = 4X^3 - AX - B$ be our cubic. Assume that $A^3 - 27B^2 \neq 0$. Then the modular invariant j_c of this cubic is not infinite, and there exists $\tau \in \mathcal{H}$ such that

$$j(\tau) = j_c.$$

Let Λ_τ be the lattice $\mathbb{Z} + \mathbb{Z}\tau$. Then the Weierstrass function $\wp$ of the lattice Λ_τ satisfies the equation

$$W: \quad \wp'(z)^2 = 4\wp(z)^3 - g_4\wp(z) - g_6.$$

Since the cubics C and W are isomorphic, there exists $\lambda \in \mathbb{C}^*$ such that $A = \lambda^4 g_4$ and $B = \lambda^6 g_6$. As

$$g_4(\lambda\Lambda_\tau) = \lambda^4 g_4(\Lambda_\tau) \quad \text{and} \quad g_6(\lambda\Lambda_\tau) = \lambda^6 g_6(\Lambda_\tau),$$

we see that if $\wp$ denotes the Weierstrass function of $\lambda^{-1}\Lambda_\tau$, then

$$\wp'(z)^2 = 4\wp^3(z) - A\wp(z) - B. \qquad \square$$

When we now consider a subgroup H of finite index μ in G, the space $\widehat{\mathcal{H}/H}$ is $\mathcal{H}/H$ compactified and equipped with an analytic structure; it is a Riemann surface, whose genus g can be computed in any of several different ways (see [Gu] Chapter I, Section 4). In particular, this genus is the dimension over $\mathbb{C}$ of the vector space $S_2(H)$ of modular cusp forms of weight 2 relative to H. When $H = \Gamma_0(N)/\{I, -I\}$, the genus g is given by the formula (see [Sh] p. 23–25)

$$g = 1 + \frac{\mu}{12} - \frac{\nu_2}{4} - \frac{\nu_3}{3} - \frac{\nu_\infty}{2}, \tag{1}$$

where

$$\begin{cases} \mu = [SL_2(\mathbb{Z}) : \Gamma_0(N)] = N\prod_{\ell|N}(1 + \ell^{-1}), \\ \nu_2 = \prod_{\substack{\ell|N \\ \ell \text{prime}}}\left(1 + \left(\frac{-1}{\ell}\right)\right) \quad \text{if 4 does not divide } N, \text{ zero otherwise}, \\ \nu_3 = \prod_{\substack{\ell|N \\ \ell \text{prime}}}\left(1 + \left(\frac{-3}{\ell}\right)\right) \quad \text{if 9 does not divide } N, \text{ zero otherwise}, \\ \nu_\infty = \sum_{\substack{d|N \\ d>0}}\varphi\left(\left(d, \frac{N}{d}\right)\right) \quad \text{where } \varphi \text{ is the Euler function.} \end{cases} \tag{2}$$

The Legendre symbols $(-1/\ell)$ and $(-3/\ell)$ can be defined as follows:

$$\left(\frac{-1}{\ell}\right) = \begin{cases} 0 & \text{if } \ell = 2 \\ 1 & \text{if } \ell \equiv 1 \bmod 4 \\ -1 & \text{if } \ell \equiv 3 \bmod 4 \end{cases} \qquad \left(\frac{-3}{\ell}\right) = \begin{cases} 0 & \text{if } \ell = 3 \\ 1 & \text{if } \ell \equiv 1 \bmod 3 \\ -1 & \text{if } \ell \equiv 2 \bmod 3. \end{cases}$$

Example 5.4.1 **(1)** If $N < 11$, we find that $g = 0$.

(2) When N is an odd prime p, the dimension of $S_2(\Gamma_0(p))$ is given by

$$\begin{cases} g = \dfrac{p+1}{12} & \text{if } p \equiv -1 \bmod 12 \\ g = \dfrac{p-5}{12} & \text{if } p \equiv 5 \bmod 12 \\ g = \dfrac{p-7}{12} & \text{if } p \equiv 7 \bmod 12 \\ g = \dfrac{p-13}{12} & \text{if } p \equiv 1 \bmod 12. \end{cases}$$

5.5 THE FIFTH OPERATION OF ARITHMETIC

In this section, we will show how the results of the preceding section provide remarkable arithmetic identities, such as for example Jacobi's identity (Section 5.1, formula (5)).

Let M_k (resp. S_k) denote the complex vector space of modular forms (resp. of cusp forms) of weight k for the group $G = PSL_2(\mathbb{Z})$. Clearly, the map $M_k \to \mathbb{C}$ which associates to a modular form its value at infinity is a linear form on M_k, whose kernel is S_k. Thus,

$$M_k = S_k \oplus \mathbb{C}G_k$$

where k is even and ≥ 4, and where G_k denotes the Eisenstein series of weight k; indeed $G_{2h}(i\infty) = 2\zeta(2h) \neq 0$ by theorem 5.3.1.

Proposition 5.5.1

(1) *M_k is zero if k is odd, negative or equal to 2.*

(2) *If $k = 0, 4, 6, 8, 10$, then M_k is a vector space of dimension 1 which has basis (respectively) $1, G_4, G_6, G_8, G_{10}$. Moreover, under these conditions, $S_k = \{0\}$.*

(3) *Multiplication by Δ defines an isomorphism $M_k \to S_{k+12}$.*

Proof. All this follows from theorem 5.4.1 and the fact that $\Delta \in S_{12}$ does not vanish on $\mathcal{H}$. □

Corollary 5.5.1 *We have*

$$\dim M_k = \begin{cases} [k/12] & \text{if } k \equiv 2 \bmod 12,\ k \geq 0 \\ [k/12] + 1 & \text{if } k \not\equiv 2 \bmod 12,\ k \geq 0. \end{cases}$$

Proof. It suffices to check these formulae for $0 \leq k < 12$, and then to see that both sides increase by 1 when k increases by 12. □

Corollary 5.5.2 *A basis of the space M_k is given by the family of monomials $G_4^{\alpha} G_6^{\beta}$, where α and β are integers ≥ 0 and $4\alpha + 6\beta = k$.*

Proof. We use induction on k, checking first that the property holds for $k \leq 6$.

If $k \geq 8$, then we see that that there exists a pair $(\alpha, \beta) \in \mathbb{N}^2$ such that $4\alpha + 6\beta = k$. As the modular form $G_4^\alpha G_6^\beta$ is non-zero at infinity, all $f \in M_k$ can be written $f = \lambda G_4^\alpha G_6^\beta + g$ with $g \in S_k$ and $\lambda \in \mathbb{C}$.

By (3) of Proposition 5.5.1, there exists $h \in M_{k-12}$ such that $g = \Delta h$. Applying the induction hypothesis to M_{k-12}, we see that the monomials $G_4^\alpha G_6^\beta$ generate M_k.

It remains to see that they are linearly independent; if we assumed the contrary, G_4^3/G_6^2 would satisfy a non-trivial algebraic equation with coefficients in $\mathbb{C}$, so it would be constant. But this is absurd. Indeed, $G_4 \in M_4$ implies $n + (n'/2) + (n''/3) = 1/3$, so $(n, n', n'') = (0, 0, 1)$, and G_4 vanishes only at ρ. Similarly, $G_6 \in M_6$ implies $n + (n'/2) + (n''/3) = 1/2$, so $(n, n', n'') = (0, 1, 0)$, and G_6 vanishes only at i. Thus we have a contradiction by considering the values of G_4^3/G_6^2 in $\mathcal{H}$. □

Some applications

(1) Since $\dim M_8 = 1$ and E_4^2 and E_8 belong to M_8, we have

$$E_8 = \lambda E_4^2 \quad \text{with } \lambda \in \mathbb{C}.$$

Considering the expansions of Section 5.3.1 (Remark 5.3.2), we obtain

$$1 = \frac{\lambda}{120}$$

for the coefficients of q, hence $\lambda = 120$.

Now considering the coefficients of q^n, we obtain the identity

$$\sigma_7(n) = \sigma_3(n) + 120 \sum_{m=1}^{n-1} \sigma_3(m)\sigma_3(n-m).$$

(2) Since $\dim M_{10} = 1$ and $E_4 E_6$ and E_{10} belong to M_{10}, we have

$$E_{10} = \lambda E_4 E_6.$$

Similarly, we obtain $\lambda = 5040/11$ and the identity

$$11\sigma_9(n) = 21\sigma_5(n) - 10\sigma_3(n) + 5040 \sum_{m=1}^{n-1} \sigma_3(m)\sigma_5(n-m).$$ □

Let us now turn our attention to the magnitude of the coefficients of a cusp form.

Theorem 5.5.1 *If f is a* **cusp form** *of weight k, then*

$$a_n = O(n^{k/2}).$$

Proof. **(1)** Since $a_0 = 0$, we have $f(\tau) = qh(\tau)$ with $|h(\tau)|$ bounded in the neighbourhood of infinity. Thus,

$$|f(\tau)| = O(q) = O(e^{-2\pi y})$$

when $q \to 0$.

(2) Now consider the function $\varphi(\tau) = |f(\tau)|y^{k/2}$. We have

$$\begin{aligned}\varphi(\gamma\tau) &= |c\tau + d|^k |f(\tau)| y'^{k/2} \\ &= |c\tau + d|^k |f(\tau)| \frac{y^{k/2}}{|c\tau + d|^k} = \varphi(\tau),\end{aligned}$$

where we set $y = \Im(\tau)$ and $y' = \Im(\gamma\tau)$.

Thus φ is invariant under G.

As φ is continuous in the fundamental domain D and tends to zero at infinity by the first part, we see that φ is bounded in D, so that

$$\varphi(\tau) = |f(\tau)|y^{k/2} \leq M$$

for a certain constant M.

Since φ is invariant under G, this inequality holds in all of $\mathcal{H}$.

By lemma 5.3.1, we have

$$f(\tau) = g(e^{2i\pi\tau}),$$

where g is holomorphic in $B = \{z \in \mathbb{C};\ |z| < 1\}$. We have

$$a_n = \frac{1}{2i\pi} \int_C \frac{g(z)}{z^{n+1}}\, dz = \int_0^1 f(x + iy) q^{-n}\, dx,$$

taking for C the circle of radius $e^{-2\pi y}$ described in a counterclockwise direction, and making the variable change $z = e^{2i\pi\tau} = q(\tau)$. Noting that $|f(\tau)| \leq My^{-k/2}$, we obtain

$$|a_n| \leq My^{-k/2} e^{2\pi ny}$$

for every $y > 0$.

Taking $y = 1/n$ (to make the exponential disappear), we obtain

$$|a_n| \leq M' n^{k/2},$$

with $M' = e^{2\pi} M$. □

Corollary 5.5.3 *If $f \in M_k \backslash S_k$, then we have*

$$|a_n| = O(n^{k-1}).$$

Proof. We can write $f = \lambda G_k + h$ with $\lambda \in \mathbb{C}^*$ and $h \in S_k$.

Since the Fourier coefficient of G_k is $O(\sigma_{k-1}(n))$, and furthermore we have

$$1 \leq \frac{\sigma_{k-1}(n)}{n^{k-1}} \leq \zeta(k-1),$$

we see that theorem 5.5.1 implies that

$$|a_n| = O(n^{k-1}).$$ □

Remark 5.5.1 Deligne proved in 1969 that for cusp forms, we have

$$|a(n)| = O(n^{(k-1)/2+\varepsilon}).$$

In particular, for the coefficients $\tau(n)$ of Δ (the Ramanujan function), we find the result

$$|\tau(n)| = O(n^{11/2+\varepsilon}).$$

But Deligne did much more since he succeeded in proving the famous *Ramanujan–Petersson conjecture*: if p is prime, then

$$|\tau(p)| \leq 2p^{11/2}.$$

5.6 THE PETERSSON HERMITIAN PRODUCT

The space of cusp forms of a given weight for a given group is equipped with a Hermitian product which is of major importance in Hecke theory. Let H be a congruence subgroup of G, and let D_H be a fundamental domain of $\mathcal{H}$ for the action of H. Recall (definition 5.3.2 of Section 5.3) that a cusp form for H is a modular form for H which vanishes at the cusps of D_H. Let $S_{2k}(H)$ be the space of cusp forms for the group H of weight $2k > 0$, and let $d\mu$ denote the invariant measure on $\mathcal{H}$ defined by

$$d\mu(\tau) := y^{-2}dx \wedge dy = \frac{y^{-2}}{-2i}(d\tau \wedge \overline{d\tau}).$$

For f and $g \in S_{2k}(H)$, we define a new measure by

$$(f, g)(\tau) := f(\tau)\overline{g(\tau)}(\Im\tau)^{2k}d\mu(\tau).$$

Proposition 5.6.1 *The exterior form (f, g) satisfies the relations*

$$(f, g) = \overline{(g, f)} \tag{1}$$

$$(f, f) \geq 0 \quad \textit{and} \quad (f, f) = 0 \ \textit{implies}\ f = 0 \tag{2}$$

$$(f, g)(\gamma\tau) = (f, g)(\tau) \quad \textit{for every}\ \gamma \in H. \tag{3}$$

Proof. (1) and (2) are obvious, so let us prove (3).

We know that if $\gamma(\tau) = (a\tau + b)/(c\tau + d)$, then

$$\Im(\gamma\tau) = \frac{\Im\tau}{|c\tau + d|^2},$$

and we easily see that

$$d\mu(\gamma\tau) = d\mu(\tau) \quad \text{(invariance of } d\mu\text{)}.$$

Furthermore, we know that if $\gamma \in H$, then

$$\begin{cases} f(\gamma\tau) = (c\tau + d)^{2k} f(\tau) \\ g(\gamma\tau) = (c\tau + d)^{2k} g(\tau), \end{cases}$$

so

$$f(\gamma\tau)\overline{g(\gamma\tau)}\Im(\gamma(\tau))^{2k} d\mu(\gamma\tau) = (f, g)(\tau). \qquad \square$$

Theorem 5.6.1 *The space $S_{2k}(H)$ of modular forms of weight $2k > 0$ for H is a Hilbert space of finite dimension for the* **Petersson Hermitian product***:*

$$(f, g) = \int_{D_H} (f, g)(\tau) = \int_{D_H} f(\tau)\overline{g(\tau)} y^{2(k-1)}\, dx \wedge dy. \tag{4}$$

Proof. We want to show that the integral converges.

(1) Since H is of finite index in G, the fundamental domain D_H can be taken to be a union of finite copies of the fundamental domain D of G which we studied earlier (at least up to a set of measure zero).

Let $T(\tau) = \tau + N$ be the "smallest" translation of H; we can also choose D_H in the vertical band $0 \leq x \leq N$. We cut D_H into two pieces in such a way that $D_H = D'_H \cup D''_H$;

$$D'_H := \{\tau \in D;\ 0 \leq x \leq N;\ y \geq 1\}$$
$$D''_H := \{\tau \in D;\ 0 \leq x \leq N;\ y \leq 1\}.$$

(2) We know (definition 5.3.7 of Section 5.3.3) that we have

$$\begin{cases} f(\tau) = \sum_{n=1}^{\infty} a_n e^{2i\pi n\tau/N} \\ g(\tau) = \sum_{n=1}^{\infty} b_n e^{2i\pi n\tau/N} \end{cases} \tag{5}$$

and that these series converge uniformly in D'_H. It follows that

$$\begin{aligned} |f(\tau)\overline{g(\tau)}y^{2(k-1)}| &\leq \sum_{m=1}^{\infty}\sum_{n=1}^{\infty} |a_m\bar{b}_n e^{2i\pi m\tau/N - 2i\pi n\bar{\tau}/N} y^{2(k-1)}| \\ &\leq \sum_{m=1}^{\infty}\sum_{n=1}^{\infty} |a_m\bar{b}_n| e^{-2\pi y(m+n)/N} y^{2(k-1)} \\ &= \sum_{\nu=2}^{\infty} c_\nu e^{-2\pi\nu y/N} y^{2(k-1)}, \end{aligned} \tag{6}$$

with $c_\nu = \sum_{m+n=\nu} |a_m\bar{b}_n|$, and this last series converges uniformly on D'_H.

It follows that

$$\left|\int_{D'} f(\tau)\overline{g(\tau)}y^{2(k-1)}\,dx\,dy\right| \leq N\int_{y=1}^{\infty} dy\Big(\sum_{\nu=2}^{\infty} c_\nu e^{-2\pi\nu y/N}y^{2(k-1)}\Big).$$

Now, if $\nu \geq 1$, we have

$$\begin{aligned}\int_{y=1}^{\infty} e^{-2\pi\nu y/N}y^{2(k-1)}\,dy &= e^{-2\pi\nu/N}\int_0^{\infty} e^{-2\pi\nu(y-1)/N}y^{2(k-1)}\,dy\\ &= e^{-2\pi\nu/N}\int_0^{\infty} e^{-2\pi(y-1)/N}y^{2(k-1)}\,dy = Ke^{-2\pi\nu/N}\end{aligned}$$

for a certain constant K.

Thus, the integral is bounded above term by term by

$$N\sum_{\nu=2}^{\infty} Kc_\nu e^{-2\pi\nu/N},$$

which is a convergent series since it is essentially the right-hand side of (6) with $y = 1$.

(3) It remains to study the convergence of the integral in D'_H.

Outside of the cusps of D''_H, there is no problem of convergence since if we remove the cusps by removing small neighbourhoods containing them, we are dealing with the integral of a continuous function in a compact set.

Because there is only a finite number of cusps, we are reduced to studying the integral in a neighbourhood V of a single cusp P. Let $S \in G$ be such that $S(P) = \infty$, and let $f_0 = f\mid_S$ and $g_0 = g\mid_S$. We have

$$\begin{cases} f(S^{-1}(\tau)) = (c\tau + d)^{2k}f_0(\tau)\\ g(S^{-1}(\tau)) = (c\tau + d)^{2k}g_0(\tau)\end{cases}$$

if $S^{-1}(\tau) = (a\tau + b)/(c\tau + d)$, hence

$$\int_V f(\tau)\overline{g(\tau)}y^{2(k-1)}\,dx \wedge dy = \int_{SV} f_0(\tau)\overline{g_0(\tau)}y^{2(k-1)}\,dx \wedge dy.$$

As f and g vanish at P, f_0 and g_0 vanish at infinity and we are reduced to the case of part (2). □

5.7 HECKE FORMS

It was noted long ago that the modular forms found in nature have a certain curious property: the Fourier coefficients $a(n)$ of their expansion in powers of q are either **multiplicative functions** or simple linear combinations of multiplicative functions.

For example the Ramanujan tau function is such that $\tau(mn) = \tau(m)\tau(n)$ if m and n are relatively prime (multiplicative function).

Consider an integer $m > 0$ and a modular form of weight k for the group $\Gamma_0(N)$, $f(\tau) = \sum_{n=0}^{\infty} a(n)q^n$. Set

$$\begin{cases} V_m f(\tau) = m^{-k/2} f|_k \begin{pmatrix} m & 0 \\ 0 & 1 \end{pmatrix} (\tau) = f(m\tau) = \sum_{n=0}^{\infty} a(n)q^{mn} \\ U_m f(\tau) = m^{-k/2-1} \sum_{j=1}^{m} f|_k \begin{pmatrix} 1 & j \\ 0 & m \end{pmatrix} (\tau) = \sum_{n=0}^{\infty} a(mn)q^n. \end{cases}$$

Then V_m and U_m are linear maps from $M_k(\Gamma_0(N))$ to $M_k(\Gamma_0(mN))$. But if m divides N, then U_m is a linear endomorphism of $M_k(\Gamma_0(N))$.

Supposing that U_m were an endomorphism of $S_{12}(SL_2(\mathbb{Z}))$ for every $m > 0$ (which is not the case!), we would have

$$U_m \Delta = \lambda_m \Delta$$

with $\lambda_m \in \mathbb{C}$.

It would follow from this that

$$\tau(m) = \lambda_m \tau(1) = \lambda_m$$

for every $m > 0$, and

$$\tau(mn) = \lambda_m \tau(n) = \tau(m)\tau(n).$$

This would prove that τ is multiplicative!

This idea can be made to work by replacing U_m and V_m by a sort of average of these operators, so as to obtain endomorphisms of the M_k and of the S_k.

5.7.1 Hecke Operators for $SL_2(\mathbb{Z})$

Let n be an integer ≥ 1.

Let $\mathbb{M}_n$ denote the set of matrices $\left(\begin{smallmatrix} a & b \\ c & d \end{smallmatrix}\right) \in \mathbb{M}_2(\mathbb{Z})$ such that $ad - bc = n$.

Clearly, $\Gamma_1 = SL_2(\mathbb{Z})$ acts on $\mathbb{M}_n$ on the left, so it is possible to decompose $\mathbb{M}_n$ into orbits; let $\Gamma_1 \backslash \mathbb{M}_n$ denote a system of representatives for these orbits. The system is obviously finite.

Then, the sum $\sum_{\mu \in \Gamma_1 \backslash \mathbb{M}_n} f|_k \mu$ does not depend on the **choice** of this system of representatives, since if $\gamma \in \Gamma_1$, then

$$f|_k \gamma\mu = (f|_k \gamma)|_k \mu = f|_k \mu,$$

because $f|_k \gamma = f$.

We now set

$$T_nf(z) = n^{k/2-1} \sum_{\mu \in \Gamma_1 \backslash \mathbb{M}_n} f \,|_k\, \mu. \tag{1}$$

Clearly, if f is holomorphic on $\mathcal{H}$, then T_nf is also holomorphic. Moreover, if $\gamma \in \Gamma_1$, then

$$\begin{aligned} T_nf \,|_k\, \gamma &= n^{k/2-1} \sum_{\mu \in \Gamma_1 \backslash \mathbb{M}_n} (f \,|_k\, \mu) \,|_k\, \gamma \\ &= n^{k/2-1} \sum_{\mu \in \Gamma_1 \backslash \mathbb{M}_n} (f \,|_k\, \mu\gamma) = T_nf, \end{aligned}$$

since the $\mu\gamma$ form another system of orbits of $\mathbb{M}_n$ modulo the left action of Γ_1. Thus, it appears that T_n, which is known as the **Hecke operator** of index n, is an endormorphism of M_k and S_k!

Theorem 5.7.1

(i) *Let n be an integer ≥ 1 and $f(\tau) = \sum_{h=0}^{\infty} a(h)q^h \in M_k$. Then we have*

$$T_nf(\tau) = \sum_{h=0}^{\infty} \left(\sum_{d|(n,h)} d^{k-1} a\left(\frac{nh}{d^2}\right)\right) q^h. \tag{2}$$

Thus, we see that M_k and S_k are stable for T_n.

(ii) *If m and n are two integers ≥ 1, then*

$$T_nT_m = \sum_{d|(n,m)} d^{k-1} T_{nm/d^2} = T_mT_n. \tag{3}$$

In particular, $T_nT_m = T_{nm}$ if n and m are relatively prime.

Proof. **(1)** Every matrix $\mu = \left(\begin{smallmatrix} a & b \\ c & d \end{smallmatrix}\right) \in \mathbb{M}_n$ can be made upper triangular (i.e. such that $c = 0$) by multiplying it on the left by $\gamma \in \Gamma_1$.

Then, multiplying it on the left by $\pm\left(\begin{smallmatrix} 1 & r \\ 0 & 1 \end{smallmatrix}\right)$ with $r \in \mathbb{Z}$, we transform it to $\pm\left(\begin{smallmatrix} a & b+dr \\ 0 & d \end{smallmatrix}\right)$. As $ad = n$, we can assume that $a > 0$ and $0 \leq b < d$. With this choice of representatives, we have

$$T_nf(\tau) = n^{k-1} \sum_{\substack{a,d>0 \\ ad=n}} \sum_{b=0}^{d-1} d^{-k} f\left(\frac{a\tau + b}{d}\right).$$

Taking into account the relation

$$\sum_{b=0}^{d-1} f\left(\frac{a\tau + b}{d}\right) = \sum_{m=0}^{\infty} da(md)q^{ma}$$

and changing the notation, we find formula (2). Note that the constant term in q is equal to $\sigma_{k-1}(n)a(0)$, so S_k is stable under T_n.

(2) Formula (3) can be deduced from formula (2) by computation. □

Special case of formula (2)
If n is a prime number p, then

$$T_p f(\tau) = \sum_{h=0}^{\infty} a(hp)q^h + p^{k-1} \sum_{h=0}^{\infty} a(h)q^{ph}, \tag{2'}$$

i.e. (over M_k)

$$T_p = U_p + p^{k-1} V_p. \tag{2''}$$

Special case of formula (3)
Knowing T_p for p prime enables us to obtain a formal expression for T_n for every n. Indeed,

(i) If p divides n exactly once, then

$$T_n = T_{n/p} T_p. \tag{3'}$$

(ii) If p^2 divides n, then

$$T_n = T_{n/p} T_p - p^{k-1} T_{n/p^2}. \tag{3''}$$

Remark 5.7.1 It can sometimes be useful to define T_0. If $h > 0$, then formula (2) gives us the coefficient of q^h; it is $\sigma_{k-1}(h)a(0)$. Thus, we can say that $T_0 f(\tau)$ should resemble $E_k(\tau)$. But in order for $T_0 f(\tau)$ to lie in M_k, we have no choice for the constant term, which must be that of $a(0)E_k(z)$, so $-(B_k/2k)a(0)$! Thus, we **set**

$$T_0 f(\tau) = a(0)E_k(\tau). \tag{1'}$$

□

We saw that if $k = 4, 6, 8, 10$ and 14 the space M_k is a line over $\mathbb{C}$; then every non-zero form of M_k is an eigenvector of every T_n, so by (2) we must have

$$a(n) = \lambda_n a(1)$$

where λ_n is the eigenvalue corresponding to T_n (i.e. $T_n f = \lambda_n f$).
If $a(1) = 0$, then $f = 0$, which is absurd. Thus, we see that $a(1) \neq 0$.

Definition 5.7.1 *A* **Hecke form** *of M_k (for $k > 0$) is an eigenfunction of all the Hecke operators T_n, such that $a(1) = 1$ (normalisation).*

The Fourier coefficients $a(n)$ of f are then eigenvalues of the T_n, and equation (3) implies that

$$a(n)a(m) = \sum_{d|(m,n)} d^{k-1} a\left(\frac{mn}{d^2}\right). \tag{3'''}$$

In particular, we see that the function $n \mapsto a(n)$ is multiplicative.

Remark 5.7.2 The same considerations apply to the space S_k when it is a line, i.e. for $k = 12, 16, 18, 20, 22$ and 26.

Lemma 5.7.1 *The Eisenstein series of even index*

$$E_{2k}(z) = -\frac{B_{2k}}{4k} + \sum_{m=1}^{\infty} \sigma_{2k-1}(m)q^m$$

are eigenvectors of all the Hecke operators.

Proof. Taking (2) into account, it suffices to satisfy ($3'''$) when n or $m \geq 0$.

This is obvious if n or m is zero.

It is also obvious if n or $m = 1$.

When $nm > 0$, we use the multiplicativity of σ_{k-1} to reduce to the case of $m = p^\mu$ and $n = p^\nu$.

We have

$$a(m)a(n) = \sigma_{k-1}(p^\mu)\sigma_{k-1}(p^\nu) = \frac{p^{(k-1)(\mu+1)} - 1}{p^{k-1} - 1} \cdot \frac{p^{(k-1)(\nu+1)} - 1}{p^{k-1} - 1}.$$

Moreover, if $\mu \leq \nu$, then

$$\sum_{d|(p^\mu, p^\nu)} d^{k-1} a\left(\frac{mn}{d^2}\right) = a(p^{\mu+\nu}) + p^{k-1}a(p^{\mu+\nu-2}) + \cdots + p^{(k-1)\mu}a(p^{\mu+\nu-2\mu}).$$

Finally, setting $p^{k-1} = r$, we easily check that

$$\frac{(r^{\mu+1} - 1)(r^{\nu+1} - 1)}{(r-1)^2} = \frac{(r^{\mu+\nu+1} - 1) + r(r^{\mu+\nu-1} - 1) + \cdots + r^\mu(r^{\nu-\mu+1} - 1)}{r - 1}.$$

□

Theorem 5.7.2 (Hecke) *For $k > 0$, the Hecke forms form a basis of M_k.*

Proof.

(1) We saw that the E_{2k} are eigenvectors of T_n. Conversely, if $a(0) \neq 0$ and if f is an eigenvector of T_0, then f is a multiple of E_{2k}.

(2) We know that $M_{2k} = \langle E_{2k} \rangle \oplus S_{2k}$, so it suffices to show that the Hecke forms form a basis of S_{2k}.

For this, we use the Petersson Hermitian product and note (Exercise 5.12) that the T_n are **self-adjoint**, i.e.

$$(T_n f, g) = (f, T_n g)$$

for all f and $g \in S_{2k}$ and for every $n > 0$. As they commute among themselves, a theorem of linear algebra asserts that they are simultaneously diagonalisable.

(3) Let us show that the coefficients $a(n)$ of the eigenvectors are **real**; indeed, we have

$$\begin{aligned} a(n)(f,f) &= (a(n)f,f) = (T_n(f),f) = (f,T_n(f)) \\ &= (f,a(n)f) = \overline{a(n)}(f,f). \end{aligned}$$

(4) Let us show that if $f \neq g$, then $(f,g) = 0$. Indeed, there exists $n \in \mathbb{N}$ such that $a(n) \neq b(n)$. But we have

$$a(n)(f,g) = (T_n f, g) = (f, T_n g) = (f, b(n)g) = \overline{b(n)}(f,g) = b(n)(f,g),$$

since $b(n) \in \mathbb{R}$. Thus $(f,g) = 0$. □

Theorem 5.7.3 *The Fourier coefficients of the Hecke forms $f \in S_k$ are real* **algebraic integers***, whose degree is* $\leq \dim S_k$.

Proof.

(1) By propositon 5.5.1 and corollary 5.5.2 of Section 5.5, the S_k are generated by forms whose Fourier coefficients are integers; thus, by (2), the $\mathbb{Z}$-module generated by these forms is stable under the T_n.

(2) Thus the matrices of the T_n in a basis of this module are matrices with integral coefficients. The eigenvalues of the T_n are thus algebraic integers, and we saw above that they are real. □

5.8 HECKE'S THEORY

Since the function $n \mapsto a(n)$ studied above has a strong tendency to be multiplicative, it is natural (at least since Euler) to associate to it the **Dirichlet series**[64] $\sum_{n=1}^{\infty} a(n)n^{-s}$.

Definition 5.8.1 *Let $f(\tau) = \sum_{m=0}^{\infty} a(m)q^m \in M_k$ be a modular form* **for** *$SL_2(\mathbb{Z})$.* The L-**function** of $f(z)$ is by definition

$$L(f,s) = \sum_{m=1}^{\infty} \frac{a(m)}{m^s}. \tag{1}$$

Because $a(n) = O(n^{k-1})$, we see that $L(f,s)$ converges in the half-plane $\mathrm{Re}(s) > k$.

Remark 5.8.1 **(1)** A Dirichlet series cannot have a term for $m = 0$, so $a(0)$ disappears in $L(f,s)$. But since $f \in M_k$, $a(0)$ is determined by the other coefficients if $k > 0$.

[64] P.G. Lejeune-Dirichlet (1805–1859).

(2) If f is a **Hecke form**, then $a(n)$ is *multiplicative* (by theorem 5.7.1 of Section 5.7), and $L(f,s)$ has an *Euler product*

$$L(f,s) = \prod_{p \text{ prime}} \left(1 + \frac{a(p)}{p^s} + \frac{a(p^2)}{p^{2s}} + \cdots\right) \tag{2}$$

But each factor of this product can be simplified using the relation

$$a(p^{r+1}) = a(p)a(p^r) - p^{k-1}a(p^{r-1}).$$

It follows that if we set

$$A_p(x) := 1 + \sum_{r=0}^{\infty} a(p^{r+1})x^{r+1},$$

then

$$A_p(x) = 1 + a(p)xA_p(x) - p^{k-1}x^2A_p(x),$$

so

$$A_p(x) = \frac{1}{1 - a(p)x + p^{k-1}x^2}; \tag{3}$$

finally, we obtain the following theorem.

Theorem 5.8.1 *If f is a* **Hecke form**, *then $L(f,s)$ is equal to the* **Euler product**

$$L(f,s) = \prod_p \frac{1}{1 - a(p)p^{-s} + p^{k-1-2s}}. \tag{4}$$

Example 5.8.1 **(1)** If $f = E_k$, then

$$a(p^r) = 1 + p^{k-1} + \cdots + p^{r(k-1)} = \frac{p^{(r+1)(k-1)} - 1}{p^{k-1} - 1}$$

and

$$A_p(x) = \frac{1}{1 - (p^{k-1}+1)x + p^{k-1}x^2} = \frac{1}{(1 - p^{k-1}x)(1-x)},$$

giving

$$L(E_k, s) = \prod_p \frac{1}{(1 - p^{k-1-s})(1 - p^{-s})} = \zeta(s-k+1)\zeta(s).$$

(2) If $f = \Delta$, then

$$L(\Delta, s) = \prod_p \frac{1}{1 - \tau(p)p^{-s} + p^{11-2s}},$$

which summarises the multiplicative properties of τ discovered by Ramanujan.

When f is modular only for the level N, we need to modify the definition of the L-functions of Hecke forms as follows.

Definition 5.8.2 *Let f be a Hecke form of weight k for $\Gamma_0(N)$, we set*

$$L(f,s) = \prod_{p|N} \frac{1}{1-a(p)p^{-s}} \prod_{p\nmid N} \frac{1}{1-a(p)p^{-s}+p^{k-1-2s}}. \tag{4'}$$

5.8.1 The Mellin Transform

The map $\varphi(t) = \sum_{n=1}^{\infty} c_n e^{-nt} \mapsto D(s) = \sum_{n=1}^{\infty} c_n/n^s$ is essentially what we call the *Mellin transform*.

Definition 5.8.3 *Let φ be a function $\mathbb{R}_+^* \to \mathbb{C}$ which is rapidly decreasing at infinity (i.e. such that $\varphi(t) = O(t^{-A})$ when $t \to \infty$, for every $A \in \mathbb{R}$) and such that $\varphi(t) = O(t^{-c})$ when $t \to 0$ for some constant c. Then $M\varphi(s) = \int_0^\infty \varphi(t)t^{s-1}\,dt$ converges absolutely and uniformly in $Re(s) \geq c+\varepsilon$ for every $\varepsilon > 0$. We call this the* **Mellin transform** *of φ.*

Example 5.8.2

(1) If $\varphi(t) = e^{-t}$, then

$$M\varphi(s) = \int_0^\infty t^{s-1}e^{-t}\,dt = \Gamma(s),$$

which is the known as **Euler's Γ function**: it interpolates $n \mapsto (n-1)!$ on $\mathbb{R}$ and can be meromorphically continued to $\mathbb{C}$ with simple poles of residue $(-1)^n/n!$ when $s = -n$, with $n \in \mathbb{N}$.

(2) If $\varphi(t) = e^{-nt}$, then

$$M\varphi(s) = \int_0^\infty t^{s-1}e^{-nt}\,dt = n^{-s}\Gamma(s).$$

(3) If $\varphi(t) = \sum_{n=1}^{\infty} c_n e^{-nt}$, then

$$M\varphi(s) = \Gamma(s)\sum_{n=1}^{\infty} \frac{c_n}{n^s}.$$

(4) If $c_n = 1$, we find the Riemann zeta function multiplied by $\Gamma(s)$.

In 1859, B. Riemann used this transform to deduce the functional equation of the function $\zeta(s)$ from that of the function $\theta(t)$. Let us present this **principle of the functional equation** which was amply used by Hecke.

Theorem 5.8.2 *If φ is sufficiently small at infinity and satisfies the equation*

$$\varphi\left(\frac{1}{t}\right) = \sum_{j=1}^{\ell} A_j t^{\lambda_j} + \varepsilon t^h \varphi(t) \quad \text{for } t > 0, \tag{5}$$

where h, A_j, λ_j are in $\mathbb{C}$, $\varepsilon = \pm 1$, then $M\varphi$ can be meromorphically continued to all of $\mathbb{C}$, and it is holomorphic everywhere except at $s = \lambda_j$ (for $j = 1, \ldots, \ell$), where it has a simple pole with residue A_j. Moreover, we have the functional equation

$$M\varphi(h - s) = \varepsilon M\varphi(s). \tag{6}$$

A proof and applications of this result can be found in Problem 1, and further interesting developments can be found in [Car].

5.8.2 Functional Equations for the Functions $L(f, s)$

Using the principle of the functional equation, we deduce a functional equation for $L(f, s)$ from the modularity of a form $f \in M_k$.

Theorem 5.8.3 *Let $f \in M_k$ be a modular form for $SL_2(\mathbb{Z})$. Then $L(f, s)$ can be meromorphically continued to the whole complex plane, and $L(f, s)$ is even holomorphic if $f \in S_k$. Otherwise $L(f, s)$ has a simple pole of residue $(2\pi i)^k a(0)/(k-1)!$ at $s = k$ and is holomorphic everywhere else.*

The meromorphic continuation of $L(f, s)$ satisfies the functional equation

$$(2\pi)^{-s}\Gamma(s)L(f, s) = (-1)^{k/2}(2\pi)^{s-k}\Gamma(k - s)L(f, k - s). \tag{7}$$

Proof. Set

$$\varphi(t) = f(it) - a(0) = \sum_{n=1}^{\infty} a(n)e^{-2\pi nt}.$$

Then φ is sufficiently small at infinity and satisfies the equation

$$\begin{aligned}\varphi\left(\frac{1}{t}\right) &= f\left(\frac{-1}{it}\right) - a(0) = (it)^k f(it) - a(0)\\ &= (-1)^{k/2}t^k\varphi(t) + (-1)^{k/2}a(0)t^k - a(0),\end{aligned}$$

and its Mellin transform is $(2\pi)^{-s}\Gamma(s)L(f, s)$. From relation (6), we obtain relation (7). Since $k > 0$ and $M\varphi$ admits a simple pole at $s = k$, $L(f, s)$ admits a simple pole at k.

For $s = 0$, $\Gamma(s)$ already has a simple pole, so $L(f, s)$ is holomorphic at $s = 0$.

The coefficient of t^k in $\varphi(1/t)$ is $(-1)^{k/2}a(0)$, so we see that the residue of $L(f, s)$ at $s = k$ is $(2\pi i)^k a(0)/\Gamma(k)$. □

Important remark: When $f \in M_k(\Gamma_0(N))$, the modified L-function of formula (4′) always converges absolutely and uniformly for $\mathrm{Re}(s) \geq k + \varepsilon$, for every $\varepsilon > 0$, and always has a meromorphic continuation to $\mathbb{C}$ with at most a simple pole at $s = k$.

However, in general, it does not have a functional equation since when we change t to $1/t, f(it)$ no longer has symmetry.

Instead of the symmetry, we have the **Fricke involution**

$$w_N : f(\tau) \longmapsto w_N f(\tau) := N^{-k/2}\tau^{-k} f\Big(\frac{-1}{N\tau}\Big),$$

which acts on $M_k(\Gamma_0(N))$ because $\left(\begin{smallmatrix} 0 & -1 \\ N & 0 \end{smallmatrix}\right)$ normalises $\Gamma_0(N)$.

Via this involution, we can decompose $M_k(\Gamma_0(N))$ into a direct sum of two eigensubspaces:

$$M_k(\Gamma_0(N)) = M_k^+(\Gamma_0(N)) \oplus M_k^-(\Gamma_0(N)).$$

If $f \in M_k^{\varepsilon}(\Gamma_0(N))$ with $\varepsilon = \pm 1$, then

$$(2\pi)^{-s} N^{s/2} \Gamma(s) L(f, s) = \varepsilon(-1)^{k/2} (2\pi)^{s-k} N^{(k-s)/2} \Gamma(k - s) L(f, k - s). \tag{7′}$$

Special case: If $N = 1$, then $w_N = id$, so $M_k^- = \{0\}$ and we recover (7).

In the general case, w_N stabilises the subspace $M_k(\Gamma_0(N))^{\text{new}}$ of **new forms** (those which do not come from a lower level dividing N, see Chapter 6, Section 6.6 and [Z] p. 262) and commutes with all the (suitably modified) Hecke operators **on this space**.

In particular, every Hecke form[65] of level N is an eigenvector of w_N, and consequently, its L-function has a functional equation. The results of Section 5.7 extend to the new forms of level $N \geq 1$.

5.9 WILES' THEOREM

Let E be an elliptic curve defined over $\mathbb{Q}$, and let

$$y^2 + a_1xy + a_3y = x^3 + a_2x^2 + a_4x + a_6$$

be a minimal Weierstrass model for E over $\mathbb{Z}$. If E has good reduction at a prime p, set

$$a_p := p + 1 - N_p,$$

where N_p denotes the number of points of the curve reduced modulo p in $\mathbb{P}_2(\mathbb{F}_p)$.

[65] A Hecke form of level N is a form of $M_k(\Gamma_0(N))^{\text{new}}$ which is an eigenvector of T_n for every n prime to N, and which is normalised.

If E has bad reduction at p, set

$$a(p) = \begin{cases} 1 & \text{if } E \text{ admits two tangents at the double rational point on } \mathbb{F}_p \\ -1 & \text{if } E \text{ admits an isolated double point in } \mathbb{F}_p. \\ 0 & \text{if } E \text{ has additive reduction.} \end{cases}$$

Recall that for $\mathrm{Re}(s) > 3/2$, we defined (Section 4.15, Chapter 4) the L-function of E by the infinite product

$$L_E(s) = \prod_{\substack{\text{bad} \\ p}} \frac{1}{1 - a(p)p^{-s}} \prod_{\substack{\text{good} \\ p}} \frac{1}{1 - a_p p^{-s} + p^{1-2s}}. \tag{1}$$

Recall also that if $\varepsilon(p) \neq 0$ for all the bad primes, the curve E is said to be semi-stable. In this case, we call the product of all the bad primes the conductor of E, and denote it by N_E.

Conjecture (Hasse) 5.9.1 *The function $L(E, s)$ can be analytic continued to all of $\mathbb{C}$. Moreover, there exists an integer N (the conductor of E) such that if we set*

$$\Lambda_E(s) = N_E^{s/2}(2\pi)^{-s}\Gamma(s)L_E(s),$$

then we have the functional equation

$$\Lambda_E(2 - s) = \pm\Lambda_E(s).$$

The following theorem shows that we can associate an elliptic curve to certain Hecke forms.

Theorem 5.9.1 (Eichler, Shimura) *Let N be an integer ≥ 1 and let $f \in S_2(\Gamma_0(N))^{\text{new}}$ be a Hecke form.*

Then there exists an elliptic curve E defined over $\mathbb{Q}$ such that the Mellin transform of f is

$$(2\pi)^{-s}\Gamma(s)L_E(s) = N^{-s/2}\Lambda_E(s). \tag{2}$$

Remark 5.9.1 This theorem is not at all obvious: one constructs E as a natural quotient of the Jacobian of the curve $X_0(N) = \widehat{\mathcal{H}/\Gamma_0(N)}$, see [Kn].

Definition 5.9.1 *An elliptic curve constructed as above is called a* **Weil curve**.

Corollary 5.9.1 *Let E be a Weil curve and let $f = \sum_{m\geq 1}^{\infty} a(n)q^n \in S_2(\Gamma_0(N))^{\text{new}}$ be the Hecke form which generates it. Then we have $f(-1/N\tau) = \varepsilon N^2 f(\tau)$, with $\varepsilon = \pm 1$.*

Moreover, E satisfies the Hasse conjecture, and the functional equation of $\Lambda_E(s)$ is given by

$$\Lambda_E(2 - s) = -\varepsilon\Lambda_E(s).$$

As long ago as 1958, Taniyama posed the converse question; here is a slightly modified version of what he asked:

"Let C be an elliptic curve defined over a field of numbers k, and let $L_C(s)$ denote the L-function from C to k, so that

$$\zeta_C(s) = \frac{\zeta_k(s)\zeta_k(1-s)}{L_C(s)}$$

is the zeta function from C to k.

If a conjecture of Hasse is true for $\zeta_C(s)$, then the Fourier series obtained from $L_C(s)$ by the inverse Mellin transform must be a modular form of weight -2 and of a special type (Hecke). If so, it is very plausible that this form is an elliptic differential for that modular function field.

The problem consists in asking whether it is possible to prove Hasse's conjecture for C going in the other direction, finding a suitable modular form from which $L_C(s)$ may be obtained".

This conjecture became much more precise following a result of Weil which is formulated in [Kn] and [Og1].

Theorem 5.9.2 (Weil) *Let N be an integer ≥ 1 and let $f(\tau) = \sum_{m\geq 1}^{\infty} a(m)q^m$ be an entire series which converges for $|q| < 1$. For every primitive Dirichlet character χ of conductor n, set*

$$\begin{cases} L(f,\chi,s) &= \sum_{n\geq 1} a(n)\chi(n)/n^s \\ \Lambda(f,\chi,s) &= |Nn^2|^{s/2}(2\pi)^{-s}\Gamma(s)L(f,\chi,s). \end{cases}$$

Assume that we have the functional equations

$$\Lambda(f,\chi,2-s) = w(\chi)\Lambda(f,\chi,s),$$

where $w(\chi)$ denotes a complex number of module 1 satisfying certain compatibility conditions. Then f is a modular form of weight 2 for $\Gamma_0(N)$.

We can now give a rather precise formulation of the Shimura–Taniyama–Weil conjecture.

Conjecture 5.9.2 *Let E be an elliptic curve defined over $\mathbb{Q}$, and let $L_E(s) = \sum_{n\geq 1} a(n)n^{-s}$ denote its L-series and $f_E(\tau) = \sum_{n\geq 1} a(n)q^n$ the inverse Mellin transform of $(2\pi)^{-s}\Gamma(s)L_E(s)$.*

Then f is in $S_2(\Gamma_0(N))$, where N denotes the conductor of E and f is a Hecke form.

Results by Shimura showed that curves with complex multiplication were Weil curves, however this gave the answer for only a very small number of curves.

On September 19, 1994, Andrew Wiles obtained a complete proof of the Taniyama–Weil conjecture for semi-stable elliptic curves defined over $\mathbb{Q}$. We will see below that that result leads to a proof of Fermat's last theorem.

Anything more than a simple statement of Wiles' result would be well beyond the scope of this book, so here it is:

Wiles' Theorem 5.9.3 *Let E be the semi-stable elliptic curve of minimal equation over $\mathbb{Z}$ given by*

$$y^2 + a_1xy + a_3y = x^3 + a_2x^2 + a_4x + a_6.$$

Let N denote the conductor of E (the product of the prime numbers for which E has bad reduction), and let $L(E, s)$ denote the L-function of E:

$$L_E(s) = \prod_{p|N} \frac{1}{1 - a(p)p^{-s}} \prod_{p\nmid N} \frac{1}{1 - a(p)p^{-s} + p^{1-2s}}$$

where $Re(s) > 3/2$ and where the $a(p)$ are given by

$$\begin{cases} 1 + a(p) = \#\{\text{solution of} X^2 + a_1X - a_2 \equiv 0 \bmod p\} & \text{if } p \mid N \\ p + 1 - a(p) = \#\widetilde{E}(\mathbb{F}_p) & \text{if } p \nmid N. \end{cases}$$

Set

$$L_E(s) = \sum_{n=1}^{\infty} \frac{a(n)}{n^s}.$$

Then if $f(\tau) = \sum_{n=1}^{\infty} a(n)e^{2i\pi n\tau}$ for $\Im(\tau) > 0$, we have

$$f\left(\frac{a\tau + b}{c\tau + d}\right) = (c\tau + d)^2 f(\tau)$$

for every $\left(\begin{smallmatrix} a & b \\ c & d \end{smallmatrix}\right) \in \Gamma_0(N)$, and for every matrix $\left(\begin{smallmatrix} a & b \\ c & d \end{smallmatrix}\right) \in SL_2(\mathbb{Z})$, the function

$$\tau \longmapsto (c\tau + d)^{-2} f\left(\frac{a\tau + b}{c\tau + d}\right)$$

admits an expansion as an integral series in the powers of $q^{1/N} = e^{2i\pi\tau/N}$.

Remark 5.9.2

(1) Given that $L_E(s)$ is defined by an Euler product, $f(\tau)$ is an eigenfunction of the Hecke operators of $\Gamma_0(N)$.

(2) Since September 19, 1994, Wiles' proof has been extended to all elliptic curves, by work of C. Breuil, B. Conrad, F. Diamond and R. Taylor [D].

Example 5.9.1 **(1)** The curve $y^2 - y = x^3 - x^2$ has conductor 11, and if we consider

$$\begin{aligned} f &= q \prod_{n \geq 1} (1 - q^n)^2 (1 - q^{11n})^2 \\ &= q - 2q^2 - q^3 + 2q^4 + q^5 + 2q^6 - 2q^7 - 2q^9 - 2q^{10} + q^{11} - 2q^{12} + \cdots \\ &= \sum a(n)q^n, \end{aligned}$$

we see that the $a(p)$ satisfy the relations of the above statement when p is prime.

(2) Recall that the dimension g of $S_2(\Gamma_0(N))$ is given by

$$g = 1 + \frac{\mu}{12} - \frac{\nu_2}{4} - \frac{\nu_3}{4} - \frac{\nu_\infty}{2},$$

so we obtain the following table.

N	1	2	3	4	5	6	7	8	9	10	11	12	13
μ	1	3	4	6	6	12	8	12	12	18	12	24	14
ν_2	1	1	0	0	2	0	0	0	0	2	0	0	2
ν_3	1	0	1	0	0	0	2	0	0	0	0	0	2
ν_∞	1	2	2	3	2	4	2	4	4	4	2	6	2
g	0	0	0	0	0	0	0	0	0	0	1	0	0

So there exists no elliptic curve of conductor 1, 2, 3, 4, 5, 6, 7, 8, 9, 10, 12, 13 which is defined over $\mathbb{Q}$.

COMMENTARY

The Introductio in Analysin Infinitorum can easily be located in an English translation [Eu], and contains (apart from reasonings in non-standard analysis) a host of identities and other fascinating results.

For more modern expositions of the theory of theta functions, see the book by Rademacher [Ra], Volume III of the *Lehrbuch der Algebra* by Weber [Web] and Igusa's book [I].

The little book *A Course in Arithmetic* by Serre contains, among other subjects, an excellent introduction to modular forms. On this topic, it is useful to note that there exists no standard vocabulary in the literature, and that, for Serre, a modular function is not necessarily of weight zero and a modular form is always holomorphic. Here, we have adopted the vocabulary of [Sh], [Ra] and [Kn], because it appears important to us that a modular function be a function on a Riemann surface. However, we wanted to avoid the introduction of this notion explicitly; to explore this aspect of the theory, the reader may consult the books by Gunning [Gu], Shimura [Sh], Reyssat [Re] and the article by Bost [Bos].

For everything concerning Hecke theory, the reader can consult the complete works of that author [Hec 2], as well as the book by Ogg [Og 1]. The Eichler–Shimura theorem, as well as a host of other subjects, are given an admirable treatment in the book by Knapp [Kn].

Naturally, the reader is invited to carefully read the foundational articles by Wiles [Wi] and [W-T], but he should not be surprised by their degree of sophistication and difficulty. Several books on this question have been published ([D-D-T] and [Co]). However, the lectures by Oesterlé and Serre ([Oe] and [Se 6]) remain valuable sources.

Finally, the works of Klein [Kl] are more historically interesting than ever, and the books by Koblitz [Ko] and Miyake [Mi] are excellent additions to the existing literature.

Exercises and Problems for Chapter 5

5.1 The goal of this exercise is to give a formal proof of **Jacobi's triple product formula**

$$1+\sum_{n=1}^{\infty}(z^n+z^{-n})x^{n^2}=\prod_{m=1}^{\infty}(1-x^{2m})(1+zx^{2m-1})(1+z^{-1}x^{2m-1}),$$

a formula whose left-hand side is often written

$$\sum_{n=-\infty}^{\infty} z^n x^{n^2}.$$

We will work in the ring $A[[x]]$ of the formal series with coefficients in $A=\mathbb{Z}[z,z^{-1}]$.

(a) For every integer $N\geq 1$, set

$$\phi_N(x;z)=\prod_{m=1}^{N}(1-x^{2m})(1+zx^{2m-1})(1+z^{-1}x^{2m-1}),$$

and let $\phi_\infty(x;z)$ denote the limit of $\phi_N(x;z)$ in $A[[x]]$ as N tends to infinity. Show that

$$\phi_N(x;z)\equiv\phi_\infty(x;z) \mod x^{2N}.$$

(b) Show by induction on N that

$$\phi_N(x;z)=C_{0,N}(x)+(z+z^{-1})C_{1,N}(x)+\cdots+(z^N+z^{-N})C_{N,N}(x),$$

where the $C_{i,N}(x)\in\mathbb{Z}[x]$ are polynomials depending on i and N. Deduce that $\phi_N(x;z)$ belongs to $\mathbb{Z}[x][z+z^{-1}]$.

(c) Show that

$$(zx+x^{2N})\phi_N(x;zx^2)=(1+zx^{2N+1})\phi_N(x;z).$$

Deduce (eliminating the index N in the notation $C_{i,N}(x)$) that for $0\leq k<N$, we have

$$x^{2k+1}(1-x^{2N-2k})C_k(x)=(1-x^{2N+2k+2})C_{k+1}(x).$$

(d) Show that $C_N(x)=x^{N^2}\prod_{m=1}^{N}(1-x^{2m})$ and deduce that

$$C_0(x)=(1-x^{2N+2})(1-x^{2N+4})\cdots(1-x^{4N}).$$

(e) To establish the triple product formula, we propose to show that for every $N\geq 1$, we have

$$\phi_N(x;z)\equiv 1+\sum_{n\geq 1}^{\infty}(z^n+z^{-n})x^{n^2} \mod n^{2N}.$$

Deduce from (c) and (d) that we have

$$\begin{cases} C_0(x)\equiv 1 \\ C_k(x)\equiv x^{1+3+\cdots+(2k-1)} \end{cases} \mod x^{2N}$$

and conclude.

5.2 We propose to deduce two classical identities involving Jacobi's triple product (Exercise 5.1).

(a) Replacing x by y^3 and z by $-y$, show that we have

$$\sum_{-\infty}^{\infty}(-1)^n y^{n(3n+1)} = \prod_{r=1}^{\infty}(1-y^{2r}).$$

(b) Replacing z by $-zx$, show that we have *Euler's formula*

$$\sum_{-\infty}^{\infty}(-1)^n z^n x^{n(n+1)} = \prod_{m=1}^{\infty}(1-x^{2m})(1-zx^{2m})(1-z^{-1}x^{2m-2}).$$

(c) Regrouping the terms which give the same exponent for x in the left-hand side and simplifying by $1-z^{-1}$, show that

$$\sum_{n=0}^{\infty}(-1)^n x^{n(n+1)} z^n (1+z^{-1}+\cdots+z^{-2n})$$
$$= \prod_{m=1}^{\infty}(1-x^{2m})(1-zx^{2m})(1-z^{-1}x^{2m}).$$

(d) Replacing z by 1, show *Jacobi's formula*:

$$\sum_{n=0}^{\infty}(-1)^n(2n+1)x^{(n(n+1))/2} = \left[\prod_{m=1}^{\infty}(1-x^m)\right]^3.$$

5.3 Using the results of Exercise 5.2, we propose to show **the first of the Ramanujan congruences**: $p(5n+4) \equiv 0 \pmod 5$.

Letting $\varphi(x)$ denote the infinite product $\prod_{m=1}^{\infty}(1-x^m)$, we have

$$\varphi(x) = \sum_{-\infty}^{\infty}(-1)^n x^{(n(3n-1))/2} \tag{1}$$

$$\varphi(x)^3 = \sum_{0}^{\infty}(-1)^n(2n+1)x^{(n(n+1))/2}, \tag{2}$$

and we propose to show that

$$\sum_{n=0}^{\infty}p(5n+4)x^n = 5\frac{\varphi(x^5)^5}{\varphi(x)^6}. \tag{3}$$

For $s \in \{0,1,2,3,4\}$, set

$$\begin{cases} G_s(x) := \displaystyle\sum_{n(3n-1)/2 \equiv s \mod 5} (-1)^n x^{n(3n-1)/2} \\ H_s(x) := \displaystyle\sum_{n(n+1)/2 \equiv s \mod 5} (-1)^n(2n+1)x^{n(n+1)/2} \\ P_s(x) := \displaystyle\sum_{n=0}^{\infty} p(5n+s)x^{5n+s}. \end{cases}$$

We have

$$\begin{cases} \varphi = G_0 + G_1 + G_2 + G_3 + G_4 \\ \varphi^3 = H_0 + H_1 + H_2 + H_3 + H_4 \\ P = P_0 + P_1 + P_2 + P_3 + P_4, \end{cases}$$

where P denotes the generating series of the function p.
Using a result of Section 5.1, show that

$$(G_0 + G_1 + G_2 + G_3 + G_4)(P_0 + P_1 + P_2 + P_3 + P_4) = 1.$$

Deduce the system of equations

$$\begin{cases} G_0P_0 + G_4P_1 + G_3P_2 + G_2P_3 + G_1P_4 = 1 \\ G_1P_0 + G_0P_1 + G_4P_2 + G_3P_3 + G_2P_4 = 0 \\ G_2P_0 + G_1P_1 + G_0P_2 + G_4P_3 + G_3P_4 = 0 \\ G_3P_0 + G_2P_1 + G_1P_2 + G_0P_3 + G_4P_4 = 0 \\ G_4P_0 + G_3P_1 + G_2P_2 + G_1P_3 + G_0P_4 = 0. \end{cases} \tag{4}$$

Show that the determinant D of the system (4), considered as a system of linear equations in P_0, P_1, P_2, P_3, P_4, is $\varphi(x)\varphi(\zeta x)\varphi(\zeta^2 x)\varphi(\zeta^3 x)\varphi(\zeta^4 x)$, where ζ denotes a primitive fifth root of unity (we will work in $(\mathbb{Z}[\zeta])[[x]]$).
Since $P_4 = D_4/D$ by Cramer's formulae, it remains to compute

$$D_4 = \begin{vmatrix} G_1 & G_0 & G_4 & G_3 \\ G_2 & G_1 & G_0 & G_4 \\ G_3 & G_2 & G_1 & G_0 \\ G_4 & G_3 & G_2 & G_1 \end{vmatrix}.$$

Prove that $G_3 = G_4 = 0 = H_2 = H_4$, by showing that these series do not contain any terms.

Deduce from (2) that we have $H_2 = 3G_0(G_0G_2 + G_1^2)$, and show that then $G_0G_2 = -G_1^2$.

Prove that $G_1(x) = -x\varphi(x^{25})$ and deduce (3) from this.

5.4 We now propose to prove the **second of the Ramanujan congruences**:

$$p(7n + 5) \equiv 0 \bmod 7.$$

(a) Let $\sum a_n x^{7n}$, $\sum b_n x^n$ and $\sum c_n x^n$ be three series in $\mathbb{F}_7[[x]]$. Assume that $a_0 = 1$ and that $(\sum a_n x^{7n})(\sum b_n x^n) = \sum c_n x^n$. Show that

$$\{b_{7n} = 0 \text{ for every } n\} \iff \{c_{7n} = 0 \text{ for every } n\}.$$

(b) Using the result (d) of Exercise 5.2, show that in $\mathbb{F}_7[[x]]$, we have

$$\prod_{m \geq 0} (1 - x^{7m}) \sum_{n=2}^{\infty} p(n-2)x^n = \sum_{\substack{h \geq 0 \\ k \geq 0}} (-1)^{h+k}(2h+1)(2k+1)x^{2+(h(h+1))/2+(k(k+1))/2}.$$

(c) Letting $\sum c_n x^n$ denote the right-hand side of this equation, show that $c_{7n} = 0$ for every n. Conclude.

5.5 Let n be an integer ≥ 1 and let π be the canonical homomorphism $\mathbb{Z} \to \bar{\mathbb{Z}}_n := \mathbb{Z}/n\mathbb{Z}$. Let $\tilde{\pi}$ denote the homomorphism $SL(2, \mathbb{Z}) \to SL(2, \bar{\mathbb{Z}}_n)$ defined by

$$\tilde{\pi}\begin{pmatrix} a & b \\ c & d \end{pmatrix} := \begin{pmatrix} \pi(a) & \pi(b) \\ \pi(c) & \pi(d) \end{pmatrix} = \begin{pmatrix} \bar{a} & \bar{b} \\ \bar{c} & \bar{d} \end{pmatrix},$$

where $\bar{x}$ denotes the element $\pi(x)$.

Recall that the kernel of $\tilde{\pi}$ is the **principal congruence subgroup** $\Gamma(n)$ of level n.

(i) Recall that $\tilde{\pi}$ is surjective. Deduce that if $\Gamma := SL(2, \mathbb{Z})$, then

$$[\Gamma : \Gamma(n)] = \# SL(2, \bar{\mathbb{Z}}_n).$$

(ii) A pair of integers (x, y) is called **primitive modulo** n if the greatest common divisor (n, x, y) is equal to 1. Let $\lambda(n)$ denote the number of pairs $(\pi(x), \pi(y))$ such that (x, y) is primitive modulo n.

Show that the function λ is multiplicative, i.e. that if n_1 and n_2 are relatively prime, then

$$\lambda(n_1 n_2) = \lambda(n_1)\lambda(n_2).$$

(iii) Show that if $\nu \geq 0$, then

$$\lambda(p^{\nu}) = p^{2\nu}\left(1 - \frac{1}{p^2}\right).$$

(iv) Show that if (c, d) is a pair of primitive integers modulo n, there exist n pairs $(\bar{a}, \bar{b}) \in \bar{\mathbb{Z}}^2$ such that $\left(\begin{smallmatrix} \bar{a} & \bar{b} \\ \bar{c} & \bar{d} \end{smallmatrix}\right) \in SL(2, \bar{\mathbb{Z}}_n)$.

(v) Deduce that

$$[\Gamma : \Gamma(n)] = n^3 \prod_{p|n}\left(1 - \frac{1}{p^2}\right).$$

5.6 Let q be a complex number such that $0 < |q| < 1$. As in Section 1.6.3 (Chapter 1), we let $r(n)$ denote the number of different ways of writing n as a sum of two squares.

(a) Show that $r(2n) = r(n)$.

(b) Show that

$$\begin{cases} \theta_3^2(q) = \sum_{n=0}^{\infty} r(n) q^n \\ \theta_4^2(q) = \sum_{n=0}^{\infty} (-1)^n r(n) q^n. \end{cases}$$

(c) Show that

$$\theta_3^2(q) + \theta_4^2(q) = 2\sum_{n=0}^{\infty} r(2n)q^{2n} = 2\theta_3^2(q^2)$$

$$\theta_3^2(q) - \theta_3^2(q^2) = \sum_{n=0}^{\infty} r(2n+1)q^{2n+1}.$$

5.7 Let $h \in]0,1[$, and let q be the unique solution in $]0,1[$ of the equation

$$k = \theta_2^2(q)\theta_3^{-2}(q).$$

(a) Show that $k' = \sqrt{1-k^2} = \theta_4^2(q)/\theta_3^2(q)$.
(b) Show that

$$\begin{cases} \dfrac{1}{2}(\theta_3^2(q) + \theta_4^2(q)) = \theta_3^2(q^2) \\ \sqrt{\theta_3^2(q)\theta_4^2(q)} = \theta_4^2(q^2). \end{cases}$$

(c) Deduce that the arithmetico-geometric mean $M(1,k')$ of 1 and k' is equal to $\theta_3^{-2}(q)$ (see Exercise 2.2 of Chapter 2).

5.8 Recall from Section 5.2 the formula

$$\left(\sum_{-\infty}^{\infty} q^{n^2}\right)^4 = 4q\left(\frac{1}{4q} - 8\sum_{m=1}^{\infty}\frac{mq^{4m-1}}{1-q^{4m}} + 2\sum_{m=1}^{\infty}\frac{mq^{m-1}}{1-q^m}\right).$$

(a) Show that

$$\left(\sum_{-\infty}^{\infty} q^{n^2}\right)^4 = 1 + 8\sum_{\substack{m\geq 1\\ m\not\equiv 0 \bmod 4}} \frac{mq^m}{1-q^m}.$$

(b) Deduce that the right-hand side is equal to

$$1 + 8\sum_{\substack{m\geq 1\\ m\not\equiv 0 \bmod 4}} \left(\sum_{k=1}^{\infty} mq^{km}\right).$$

(c) Deduce formula (5) of Section 5.1.

5.9 Set

$$Q_0 := \prod_{n=1}^{\infty}(1-q^{2n}), \qquad Q_1 := \prod_{n=1}^{\infty}(1+q^{2n}),$$

$$Q_2 := \prod_{n=1}^{\infty}(1+q^{2n-1}), \qquad Q_3 := \prod_{n=1}^{\infty}(1-q^{2n-1}).$$

(a) Show that $Q_1Q_2Q_3 = 1$.

(b) Prove the relations

$$Q_0Q_1 = Q_0(q^2), \qquad Q_0Q_3 = Q_0(q^{1/2}),$$
$$Q_2Q_3 = Q_3(q^2), \qquad Q_1Q_2 = Q_1(q^{1/2}).$$

5.10 Let q be a complex number such that $0 < |q| < 1$.

(a) Show that $\theta_2(q)$, $\theta_3(q)$ and $\theta_4(q)$ are never zero.

(b) Show that if $q \in]0, 1[$, then θ_4 is a strictly decreasing function of q and θ_2 and θ_3 are strictly increasing functions of q.

5.11 Let $n > 0$ be an integer, and let $T \in \Gamma(n) \subset SL_2(\mathbb{Z})$ be a matrix. We write

$$T = \begin{pmatrix} a & b \\ c & d \end{pmatrix}, \qquad Tz = T(z) = \frac{az+b}{cz+d}.$$

(i) Show that

$$\frac{dT(z)}{dz} = (cz+d)^{-2}.$$

(ii) Setting $J_T(z) = (cz+d)^{-2}$, show that

$$J_{TS}(z) = J_S(Tz)J_S(z),$$

whenever S is another matrix of $\Gamma(n)$.

(iii) For every $T \in \Gamma(n)$, take a function $\mu_T(z)$ which is holomorphic on $\mathcal{H}$ and never vanishes. Assume that these functions satisfy the formula

$$\mu_{ST}(z) = \mu_S(Tz)\mu_T(z).$$

We propose to construct a function $f(z)$, holomorphic on $\mathcal{H}$, such that

$$f(Tz) = \mu_T(z)f(z), \qquad \text{for every } T \in \Gamma(n). \tag{1}$$

Show that if $h(z)$ is a holomorphic function on $\mathcal{H}$ such that the series

$$f(z) = \sum_{T \in \Gamma(n)} \frac{h(Tz)}{\mu_T(z)}$$

converges normally on every compact subset of $\mathcal{H}$, then f satisfies (1).

(iv) Set $\Gamma_0 = \{T \in \Gamma(n);\ \mu_T(z) \equiv 1\}$. Show that Γ_0 is a subgroup of $\Gamma(n)$.

(v) Let $\mathcal{R}$ be a system of representatives of left cosets of $\Gamma(n)$ modulo Γ_0, and let h be a holomorphic function on $\mathcal{H}$ such that $h(Sz) \equiv h(z)$ for every $S \in \Gamma_0$. Set

$$g_{\mathcal{R}}(z) = \sum_{T \in \mathcal{R}} \frac{h(Tz)}{\mu_T(z)},$$

and assume that the series is convergent on every compact subset of $\mathcal{H}$. Show that $g_{\mathcal{R}}(z)$ does not depend on the choice of $\mathcal{R}$.

(vi) Set $g = g_{\mathcal{R}}$. Show that g satisfies condition (1).

(vii) Let $k \in \mathbb{N}^*$ and $\mu_T(z) = J_T(z)^{-k} = (cz+d)^{2k}$. Determine Γ_0 and show that we can take $h = e^{2i\pi\nu z/n}$, with arbitrary $\nu \in \mathbb{N}$.

The function g constructed in (vi) is called the *Poincaré series* of weight $2k$ and of character ν of the group $\Gamma(n)$.

(viii) By studying the behaviour of g at the cusps of $\mathcal{H}/\Gamma(n)$, show that g is a modular form of weight $2k$ for $\Gamma(n)$ (this is harder).

5.12 Let n be an integer ≥ 1.

Using the definition of the Hecke operator T_n (Section 5.7, formula (1)) and the invariance of the exterior form $d\mu$ (see Section 5.6), show that T_n is self-adjoint for the Petersson Hermitian product, which means that if f and g are in S_{2k}, then

$$(T_n f, g) = (f, T_n g).$$

Problem I

Hardy's Theorem

Preamble: *The following properties of the Γ function will be used without proof; they are not involved in the first part of the problem.*

For every complex number s, let $Re(s)$ denote its real part and $\Im(s)$ its imaginary part. For $Re(s) > 0$, set

$$\Gamma(s) = \int_0^{+\infty} e^{-t} t^{s-1}\, dt.$$

The function Γ is holomorphic in the half-plane $Re(s) > 0$. It extends to a meromorphic function on $\mathbb{C}$ whose poles are the negative integers and zero. These poles are simple, and the residue of Γ at the point $s = -p$, $(p \in \mathbb{N})$ is $(-1)^p/p!$.

If s is not a pole, we have $\Gamma(s+1) = s\Gamma(s)$, and $\Gamma(s) \neq 0$.

Let σ_1, σ_2 be real numbers such that $\sigma_1 \leq \sigma_2$, and let m be a positive integer; we have

$$\lim_{|t| \longrightarrow +\infty} |t^m \Gamma(\sigma + it)| = 0$$

uniformly, for an element σ of $[\sigma_1, \sigma_2]$.

Finally, if c and x are strictly positive real numbers, we have

$$e^{-x} = \frac{1}{2i\pi} \int_{Re(s)=c} x^{-s}\Gamma(s)\, ds,$$

where the line $Re(s) = c$ is oriented by increasing ordinates. (This convention for the orientation holds for all the analogous integrals appearing in this problem.)

If z is a non-zero complex number, let $\operatorname{Arg}(z)$ denote the unique determination of the argument of z which lies in $[-\pi, \pi[$, and set $\log(z) = \log|z| + i\operatorname{Arg}(z)$. Then, for every complex number a, we have $z^a = e^{a\log(z)}$.

Throughout this problem, $\mathcal{P}$ denotes the set of complex numbers with strictly positive imaginary part.

Let λ be a strictly positive real number; a function f defined in $\mathcal{P}$ is said to be periodic of period λ if for every $z \in \mathcal{P}$, we have $f(z+\lambda) = f(z)$.

First Part

(1) Let f be a function defined in $\mathcal{P}$, holomorphic and periodic of period λ.

(a) Prove that there exists a function g, defined and holomorphic in the open set

$$\{z;\ z \in \mathbb{C} \text{ and } 0 < |z| < 1\},$$

such that

$$g(e^{2i\pi z/\lambda}) = f(z).$$

(b) Let $z_0 = x_0 + iy_0 \in \mathcal{P}$. For $n \in \mathbb{Z}$, set

$$a_n = \frac{1}{\lambda}\int_{x_0}^{x_0+\lambda} f(t + iy_0)e^{-2i\pi n(t+iy_0)/\lambda}\, dt. \tag{1}$$

Prove that a_n is independent of z_0, and that

$$f(z) = \sum_{n=-\infty}^{+\infty} a_n e^{2i\pi nz/\lambda}, \tag{2}$$

where this series converges uniformly on every compact subset of $\mathcal{P}$. The function f is said to be holomorphic (resp. meromorphic) at infinity if the function g is holomorphic (resp. meromorphic) at zero; give the conditions on the a_n for this to hold. In what follows, we will say that the a_n are the Fourier coefficients of f.

(c) Assume that there exist two positive constants c and ρ such that for every $z = x + iy \in \mathcal{P}$, with $y \leq 1$, we have

$$|f(x + iy)| \leq cy^{-1-\rho}. \tag{3}$$

Prove that

$$\sup_{n\in\mathbb{Z}*} |a_n|\ |n|^{-\rho-1} < +\infty. \tag{4}$$

(2) **(a)** Let $\rho > 0$. Show that the sequence u defined, for $n \geq 1$, by

$$u_n = n^\rho \frac{n!}{(\rho + 1)(\rho + 2)\cdots(\rho + h)\cdots(\rho + n)},$$

is bounded. (Use the series of general term $\log(u_{n+1}/u_n)$.)

(b) Let $(a_n)_{n\geq 0}$ be a sequence of complex numbers. Assume that there exists a strictly positive real number ρ such that

$$\sup_{n\in\mathbb{N}*} |a_n|n^{-\rho} < +\infty, \tag{5}$$

and consider the map f, of $\mathcal{P}$ in $\mathbb{C}$, defined by

$$f(z) = \sum_{n=0}^{+\infty} a_n e^{2i\pi nz/\lambda}. \tag{6}$$

Show that f is holomorphic and that (3) is satisfied for a suitable value of the positive constant c.

Show that for every strictly positive real γ, we have

$$\lim_{t\longrightarrow+\infty} t^\gamma |f(it) - a_0| = 0.$$

Second Part

Let λ be a strictly positive real number and $(a_n)_{n\geq 0}$ a sequence of complex numbers. Assume that there exists $\rho > 0$ such that (5) is satisfied. Define f by (6), and for $Re(s) > \rho + 1$, set

$$\varphi(s) = \sum_{n=1}^{+\infty} a_n n^{-s}, \qquad \Phi(s) = \left(\frac{2\pi}{\lambda}\right)^{-s} \Gamma(s)\varphi(s).$$

(1) **(a)** Show that φ is holomorphic for $Re(s) > \rho + 1$.

(b) Show, carefully, that

$$\Phi(s) = \int_0^{+\infty} t^{s-1}(f(it) - a_0)\, dt, \qquad \text{for } Re(s) > \rho + 1,$$

and conversely, that for $\alpha > \rho + 1$ and $y > 0$, we have

$$f(iy) - a_0 = \frac{1}{2i\pi} \int_{Re(s)=\alpha} y^{-s}\Phi(s)\, ds.$$

(c) Show that $s^2\Phi(s)$ is bounded on every vertical strip of the half-plane $Re(s) > \rho + 1$.

(2) Let ε and k be real numbers such that $\varepsilon \in \{1, -1\}$ and $k > 0$. Assume that Φ has the following properties (A) and (B):

(A) Let Ω be the set of complex numbers not equal to 0 or k. The function Φ admits a holomorphic continuation to Ω, and this continuation, which we also call Φ, satisfies $(\forall s \in \Omega)(\Phi(s) = \varepsilon\Phi(k - s))$.

(B) The function $s \mapsto \Phi(s) + a_0(1/s + (\varepsilon/(k - s))$ extends to an entire function of s, and is bounded on every vertical strip.

(a) Let α be a real number such that $\alpha > \rho + 1$ and $\alpha > k$. Let U be the subset of $\mathbb{C}$ consisting of complex numbers s such that

$$k - \alpha \leq Re(s) \leq \alpha \quad \text{and} \quad |Im(s)| \geq 1.$$

Show that $s^2\Phi(s)$ is bounded on the boundary of U, and then that $s^2\Phi(s)$ is bounded in U.

[Use the preceding result, and consider for every $a > 0$ the function $s \mapsto e^{as^2}s^2\Phi(s)$; recall also the statement of the *Maximum Principle:* Let V be a bounded open subset of $\mathbb{C}$. Let g be a function which is defined and continuous in the closure of V and holomorphic in V. If ∂V denotes the boundary of V, then $\sup_{z\in V} |g(z)| = \sup_{z\in\partial V} |g(z)|$].

(b) For every strictly positive real y, set

$$I(y) = \int_{Re(s)=k-\alpha} y^{-s}\Phi(s)\, ds \quad \text{and} \quad J(y) = \int_{Re(s)=\alpha} y^{-s}\Phi(s)\, ds.$$

Show that $I(y) = \varepsilon y^{-k} J(1/y)$ and $J(y) - I(y) = 2i\pi a_0(\varepsilon y^{-k} - 1)$.

(c) Deduce from (b) that f has the following property:

(C) $$f(z) = \varepsilon\left(\frac{z}{i}\right)^{-k} f\left(-\frac{1}{z}\right).$$

(3) With the notation of the preceding paragraph, show that if f has the property (C), then Φ has the properties (A) and (B). (Use the expression of $\Phi(s)$ obtained in (Part 2,1,b) and also make use of (1) in the integration interval).

(4) For every element z of $\mathcal{P}$, set

$$\theta(z) = \sum_{n\in\mathbb{Z}} e^{i\pi n^2 z}.$$

(a) Show that, for t strictly positive real and y real, we have

$$\int_{-\infty}^{+\infty} e^{-\pi x^2 t} e^{-2i\pi xy}\, dx = \frac{1}{\sqrt{t}} e^{-\pi y^2/t}.$$

(You may use the equality $\Gamma(1/2) = \sqrt{\pi}$ without proof).

(b) For t strictly positive real and x real, set

$$\psi(x) = \sum_{n\in\mathbb{Z}} e^{-\pi(x+n)^2 t}.$$

The function ψ is a periodic function of the real variable x. Give its Fourier series and show that this series converges to ψ. Deduce the equality $\theta(it) = 1/\sqrt{t}\theta(-1/it)$.

(c) Under the hypothesis ($\lambda = 2, k = 1/2, \varepsilon = 1$), and choosing a suitable sequence $(a_n)_{n\geq 0}$, show that θ has the property (C).

(d) For every complex number s with $Re(s) > 1$, set

$$\zeta(s) = \sum_{n=1}^{+\infty} n^{-s}.$$

Deduce certain properties of the ζ function from the preceding question (show in particular that ζ admits a simple pole at $s = 1$, of residue 1).

Third Part

Use the following result without proof: when $|t|$ tends to infinity (t real), we have

$$\lim |\Gamma(\sigma + it)|(2\pi)^{-1/2} e^{(\pi|t|/2)} |t|^{1/2-\sigma} = 1$$

uniformly for σ belonging to a compact subset of $\mathbb{R}$.

(1) Let σ_1, σ_2 be real numbers satisfying $\sigma_1 < \sigma_2$, U (resp. V) the subset of $\mathbb{C}$ defined by the inequalities $\sigma_1 \leq Re(s) \leq \sigma_2$ and $|\Im(s)| \geq 1$ (resp. $\sigma_1 \leq Re(s) \leq \sigma_2$ and $\Im(s) \geq 1$).

Let h (resp. ℓ) be a function which is defined and holomorphic in the neighbourhood of U (resp. V). Assume that there exist real positive numbers α, β_1, β_2 such that

$$\begin{cases} \sup_{s\in U} |h(s)|e^{-\alpha|s|} < +\infty \\ \sup_{|t|>1} |t|^{-\beta_j} |h(\sigma_j + it)| < +\infty \quad (j = 1, 2). \end{cases}$$

$$\left(\text{resp.} \begin{cases} \sup_{s\in V} |\ell(s)|e^{-\alpha|s|} < +\infty \\ \sup_{t\geq 1} t^{-\beta_j} |\ell(\sigma_j + it)| < +\infty \ (j = 1, 2). \end{cases}\right)$$

Let L be the affine function such that $L(\sigma_j) = \beta_j$, $(j = 1, 2)$.

Prove that there exists a real M such that for every $\sigma \in [\sigma_1, \sigma_2]$, we have

$$\sup_{|t|\geq 1} |t|^{-L(\sigma)} |h(\sigma + it)| \leq M \ \left(\text{resp.} \sup_{t\geq 1} t^{-L(\sigma)} |\ell(\sigma + it)| \leq M.\right)$$

(Reduce this to proving the result concerning V and ℓ, and then, dividing ℓ by the function $(s/i)^{L(s)}$, reduce to the case $\beta_1 = \beta_2 = 0$.)

Now, we return to the notation and hypotheses of the second part.

The function f satisfies (C) and is not constant. Let m be a strictly positive integer such that $a_m \neq 0$. Let Z be an integral of $s^{(k-1)/2}m^2\varphi(s)$, in the quadrant

$$Re(s) > 0, \qquad \mathrm{Im}(s) > 0.$$

(2) Let σ_1, σ_2 be real numbers satisfying $0 < \sigma_1 < \sigma_2$, and let V denote the subset of $\mathbb{C}$ defined by $\sigma_1 \leq Re(s) \leq \sigma_2$ and $\mathrm{Im}(s) \geq 1$. Show that there exists $\alpha > 0$ such that $Z(s)e^{-\alpha|s|}$ is bounded on V.

(3) Let σ be a real number such that $\sigma > \rho + 1$. For every real a, prove that $\sup_{t\geq 1} t^{-a}|Z(\sigma + it)| < +\infty$ if and only if $a \geq (k+1)/2$.

(4) **(a)** For every real σ, prove that there exists $a > 0$ such that

$$\sup_{|t|\geq 1} |t|^{-a}|\varphi(\sigma + it)| < +\infty.$$

(Use question 1, taking σ_2 strictly greater than $\rho + 1$, and $\sigma_1 = k - \sigma_2$.)

(b) For every real σ with $\sigma > \rho + 1$ and every element $z \in \mathcal{P}$, show that

$$f(z) - a_0 = \frac{1}{2i\pi}\int_{Re(s)=\sigma}\left(\frac{z}{i}\right)^{-s}\Phi(s)\,ds.$$

(c) Evaluate the following integral, where z is an element of $\mathcal{P}$:

$$\frac{1}{2i\pi}\int_{Re(s)=k/2}\left(\frac{z}{i}\right)^{-s}\Phi(s)\,ds.$$

Assume from now on that the Fourier coefficients of f (Part 1,1,b) are real, that there exists $\beta \in [0, (k+1)/2[$ such that, when u tends to 0 by strictly positive values, $u^\beta|f(e^{iu})|$ is bounded, and finally, that the function φ has only a finite number of zeros on the line

$$\left\{s \in \mathbb{C};\ Re(s) = \frac{k}{2}\right\}.$$

(5) **(a)** Prove that $i^{(1-\varepsilon)/2}\Phi(s)$ is real for $Re(s) = k/2$, and that when u tends to 0 by strictly positive values,

$$u^\beta \int_{-x}^{+\infty} e^{t(\pi/2-u)}\left|\Phi\left(\frac{k}{2} + it\right)\right| dt$$

is bounded.

(b) Deduce that

$$T^{-\beta}\int_0^T t^{(k-1)/2}\left|\Phi\left(\frac{k}{2} + it\right)\right| dt$$

is bounded as the real number T tends to $+\infty$.

(c) Prove that $\sup_{t\geq 1} t^{-\beta}|Z(k/2 + it)| < +\infty$.

What conclusion can we draw from these computations?

(6) Let the notation be as in the last question of the second part.

(a) For every element z of $\mathcal{P}$, prove the equality

$$\theta\left(1-\frac{1}{z}\right)=\left(\frac{z}{i}\right)^{1/2}\sum_{n=-\infty}^{+\infty}e^{i\pi(n+1/2)^2 z}.$$

(b) Prove that the ζ function has an infinity of zeros on the line $Re(s)=1/2$ (Hardy's theorem).

6

NEW PARADIGMS, NEW ENIGMAS

The title of this last chapter is a reference to a book by T. Kuhn devoted to the structure of scientific revolutions [Kuh]. It is far too early, of course, to tell if the work of Wiles (and others) should be considered within the framework of a scientific revolution, but we can certainly note some intriguing facts which appear to confirm this hypothesis.

The **anomaly**: Taniyama's original conjecture, followed by the construction, in 1969, of curves related to a non-trivial solution of Fermat's equation, and the "impossible" properties of their p-torsion – all this appeared to be no more than an amiable joke, since the most common opinion twenty-five years ago was that the (second case of) Fermat's last theorem was probably false. Indeed, had not the logician Zinoviev "proved", in 1977, that Fermat's assertion was unprovable?

Kuhn declares that "The rise of consciousness of an anomaly opens a period during which conceptual categories are readjusted until that which was originally anomalous actually become the expected result" [Kuh]. In the case we are considering here, this was the period from 1969 to 1985; 1985 was the year in which Frey and Serre made their conjectures public (Frey did so orally, while Serre did it in a course taught at the Collège de France and a famous article).

Then came Wiles' long and solitary voyage (1986–1994), which ensured the definitive success of the **paradigm**. The scientific landscape changed radically: "a new theory breaks a tradition of scientific research, and introduces a new one, constructed according to different rules, in the framework of a different discursive universe" continues T. Kuhn [Kuh]. One brief look at a book like [Rbb] is more convincing than any commentary.

In this last chapter of the present book, we first give some necessary preliminaries, and then proceed to the description of those amazing "Orlando's mares"[66], the curves $E_{A,B,C}$ associated to assumed non-trivial solutions of Fermat's equation. This is the object of Section 6.5. In Section 6.6, we explain how the awareness of this anomaly finally occurred after fifteen years. As the content of Wiles' theorem and proof lies well beyond the scope of this book, we end the chapter (Section 6.8) with a discussion of what, exactly, the "new tradition of scientific research" in the domain may represent.

[66] In the legend of "Orlando Furioso", Orlando's mare possessed every desirable virtue except that of existence.

6.1 A SECOND DEFINITION OF THE RING $\mathbb{Z}_p$ OF p-ADIC INTEGERS

In Chapter 3, we considered the reduction of an elliptic curve modulo a prime number.

However, very often, such reductions do not give the desired results, and one must go further, and work modulo a sufficiently large power of p.

Example 6.1.1

(1) Assume that $p = 3$, or 5. Then, if we want to show that the **first case** of Fermat's equation

$$x_1^p + x_2^p + x_3^p = 0, \qquad x_1x_2x_3 \not\equiv 0 \bmod p$$

is impossible, it is no use to consider the congruence modulo p, since

$$x_1^p + x_2^p + x_3^p \equiv (x_1 + x_2 + x_3)^p \bmod p.$$

However, considering the congruence modulo p^2 gives the result.

(2) Conversely, one can check that if $p = 7$, there exists no h-th power of p which gives this result working modulo p^h.

But how about working modulo "p^∞"? Can we do this? what does it mean? and in what ring should we work?

From now on, let p denote a fixed prime number. Let us now give a second construction of the ring $\mathbb{Z}_p$ which is given in [Se 1] and uses the motion of **projective limit**.

For every integer $n \geq 1$, let A_n denote the ring $\mathbb{Z}/p^n\mathbb{Z}$. As every congruence modulo p^n implies a congruence modulo p^{n-1} (when $n \geq 1$), we see that we have canonical homomorphisms φ_n:

$$\cdots \longrightarrow A_n \xrightarrow{\varphi_n} A_{n-1} \cdots \longrightarrow A_2 \xrightarrow{\varphi_2} A_1 = \mathbb{Z}/p\mathbb{Z},$$

and the object we want to construct is the object "at infinity" on the left, i.e. a hypothetical ring equipped with hypothetical homomorphisms

$$\cdots \circ \varphi_{n+2} \circ \varphi_{n+1} = \pi_n \colon \; X \longrightarrow A_n$$

such that the following diagram is commutative:

$$\begin{array}{ccc} X & \xrightarrow{\pi_n} & A_n \\ & \searrow^{\pi_{n-1}} & \downarrow \varphi_n \\ & & A_{n-1}. \end{array}$$

Thus, we want X in $\prod_{n\geq 1} A_n$, and taking into account the above arguments, we take

$$X = \left\{x \in \prod_{n\geq 1} A_n;\ \forall n > 1,\ \varphi_n \circ \varepsilon_n(x) = \varepsilon_{n-1}(x)\right\},$$

where ε_n denotes the projection of $\prod_{n\geq 1} A_n$ onto the n-th coordinates.

We easily check that X is a subring of the product ring $\prod_{n\geq 1} A_n$. Finally, we also note that X is equipped with a natural topology which is the topology induced by that of $\prod_{n\geq 1} A_n$ where A_n is equipped with its discrete topology.

Because each A_n is finite, A_n is compact, so **Tychonoff's theorem** shows that $\prod_{n\geq 1} A_n$ is compact.

As X is closed in that product, X is compact. One can show (see [Se 1]) that the set of X such that $\varepsilon_n(x) = 0$ is an ideal of X equal to $p^n X$, and that

$$X/p^n X \cong A_n = \mathbb{Z}/p^n\mathbb{Z}.$$

Since $\varepsilon_1(X) = \mathbb{Z}/p\mathbb{Z}$ is a field, we see that the kernel of ε_1, i.e. pX, is a maximal ideal of X.

We show that if $x \in X$ does not belong to this ideal, then x is invertible. It follows that X is a **local ring** whose unique maximal ideal is pX.

Now, let $x \in X$. If there exists a greatest integer $n \geq 1$ such that $\varepsilon_n(x) = 0$, we let the "p-adic valuation of x" be the number

$$v_p(x) := n.$$

We set $v_p(x) = 0$ when $\varepsilon_1(x) \neq 0$, and $v_p(0) = +\infty$ when all the $\varepsilon_n(x)$ are zero.

Example 6.1.2

(1) Let 1_n be the class of unity in A_n. Then unity in X is given by the sequence $1 = (1_1, 1_2, 1_3, \ldots)$. Thus, we have $v_p(1) = 0$.

(2) It follows that

$$p := p.1 = (0, p1_2, p1_3, \ldots).$$

Thus, $v_p(p) = 1$.

(3) Similarly, for every $h \in \mathbb{N}$,

$$v_p(p^h) = h.$$

(4) An invertible element u of $\mathbb{Z}_p$ can be written

$$u = (u_1, u_2, u_3, \ldots)$$

with $u_n \neq 0$ for every n. Thus $v_p(u) = 0$.

Let us summarise all these remarks in theorem 6.1.1.

Theorem 6.1.1

(1) *X is a ring.*

(2) *An element of X is invertible if and only if it is not divisible by p.*

(3) *Every non-zero element x of X can be written in a unique way in the form $p^n u$, with $n \in \mathbb{N}$ and u invertible. Furthermore, we have $v_p(x) = n$.*

(4) *The p-adic v_p valuation has the following properties:*

$$\begin{cases} v_p(xy) = v_p(x) + v_p(y) \\ v_p(x+y) \geq v_p(x) + v_p(y) \end{cases}$$

Remark 6.1.1 It can also be shown [Se 1] [p. 25] that the topology of X can be defined by the p-adic distance:

$$d_p(x, y) = p^{-v_p(x-y)},$$

and that the ring X is a complete space in which $\mathbb{Z}$ is dense.

Since X is an integral domain (this follows from $v_p(xy) = v_p(x)+v_p(y)$), we can consider its fraction field K.

Clearly, every non-zero number x in K can be written uniquely in the form

$$x = p^n u,$$

with $n \in \mathbb{Z}$ and $u \in X$ invertible.

Now, using the notation of Chapter 3, we can speak of the p-adic absolute value on K.

Notation 6.1.1

$$v_p(x) := n,\ v_p(0) = +\infty$$
$$|x|_p = p^{-v_p(x)}$$
$$d_p(x) = |x - y|_p.$$

Lemma 6.1.1 *The field K equipped with the distance d_p is locally compact (so it is complete), and the field $\mathbb{Q}$ of rational numbers is dense in K.*

Summary 6.1.1 We have defined a field K, complete for a "p-adic" distance d_p, in which the prime subfield isomorphic to $\mathbb{Q}$ is *dense.* The field K is thus a *completion* of $\mathbb{Q}$ for its p-adic distance d_p, and theorem 3.2.2 of Section 3.2 (Chapter 3) shows that the field K and $\mathbb{Q}_p$ are isomorphic via an isomorphism φ which respects the p-adic absolute value. It follows in particular that φ induces an isomorphism of local rings of X over $\mathbb{Z}_p$ (defined in theorem 3.3.3 of Section 3.3, Chapter 3).

Remark 6.1.2 When we want to do analysis and use the (limited) analogy between $\mathbb{Q}_p$ and $\mathbb{R}$, it is actually easier to consider that $\mathbb{Q}_p$ is obtained from $\mathbb{Q}$ by completion for the p-adic distance d_p defined over $\mathbb{Q}$ as in Chapter 3.

6.2 THE TATE MODULE $T_\ell(E)$

Assume that E is an elliptic curve defined over a field K, and let $\bar{K}$ denote an algebraic closure of K (think $K = \mathbb{Q}$ and $\bar{K} = \bar{\mathbb{Q}}$).

Take a prime number ℓ different from the characteristic of K (ℓ can be an arbitrary prime if $K = \mathbb{Q}$), and let $n \geq 1$. We know that

$$E[\ell^n] \cong \mathbb{Z}/\ell^n\mathbb{Z} \oplus \mathbb{Z}/\ell^n\mathbb{Z},$$

and that if $[\ell]$ denotes multiplication by ℓ, we have

$$\cdots \longrightarrow E[\ell^n] \xrightarrow{[\ell]} E[\ell^{n-1}] \xrightarrow{[\ell]} \cdots \xrightarrow{[\ell]} E[\ell].$$

Thus, we are in a situation which is completely analogous to the one we encountered in the first section, and we let $T_\ell(E)$ denote the object whose presence is felt on the left at infinity.

Here is a definition analogous to that of the first section (whose notation we use).

Definition 6.2.1 *Let ℓ be a prime, and E an elliptic curve. The* **Tate module** *of E at ℓ, denoted by $T_\ell(E)$, is the subgroup of $\prod_{n\geq 1} E[\ell^n]$ defined by*

$$T_\ell(E) = \left\{x \in \prod_{n\geq 1} E[\ell^n];\ \forall n \geq 1,\ [\ell] \circ \varepsilon_n(x) = \varepsilon_{n-1}(x)\right\},$$

where ε_n denotes the projection $\prod_{m\geq 1} E[\ell^m] \to E[\ell^n]$.

The remark above thus implies that if ℓ is different from the characteristic of K, then

$$E[\ell^n] \cong \frac{1}{\ell^n}\mathbb{Z}/\mathbb{Z} \oplus \frac{1}{\ell^n}\mathbb{Z}/\mathbb{Z} \cong \mathbb{Z}/\ell^n\mathbb{Z} \oplus \mathbb{Z}/\ell^n\mathbb{Z}.$$

Multiplication by ℓ on $E[\ell^n]$ induces the canonical projection

$$\mathbb{Z}/\ell^n\mathbb{Z} \oplus \mathbb{Z}/\ell^n\mathbb{Z} \longrightarrow \mathbb{Z}/\ell^{n-1}\mathbb{Z} \oplus \mathbb{Z}/\ell^{n-1}\mathbb{Z}.$$

Thus, by passage to the "projective limit", we see that $T_\ell(E)$ is a free $\mathbb{Z}_\ell$-module of rank 2:

$$T_\ell(E) \cong \mathbb{Z}_\ell \oplus \mathbb{Z}_\ell.$$

Moreover, the action of the Galois group $G := \mathrm{Gal}(\bar{K}/K)$ commutes with multiplication by ℓ, so G acts on the module $T_\ell(E)$, which is thus a G-module over $\mathbb{Z}_\ell$ (we also call it a $\mathbb{Z}_\ell[G]$-module, as in Chapter 3).

Finally, we note that the definitions of the topologies of G and of $\mathbb{Z}_\ell$ ensure the *continuity* of the action of G on $T_\ell(E)$, see Section 3.6 (Chapter 3).

To summarise, we have the following theorem.

Theorem 6.2.1 *Let E be an elliptic curve defined over K. For every prime ℓ different from the characteristic of K, the Tate module $T_\ell(E)$ is a G-module over $\mathbb{Z}_\ell$ (here G denotes $\mathrm{Gal}(\bar{K}/K)$), and it is a free $\mathbb{Z}_\ell$-module of rank 2. Furthermore, the representation $\rho_\ell : G \to GL_{\mathbb{Z}_\ell}[T_\ell(E)]$ is continuous.*

Remark 6.2.1 If K is a number field, and if E/K does not have complex multiplication, a version of Serre's theorem says that $\rho_\ell(G) = GL_{\mathbb{Z}_\ell}[T_\ell(E)]$ for all but a finite number of primes ℓ (see [Se 3]).

6.3 A MARVELLOUS RESULT

We can use a prime $\pi \in \mathbb{N}$ in two different ways:

(1) We can consider it as the true substance of the homomorphism $[\pi]$, and call it ℓ; then we construct the Tate module $T_\ell(E)$.

(2) We can use it to reduce modulo π; then we call it p and associate to it the finite field $\mathbb{F}_p$.

These two objects appear quite different: $T_\ell(E)$ belongs to a category of modules defined over a ring of infinite characteristic, while $\mathbb{F}_p$ is in the category of fields of characteristic p.

If moreover we vary the prime π in $\mathbb{N}$, we can expect to come upon a whole jungle of modules and fields, behaving in a most anarchic fashion, since we know that primes have extremely individualistic natures (for example, the p-adic topologies of $\mathbb{Q}$ are not mutually comparable (see Problem 1 of Chapter 3)).

However, the presence of an elliptic curve E defined over $\mathbb{Q}$ makes it possible to put a great deal of order in this chaos.

This results from the fact that the modules $T_\ell(E)$ are not only modules over $\mathbb{Z}_\ell$, but also, as we said in the preceding section, G-modules (with $G = \operatorname{Aut}\bar{\mathbb{Q}}$).

Now, to each prime p we associate a Frobenius element (modulo an inertia group which is not important here) in G (see [L] p. 17). We write σ_p for this element, and call it the **Frobenius at p.**

Then, if $p \neq \ell$ and if the reduction $\tilde{E}_p$ of E modulo p is a smooth curve (i.e. has "good reduction"), then

$$\begin{cases} \det \rho_\ell(\sigma_p) = & p \in \mathbb{Z} \\ \operatorname{trace} \rho_\ell(\sigma_p) = & a_p \in \mathbb{Z}, \end{cases}$$

where $a_p = p + 1 - \#(\widetilde{E}_p(\mathbb{F}_p))$, see Section 4.9 (Chapter 4).

Thus, we see that for fixed p and for an arbitrary prime number ℓ outside of a finite set, the characteristic polynomial of $\rho_\ell(\sigma_p)$ is given by

$$X^2 - \big(p + 1 - \#(\widetilde{E}_p(\mathbb{F}_p))\big)X + p.$$

This means that **it does not depend** on ℓ, but only on E and p.

The different representations ρ_ℓ are thus closely related to each other.

Moreover, if for fixed ℓ we know the character of the representation ρ_ℓ, then for every p outside of a finite set, the cardinal of the group of points of the curve reduced $\tilde{E}_p$ in $\mathbb{F}_p$ is known.

Remark 6.3.1 Hasse's theorem (see Section 4.9 of Chapter 4) shows that

$$|a_p| \leq 2\sqrt{p}.$$

This comes down to saying that the **Riemann hypothesis** is satisfied by the zeta function of the curve $\tilde{E}_p$ (see Section 4.5 of Chapter 4).

In what follows, we will use the letters ℓ and p interchangeably for prime numbers.

6.4 TATE LOXODROMIC FUNCTIONS

In this section, we will do a little p-adic analysis, considering $\mathbb{Q}_p$ more as the analogue of the field of real numbers $\mathbb{R}$ than as the fraction field of the local ring $\mathbb{Z}_p$.

In fact, we propose to transpose the loxodromic parametrisation of elliptic curves given in Section 2.10 (Chapter 2) into this context.

Here, q denotes a p-adic number whose valuation $|q|_p$ is strictly less than 1, and $\langle q \rangle$ is the subgroup of $\mathbb{Q}_p^*$ generated by q.

For every $z \in \mathbb{Q}_p^* \backslash \langle q \rangle$ (i.e. $z \in \mathbb{Q}_p^*$, $z \notin \langle q \rangle$), the series which defines the function ρ is absolutely convergent. In fact, the elements of $\langle q \rangle$ are poles of order 2 of the "p-adic meromorphic function" $z \mapsto \rho(z)$ defined by

$$\rho(z) := \sum_{n \in \mathbb{Z}} \frac{q^n z}{(1 - q^n z)^2}.$$

As in Chapter 2, we can construct a **Tate cubic**:

$$E_q : \qquad y^2 + xy = x^3 + a_4 x + a_6,$$

with

$$\begin{cases} a_4 = -5 \displaystyle\sum_{n \geq 1} \frac{n^3 q^n}{1 - q^n} \in \mathbb{Z}_p \\ a_6 = -\dfrac{1}{12} \displaystyle\sum_{n \geq 1} \frac{(7n^5 + 5n^3) q^n}{1 - q^n} \in \mathbb{Z}_p, \end{cases}$$

and we can parametrise it by the loxodromic functions

$$\begin{cases} x(z) = \displaystyle\sum_{n \in \mathbb{Z}} \frac{q^n z}{(1 - q^n z)^2} - 2 \sum_{n \geq 1} \frac{n q^n}{1 - q^n} \\ y(z) = \displaystyle\sum_{n \in \mathbb{Z}} \frac{q^{2n} z^2}{(1 - q^n z)^3} + \sum_{n \geq 1} \frac{n q^n}{1 - q^n}. \end{cases}$$

The following map is an isomorphism of analytic groups:

$$\varphi \begin{cases} z \longmapsto (x(z), y(z)) \\ \mathbb{Q}_p^* / \langle q \rangle \longrightarrow E_q(\mathbb{Q}_p). \end{cases}$$

The modular invariant of the cubic E_q is given by

$$j = \frac{1}{q} + 744 + 196884q + \cdots$$

Thus, we see that $|j|_p > 1$ so that $j \notin \mathbb{Z}_p$. Moreover, the reduction modulo p of the curve E_q is given by

$$y^2 + xy = x^3,$$

which in homogeneous coordinates gives

$$Z(Y^2 + XY) - X^3 = 0.$$

We easily see that the reduced curve admits the double point $(0, 0, 1)$ and that the tangents at this point are $y = 0$ and $y + x = 0$. The curve E_q thus has **semi-stable** reduction at p, with **rational** tangents to the double point. Conversely, we show that every elliptic curve defined over $\mathbb{Q}_p$ whose modular invariant j is such that $|j|_p > 1$ and which has semi-stable reduction at p with rational tangents at the double point is isomorphic over $\mathbb{Q}_p$ to a Tate cubic E_q.

Remark 6.4.1 The analytic isomorphism φ extends to "the" algebraic closure $\bar{\mathbb{Q}}_p$ of $\mathbb{Q}_p$:

$$\varphi : \bar{\mathbb{Q}}_p/\langle q\rangle \twoheadrightarrow E(\bar{\mathbb{Q}}_p),$$

and furthermore, one can show that it is compatible with the action of the Galois group

$$\mathrm{Gal}(\bar{\mathbb{Q}}_p/\mathbb{Q}_p) = \mathrm{Aut}(\bar{\mathbb{Q}}_p/\mathbb{Q}_p).$$

For more details on all this, see [Sil] and [Ro].

6.5 CURVES $E_{A,B,C}$

Although the role of elliptic functions in the theory of Fermat's equation (relation $\theta_{10}^4 + \theta_{01}^4 = \theta_{00}^4$) can be traced back to Jacobi, it was only from the end of the 1960s that this role took on a geometric aspect, via the construction of the curves $E_{A,B,C}$.

The motivations behind this construction (Hellegouarch in 1969, taken up again by Kubert-Lang and Frey) are indicated in the Appendix at the end of this volume.

Let us proceed pragmatically: to every primitive point (a, b, c) of the Fermat curve

$$x^p + y^p + z^p = 0,$$

we want to associate a cubic $E_{A,B,C}$ which is **smooth** (i.e. an elliptic curve) if and only if the point (a, b, c) is not a **trivial solution** of Fermat's equation (i.e. $abc \neq 0$). This simple condition led me to search for curves $E_{A,B,C}$ of the form

$$y^2 = (x - \alpha)(x - \beta)(x - \gamma),$$

with the following conditions (since α, β, γ can only be defined up to translation):

$$\beta - \gamma = a^p, \qquad \gamma - \alpha = b^p, \qquad \alpha - \beta = c^p,$$

in such a way that the discriminant of the right-hand side of the equation is $(a, b, c)^{2p}$, which is $\neq 0$.

Naturally, this system of equations has an infinity of solutions, but if we take $\gamma = 0$, we have

$$E_{A,B,C}: \quad y^2 = x(x - a^p)(x + b^p).$$

Definition 6.5.1

(1) *Let A, B, C be three relatively prime non-zero integers. We say that the equation $A + B + C = 0$ is an* **ABC relation**.

(2) *Given an ABC relation, we set*

$$E_{A,B,C}: \qquad y^2 = x(x - A)(x + B).$$

Remark 6.5.1

(1) Saying that $ABC \neq 0$ is equivalent to saying that the cubic $E_{A,B,C}$ is smooth.

(2) If we take a circular permutation of (A, B, C), the new curve $E_{B,C,A}$ is isomorphic to $E_{A,B,C}$ over $\mathbb{Q}$.

(3) If we take an odd permutation of (A, B, C), then in general, the new curve (for instance $E_{B,A,C}$) is only isomorphic to $E_{A,B,C}$ over a field containing $\sqrt{-1}$.

(4) The relation $A+B+C = 0$ is equivalent to $-A-B-C = 0$, and the curves $E_{-A,-B,-C}$ and $E_{A,C,B}$ are isomorphic over $\mathbb{Q}$.

(5) If $A = B$, then $E_{A,B,C} = E_{B,A,C}$ and the isomorphism noted in (3) is exactly a complex multiplication by $\sqrt{-1}$.

(6) Note that although Fermat's equation has no non-trivial solutions, there are still a great many ABC relations, which makes it possible to test the theory.

(7) The curve $E_{A,B,C}$ is also known as a "Hellegouarch–Frey" or "Frey–Hellegouarch" curve associated to $A + B + C = 0$.

6.5.1 Reduction of Certain Curves $E_{A,B,C}$

We restrict ourselves here to studying the curves $E_{A,B,C}$ for which (A, B, C) is a triple of relatively prime integers such that

$$A \equiv 3 \bmod 4 \qquad B \equiv 0 \bmod 32.$$

Let us begin by studying the reduction modulo a prime number ℓ.

Proposition 6.5.1 *Let ℓ be a prime.*

(1) *If ℓ does not divide ABC, the curve $E_{A,B,C}$ has good reduction modulo ℓ (it remains smooth).*

(2) *If $\ell \neq 2$ divides ABC, the reduction of $E_{A,B,C}$ modulo ℓ is a curve of genus zero and multiplicative type.*

(3) *If $\ell = 2$, and if 2 divides ABC the reduction modulo ℓ of a minimal model of $E_{A,B,C}$ is a curve of genus zero and multiplicative type.*

Proof. **(1)** If ℓ does not divide ABC, the curve reduced modulo ℓ is obviously smooth.

(2) If an odd prime ℓ divides A, for example, the equation of the reduced curve is

$$Y^2 = X^2(X + \bar{B}),$$

and as $\bar{B} \neq \bar{0}$, we see that the tangents at the double point are distinct (in $\bar{\mathbb{F}}_\ell$); thus the reduction is of multiplicative type.

(3) For the case $\ell = 2$, we make the variable change

$$x = 4\xi, \qquad y = 8\eta + 4\xi,$$

and the equation of $E_{A,B,C}$ becomes

$$\eta^2 + \xi\eta = \xi^3 + c\xi^2 + d\xi \tag{1}$$

with

$$c = \frac{B - 1 - A}{4} \quad \text{and} \quad d = \frac{-AB}{16}.$$

It follows that (1) has reduction modulo 2 given by

$$\eta^2 + \xi\eta = \begin{cases} \xi^3, & \text{if } A \equiv 7 \bmod 8 \\ \xi^3 + \xi^2, & \text{if } A \equiv 3 \bmod 8, \end{cases}$$

and we see that it has a double point at $(0, 0)$ with tangents given by

$$\begin{cases} \eta(\eta + \xi) = 0, & \text{if } A \equiv 7 \bmod 8 \\ \eta^2 + \eta\xi + \xi^2 = 0, & \text{if } A \equiv 3 \bmod 8. \end{cases}$$

□

Vocabulary

We saw in Chapter 4 that an elliptic curve such as $E_{A,B,C}$, which admits either good reduction or bad reduction of multiplicative type at each prime ℓ, is called a *semi-stable elliptic curve;* we define the **conductor** of a semi-stable curve to be the product of the primes where it has bad reduction.

Let ABC be a relation.

Corollary 6.5.1 *When $A \equiv 3 \bmod 4$ and $B \equiv 0 \bmod 32$, the conductor N of the* **semi-stable** *curve*

$$E_{A,B,C}: \quad y^2 = x(x - A)(x + B)$$

is the **radical** *of ABC, i.e. the product of the primes dividing ABC. Thus, we have*

$$N = rad(ABC).$$

Remark 6.5.2

(1) The adjective "semi-stable" comes from the fact that a semi-stable elliptic curve over $\mathbb{Q}$ remains semi-stable over a finite extension of $\mathbb{Q}$.

(2) Although equation (1) is not the most natural equation for $E_{A,B,C}$, it has the advantage of being a globally minimal equation (up to $\mathbb{Q}$-isomorphism) for this curve over $\mathrm{Spec}(\mathbb{Z})$, in the sense of Chapter 4, Section 4.14.

(3) If $ABC \equiv 0 \pmod{32}$, then one and only one of the curves $E_{A,B,C}$ is semi-stable at 2. To find it, we note that it is the curve whose equation is congruent to $y^2 = x^2(x+1) \bmod 4$.

6.5.2 Property of the Field K_p Associated to E_{a^p,b^p,c^p}

Another, more technical, motivation for the construction of the curves E_{a^p,b^p,c^p} related to a hypothetical solution of degree p Fermat's equation was the intuition that the group $E_{a^p,b^p,c^p}[p]$ of p-division points of the curve E_{a^p,b^p,c^p} should possess remarkable properties and – why not? – maybe properties in contradiction with what was known about elliptic curves.

Let K_p be the field generated by the coordinates of all the p-division points of E_{a^p,b^p,c^p}. We already saw in Chapter 4 that the field K_p is a Galois extension of $\mathbb{Q}$; we show that it always contains a primitive p^i-th root of unity ζ_p (see Section 4.6.3 of Chapter 4). We will now see that it is not "ramified" over $\mathbb{Q}_\ell(\zeta_p)$ when ℓ is an odd prime dividing abc.

Theorem 6.5.1 (Hellegouarch 1969) *Let ℓ be a prime dividing abc. Then the field K_p associated to the curve E_{a^p,b^p,c^p} can be considered as a subfield of $\mathbb{Q}_\ell(\zeta_p, 2^{1/p})$.*

Proof.

(1) A computation shows that the j-invariant of a curve E_{a^p,b^p,c^p} satisfies the equation

$$2^{-8}(abc)^{2p}j = (a^{2p} - b^p c^p)^3 = (b^{2p} - c^p a^p)^3 = (c^{2p} - a^p b^p)^3.$$

Thus, we see that up to a power of 2, j is the $2p$-th power of an element of $\mathbb{Q}_\ell$ of ℓ-adic absolute value >1.

(2) By what we saw in Section 6.5, the curve E_{a^p,b^p,c^p} is equivalent to the curve E_q over the field $\mathbb{Q}_\ell$ or over an unramified quadratic extension of this field.

(3) The field $\mathbb{Q}_\ell(2^{1/p}, \zeta_p)$ contains the field K_p. Indeed, if we let L denote this field, the group $E_q(L)$ is isomorphic to $L^*/\langle q\rangle$. Since j is a p-th power in L, then the same holds for q, and there exists q' in L such that $q = (q')^p$. Thus, we see that $L^*/\langle q\rangle$ contains the group $\langle q', \zeta_p\rangle/\langle q\rangle$ which is isomorphic to $\mathbb{Z}/p\mathbb{Z} \times \mathbb{Z}/p\mathbb{Z}$. Thus $E_q(L)$ contain $E_q[p]$. □

6.5.3 Summary of the Properties of E_{a^p,b^p,c^p}

In the above, we associated to a *rational point* (a, b, c) of the Fermat curve given by

$$F_p: \quad x^p + y^p + z^p = 0, \quad p \geq 5,$$

the *cubic* E_{a^p,b^p,c^p} i.e. a certain algebraic function field of genus ≤ 1, defined over $\mathbb{Q}$, satisfying the following properties.

(1) If the point (a, b, c) is non-trivial (i.e. $abc \neq 0$), the cubic is smooth (function field of genus 1).
(2) If the point is non-trivial, the corresponding elliptic curve is semi-stable.
(3) If the point is non-trivial, the Galois representation of $\mathrm{Aut}(\bar{\mathbb{Q}})$ in the Tate module at p of E_{a^p,b^p,c^p} has **very little ramification** (it is unramified outside of $2p$, and has little ramification at p; essentially it is ramified only at 2). Indeed, if ℓ does not divide $2p$ and **if ℓ divides abc**, we saw in the theorem that K_p is unramified over $\mathbb{Q}_\ell$. **If ℓ does not divide abc,** this follows from the Néron–Ogg–Shafarevitch criterion ([Sil] p. 179).

Examples of curves $E_{A,B,C}$

Since Wiles' theorem shows that Fermat's equation of degree $p > 2$ has no non-trivial solutions, we use equations which do have non-trivial solutions:

(1) $X^2 + Y^2 - W^2 = 0$
(2) $X^p + Y^p + 2Z^p = 0$, p a prime > 2, which always has the solution $(1, 1, -1)$.

Note that this last equation is that of the curve (G_p) of the preamble.

First example
We want a triple $\pm(a^2, b^2, -c^2)$ such that $b^2 \equiv 1 \bmod 8$ and $c^2 \equiv 0 \bmod 32$. If we start from the relation $4^2 + 3^2 = 5^2$, which we write $4^2 = (5 + 3\varepsilon)(5 - 3\varepsilon) = \mathrm{Norm}(5 + 3\varepsilon)$ with $\varepsilon^2 = 1$, we are led to compute

$$(5 + 3\varepsilon)^2 = 2(17 + 15\varepsilon),$$

and we find that $15^2 - 17^2 + 8^2 = 0$. Thus, the curve $E_{15^2,-17^2,8^2}$ is a smooth semi-stable cubic.

Second example
The curve $E_{A,B,C}$ associated to the obvious solution $(X, Y, Z) = (1, 1, 1)$ can be written

$$Y^2 = X(X^2 - 1).$$

It is an elliptic curve with complex multiplication whose reduction modulo 2 is not multiplicative. A theorem of Weil and Shimura asserts that every elliptic curve with complex multiplication is a Weil curve. Thus, this curve is a Weil curve of conductor 32 (see [Og 2] for the computation of the conductor).

6.6 THE SERRE CONJECTURES

We saw in Chapter 5 that the coefficients of the expansion of a Hecke form f in powers of q (except for a_0) are algebraic integers belonging to a finite extension K_f of $\mathbb{Q}$ contained in $\mathbb{C}$.

Let $\bar{\mathbb{Z}}$ denote the ring of algebraic integers contained in $\mathbb{C}$, i.e. the ring of elements z of $\mathbb{C}$ which are roots of a monic polynomial with coefficients in $\mathbb{Z}$. Clearly, $\bar{\mathbb{Z}}$ is contained in the algebraic closure $\bar{\mathbb{Q}}$ of $\mathbb{Q}$ in $\mathbb{C}$. Below, we use both $\mathrm{Aut}(\bar{\mathbb{Q}})$ and $G_{\mathbb{Q}}$ to denote the Galois group of the extension $\mathbb{Q} \subset \bar{\mathbb{Q}}$ (see Chapter 3).

Let $p \in \mathbb{N}$ be a prime and let $| \ |_p$ be the p-adic absolute value of $\mathbb{Q}$ attached to p. In Section 4.10 (Chapter 4), we defined the ring $\mathbb{Z}_{(p)} := \{x \in \mathbb{Q};\ |x|_p \leq 1\}$, and we know that there exists a canonical homomorphism of **reduction modulo** p:

$$\begin{cases} \mathbb{Z}_{(p)} \longrightarrow \mathbb{F}_p \\ x \longmapsto \bar{x}. \end{cases}$$

This homomorphism extends to a *place* of $\mathbb{Q}$, i.e. a map

$$\varphi : \mathbb{Q} \longrightarrow \mathbb{F}_p \cup \{\infty\}$$

defined by

$$\begin{cases} \varphi(x) = \bar{x} \in \mathbb{F}_p & \text{if } x \in \mathbb{Z}_{(p)} \\ \varphi(x) = \infty & \text{if } x \in \mathbb{Q} \setminus \mathbb{Z}_{(p)}. \end{cases}$$

An application of Zorn's lemma (see Problem 4 of Chapter 3) shows that we can extend every place φ of $\mathbb{Q}$ to a place $\bar{\varphi} : \bar{\mathbb{Q}} \to \bar{\mathbb{F}}_p \cup \{\infty\}$. Note, however, that this extension is not unique, although this is not very important in what follows.

This place induces a homomorphism $\bar{\mathbb{Z}} \to \bar{\mathbb{F}}_p$ whose kernel $\mathfrak{p}$ is a maximal ideal of $\bar{\mathbb{Z}}$; it is defined by

$$\mathfrak{p} := \{z \in \bar{\mathbb{Z}};\ \bar{\varphi}(z) = \bar{0} \in \bar{\mathbb{F}}_p\}.$$

We say that the place $\bar{\varphi}$ lies **above** φ, and that $\mathfrak{p}$ lies **above** the ideal (p) generated by p in $\mathbb{Z}$.

Thus, we see that to every Hecke cusp form

$$f = \sum_{n \geq 1}^{\infty} a_n q^n, \quad a_n \in \bar{\mathbb{Z}},$$

we can associate a "**form reduced modulo** p" (in fact $\mathfrak{p}$!):

$$\bar{f} = \sum_{n \geq 1}^{\infty} \bar{a}_n q^n \in \bar{\mathbb{F}}_p[[q]].$$

Remark 6.6.1 Since the coefficients a_n of f belong to a finite extension K_f of $\mathbb{Q}$ (see theorem 5.7.3 of Section 5.7, Chapter 5), we see that there exists an exponent h such that

$$\bar{f} \in \mathbb{F}_{p^h}[[q]].$$

Definition 6.6.1 *Take a natural integer $N \geq 1$, a homomorphism $\varepsilon : (\mathbb{Z}/N\mathbb{Z})^\times \to \mathbb{C}^*$ and a cusp form f of weight k for the level $N : f \in S_k(N)$. We say that f is a* **new form of type** (N, k, ε) *if*

(i) *the series $f = \sum_{n\geq 1} a_n q^n$ with $q = e^{2i\pi\tau}$ converges in $\mathcal{H} = \{\tau \in \mathbb{C}; \Im\tau > 0\}$ and satisfies the modular relation*

$$f\Big(\frac{a\tau + b}{c\tau + d}\Big) = \varepsilon(d)(c\tau + d)^k f(z). \tag{1}$$

(ii) *f vanishes at the* **cusps**, *i.e. for every $\left(\begin{smallmatrix} a & b \\ c & d \end{smallmatrix}\right) \in SL_2(\mathbb{Z})$, the function*

$$\tau \longmapsto (c\tau + d)^{-k} f\Big(\frac{a\tau + b}{c\tau + d}\Big)$$

has a series expansion of type (1) with q replaced by $q^{1/N}$ and in which the constant term is zero.

(iii) *f is a normalised eigenvector of all the Hecke operators attached to the level N (in particular $a_1 = 1$).*

(iv) *f is "***new***", i.e. it does not come from a Hecke form of weight k whose level N' strictly divides N.*

Given this, we have the following result (due to Deligne), which we do not prove here.

Theorem 6.6.1 *Let $f = \sum_{n\geq 1} a_n q^n$ be a* **new form** *of type (N, k, ε), and let K be the subfield of $\mathbb{C}$ generated by the a_n and the image of ε.*

Then K is a finite extension of $\mathbb{Q}$, and the a_n are contained in $\mathcal{O}_K := K \cap \bar{\mathbb{Z}}$.

For every prime number $p \in \mathbb{N}$, there exists a continuous representation $\rho_p : G_\mathbb{Q} \to GL_2(\mathcal{O}_K/p\mathcal{O}_K)$, unramified at every ℓ prime to N and such that, for every ℓ as above, we have

$$\begin{cases} Tr\rho_p(Frob_\ell) = \bar{a}_\ell \in \mathcal{O}_K/p\mathcal{O}_K \\ \det \rho_p(Frob_\ell) = \overline{\varepsilon(\ell)\ell^{k-1}} \in \mathcal{O}_K/p\mathcal{O}_K. \end{cases}$$

To give a rigorous meaning to this statement, we need to define what we mean by a "Galois representation unramified at ℓ".

Let us admit that a Galois representation

$$\rho : G_\mathbb{Q} \longrightarrow GL(V)$$

has an **Artin conductor** N_ρ defined by the usual formula (see [Se 5]) even when V is defined over a field of characteristic p.

The representation ρ is not ramified at ℓ if and only if ℓ does not divide N_ρ (i.e. if ρ is trivial on the inertia group of ℓ).

In the 1970s, J-P. Serre asked the following two fundamental questions:

(1) Under what hypotheses does an **irreducible** continuous representation $\rho : G_{\mathbb{Q}} \to GL_2(\bar{\mathbb{F}}_p)$ come from a modular form, via the preceding theorem?
(2) In the affirmative, how can we find the **minimal type** (N, k, ε) of this form?

In 1973, Serre conjectured that the only condition for the question (1) is that $\rho(c) = -1$, where c denotes the trace of complex conjugation on $\bar{\mathbb{Q}}$. In 1987, he published (see [Se 4]) the recipes which produce the conjectured **minimal type** of the form starting from the properties of the representation.

Example 6.6.1 Suppose we are given an abc relation for which a curve $E_{A,B,C}$ is semi-simple. Then for every odd prime p, we obtain a continuous Galois representation (see Chapter 4, Section 4.7):

$$\rho : G_{\mathbb{Q}} \longrightarrow E_{a,b,c}[p],$$

and we deduce from **Mazur's theorem** (theorem 4.10.3, Section 4.10, Chapter 4) that it is **irreducible** if p is sufficiently large. The Artin conductor N_ρ of this representation is

$$N_\rho = \prod_{\substack{\ell \neq p \\ v_\ell(j_E) \not\equiv 0 \bmod p}} \ell = \prod_{\substack{\ell \neq p \\ v_\ell(ABC/2^4) \not\equiv 0 \bmod p}} \ell;$$

it is the **minimal** level of the modular form f predicted by the Serre conjectures. □

At this time, the first Serre conjecture has been proven in the case where the image of ρ is a **solvable group**: this follows from works of Hecke, Langlands and Tunnell. But except for a finite number of special cases, it has not been proven in general.

However, the second conjecture is practically proven in the case where p is different from 2 or 3. It follows from the combined work of a large number of mathematicians, among whom we must mention B. Mazur and K. Ribet.

The "philosophy" which makes these conjectures so precious is based on the fact that the representation ρ_p associated to a new form f of level N can be much simpler than what one might expect; in particular, its Artin conductor N_ρ can be much smaller than N.

The form f is then congruent modulo p to a form whose level is a very small divisor of N, which leads to astonishing consequences. This is what we now consider for Fermat's equation F_p (and also for its false twin G_p!).

6.7 MAZUR–RIBET'S THEOREM

The strategy used by Wiles in his proof of Fermat's last theorem is based on a brilliant intuition of G. Frey according to which the curves E_{a^p,b^p,c^p}, associated to non-trivial primitive solutions of Fermat's equation cannot be Weil curves (at least for sufficiently large p).

This essential point was proven by Mazur and Ribet (see [Oe]), who in fact proved the much more general result conjectured by J.-P. Serre in 1985 [Se 4], in which the fundamental object is more the **representation** ρ_p than the curve E_{a^p,b^p,c^p}.

6.7.1 Mazur–Ribet's Theorem

Contrarily to "Fermat's last theorem", this result has no simple statement: it states a property of "irreducible, modular, finite representations of level N".

As we want to apply this to representations attached to certain curves $E_{A,B,C}$, we restrict ourselves to giving an idea of the meaning of these hypotheses in that particular context.

Thus, let us take a suitable curve $E = E_{A,B,C}$, which is assumed to be a **Weil curve.** Recall that this means that if N denotes the conductor of E, there exists a new form $f \in S_2(\Gamma_0(N))$ which we can write

$$f(z) = \sum_{n=1}^{\infty} a_n q^n, \quad a_1 = 1, \quad q = e^{2i\pi z},$$

which is an eigenfunction of the Hecke operators, and when ℓ does not divide N, satisfies the relation

$$\#\widetilde{E}(\mathbb{F}_\ell) = \ell + 1 - a_\ell.$$

Let p be a fixed prime (greater than or equal to 5). As in Chapter 4, we are interested in the representation

$$\rho\colon \quad G_{\mathbb{Q}} := \mathrm{Aut}(\mathbb{Q}) \longrightarrow GL_2(\mathbb{F}_p)$$

associated to the $G_{\mathbb{Q}}$-module $E[p]$.

We show that if ℓ is a prime number not dividing pN, and if $\sigma_\ell \in G_{\mathbb{Q}}$ is a "Frobenius element" at ℓ, we have

$$\begin{cases} \text{trace } \rho(\sigma_\ell) \equiv a_\ell \bmod p \\ \det \rho(\sigma_\ell) \equiv \ell \bmod p. \end{cases}$$

We then say that the representation ρ is "modular", because it is attached to the modular form f modulo p.

Mazur's theorem (theorem 4.10.3, Section 4.10, Chapter 4) allows us to see that this representation is **irreducible.**

We now assume that (a, b, c) denotes a "primitive" solution (i.e. a, b, c are relatively prime) of Fermat's equation

$$x^p + y^p + z^p = 0,$$

and we take

$$\{A, B, C\} = \{a^p, b^p, c^p\},$$

arranging for $E_{A,B,C}$ to be semi-stable.

Then, we saw (Section 6.5.1) that the conductor N of $E_{A,B,C}$ is $\mathrm{rad}(abc)$. If now the **odd** prime ℓ divides N, then the proof of theorem 6.5.1 shows that the exponent of ℓ in the minimal discriminant of $E_{A,B,C}$ is always divisible by p; we will say that the representation ρ is *finite* at every prime number $\neq 2$.

Remark 6.7.1 Other important properties of ρ follow from the definitions:

(1) The representation ρ is **continuous**, i.e. it factorises through the homomorphism

$$\text{Aut}\,\bar{\mathbb{Q}} \longrightarrow \text{Aut}\,K_p.$$

(2) The determinant of ρ is a continuous representation of dimension 1:

$$\text{Aut}(\bar{\mathbb{Q}}) \longrightarrow \mathbb{F}_p^{\times},$$

which is such that complex conjugation gives $-1 \pmod p$). We say that the representation is **odd**.

Mazur–Ribet's Theorem 6.7.1 *Let $\rho : G_{\mathbb{Q}} \to GL_2(F)$ be a representation of $G_{\mathbb{Q}}$,* **irreducible** *over $\bar{F}$ and* **modular** *of* **level N** *in the linear group of a 2-dimensional vector space over a field F of characteristic $p \geq 3$.*

If ρ is **finite** *at p and of* **weight 2**, *then ρ satisfies Serre's second conjecture, i.e. we can take N to be the* **Artin conductor of the representation.**

Corollary 6.7.1 *Wiles' theorem implies Fermat's "last theorem".*

Proof. There is no question of giving the proof of the Mazur–Ribet theorem here; it is extremely difficult [Ri 1], but we can easily prove the corollary.

Let (a, b, c) be a non-trivial primitive solution of Fermat's equation of exponent $p \geq 5$.

Since the curve E_{a^p,b^p,c^p} is semi-stable, Wiles' theorem shows that it is a Weil curve.

Since the representation $\rho_p : G_{\mathbb{Q}} \to E_{a^p,b^p,c^p}[p]$ satisfies the hypotheses of the Mazur–Ribet theorem, we can take N to be the Artin conductor of ρ_p.

The theorem of Section 6.5.2 comes down to saying that this conductor is equal to 2; thus we return to a modular representation of level 2. Serre's second conjecture (now proven in the theorem) shows that the minimal type of the modular form that this representation comes from is (2,2, trivial), which implies the existence of a non-zero modular form $f \in S_2(\Gamma_0(2))$. But we saw (Section 5.9, Chapter 5) that $S_2(\Gamma_0(2)) = \{0\}$, so we obtain a contradiction. □

Remark 6.7.2 The identical method cannot be applied to the equation

$$a^p + b^p + 2c^p = 0, \quad (p \geq 5), \qquad (G_p)$$

because the curve E_{a^p,b^p,c^p} **is not semi-stable**. However, we will see below what Ribet, Darmon and Merel made of the method in this case.

6.7.2 Other Applications

One can prove the following result [Se 4, p. 202–205] by an analogous procedure.

Theorem 6.7.2 (Serre) *Let p be a prime ≥ 11. Let L be a prime $\neq p$ belonging to the set*

$$S = \{3, 5, 7, 11, 13, 17, 19, 23, 29, 53, 59\},$$

and let α be an integer ≥ 0. Then the equation

$$a^p + b^p + L^\alpha c^p = 0$$

has no solutions with $a, b, c \in \mathbb{Z}$, $abc \neq 0$.

Proof. We easily reduce to the case where a, b, c are relatively prime. As above, we consider the suitable semi-stable curve $E_{a^p, b^p, L^\alpha c^p}$, and we show that the representation ρ_p is associated to a modular form $f \in S_2(\Gamma_0(2L))$ which is an eigenvalue of the Hecke operators T_p for $p \nmid 2L$.

(1) Assume that $L \in \{3, 5\}$. We saw that $S_2(\Gamma_0(2L)) = \{0\}$, so we have a contradiction as above.

(2) If $L \in \{7, 11, 13, 17, 19, 23, 29, 53, 59\}$, the situation is more difficult: we refer to [Se 4, p. 202–205] for details. □

Remark 6.7.3 A limit to the possible improvements on these results is reached for $\ell = 5$ and $P = 31$, since we have

$$1 + 31 + (-2)^5 = 0.$$

The equation $x^p + y^p + 2z^p = 0$

Naturally, a host of Diophantine equations remains to be studied, but the simplest case appears to be our **countersubject**, i.e. the following conjecture stated by Dénes in 1952 [Den].

Dénes' Conjecture 6.7.1 *Let p be an odd prime. If the three natural non-zero integers x^p, y^p and z^p lie in an arithmetic progression, then $x = y = z$.*

The reader can easily check that this conjecture is equivalent to our countersubject (a countersubject which arises as naturally as Fermat's equation, in the study of the p^2-division points of elliptic curves, see the Appendix). In [Den, Satz 9] Dénes proves the following result (which proves his conjecture for $p < 31$).

Theorem 6.7.3 (Dénes) *Let p be a regular odd prime such that the order of 2 in $\mathbb{F}_p^*$ is either even or equal to $(p-1)/2$. If $2^{p-1} \not\equiv 1 \bmod p^2$, then the conjecture holds for p.*

Remark 6.7.4 The congruence $2^{p-1} \equiv 1 \bmod p^2$ is very exceptional, but we know that it happens for $p = 1093$ and $p = 3511$ (**Wieferich numbers**). However, we do not know if it can happen for an infinity of values of p.

Very recently (see [Ri 2]), K. Ribet showed the following result.

Theorem 6.7.4 (Ribet) *There exists no solution of $x^p + y^p + 2z^p = 0$ in non-zero relatively prime integers when 2 divides the product xyz.*

To prove this theorem, Ribet notes, following K. Rubin and A. Silverberg, that Wiles' theorem extends to **all** the curves $E_{a,b,c}$ (not only the semi-simple ones), and then applies the method of Section 6.7 (noting that the Artin conductor N of ρ_p strictly divides 32, and that $g = \dim S_2(\Gamma_0(N))$ is zero when N divides 16).

But what happens when xyz is odd? Using techniques due to Darmon, Ribet was able to prove the following result.

Theorem 6.7.5 (Ribet) *If the prime number p is congruent to 1 modulo 4, then Dénes' conjecture holds.*

It remained to prove this conjecture for $p \equiv 3$ modulo 4, and this has just been accomplished by Darmon and Merel ([D-M])!

Theorem 6.7.6 (Darmon–Merel) *Dénes' conjecture holds.*

Darmon and Merel's proof (for $p > 3$) uses **new arguments** which establish the **surjectivity** of the Galois representation ρ_p associated to the curve $E_{a^p,b^p,2c^p}$ when the non-trivial solution (a, b, c) is different from $(1, 1, -1)$.

Remark 6.7.5 In fact, Darmon and Merel showed that this property is valid for every curve $E_{A,B,C}$ not equivalent to $E_{1,1,-2}$, when $p > 3$.

6.8 SZPIRO'S CONJECTURE AND THE *abc* CONJECTURE

We conclude this discussion by presentating two conjectures which will perhaps replace Fermat's last theorem for our mathematical grandchildren.

6.8.1 Szpiro's Conjecture

In a talk given in Hanover in 1983, Szpiro stated the following conjecture:

Weak Conjecture 6.8.1 *There exists $\alpha > 0$ and $\beta > 0$ such that for every semi-stable elliptic curve E over $\mathbb{Q}$, we have*

$$|\Delta_E| \leq \alpha N_E^{\beta}$$

where Δ_E is the minimal discriminant of E and N_E is its conductor.

The most optimistic form of Szpiro's conjecture (see Exercise 6.20) is given in the following statement.

Strong Conjecture 6.8.2 *For every $\varepsilon > 0$, there exists a constant $C(\varepsilon) > 0$ such that for every semi-stable elliptic curve E over $\mathbb{Q}$, we have*

$$|\Delta_E| \leq C(\varepsilon) \cdot N_E^{6+\varepsilon}.$$

Arithmetic consequence 6.8.1 *For every* $\varepsilon > 0$, *there exists a constant* $C_1(\varepsilon) > 0$ *such that for every triple* (a, b, c) *of relatively prime integers with sum equal to zero such that* $a \equiv -1 \bmod 4$ *and* $b \equiv 0 \bmod 32$, *we have*

$$|abc| \leq C_1(\varepsilon)\,\mathrm{rad}(abc)^{3+\varepsilon}.$$

Proof. We saw in Section 6.5 that the discriminant Δ of the curve $E_{a,b,c}$ is

$$\Delta = \left(\frac{abc}{16}\right)^2$$

and that $E_{a,b,c}$ has a semi-stable minimal model E. We saw also that the conductor of E is $\mathrm{rad}(abc)$. We deduce immediately from the strong version of Szpiro's conjecture that

$$\left(\frac{abc}{16}\right)^2 \leq C(2\varepsilon)\,\mathrm{rad}(abc)^{6+2\varepsilon},$$

hence

$$|abc| \leq 16\sqrt{C(2\varepsilon)}\mathrm{rad}(abc)^{3+\varepsilon}.$$

This gives the result with $C_1(\varepsilon) = 16\sqrt{C(2\varepsilon)}$. □

Generalised Szpiro's conjecture 6.8.3 *For every* $\varepsilon > 0$, *there exists a constant* $C_1(\varepsilon)$ *such that for every triple* (a, b, c) *of relatively prime integers with sum equal to zero, we have*

$$|abc| \leq C_1(\varepsilon)\,\mathrm{rad}(abc)^{3+\varepsilon}.$$

6.8.2 *abc* Conjecture

The *abc* conjecture was born during a discussion between J. Oesterlé and D.W. Masser in 1985.

***abc* Conjecture 6.8.4** *For every* $\varepsilon > 0$, *there exists a constant* $C_2(\varepsilon)$ *such that for every triple* (a, b, c) *of relatively prime integers with sum equal to zero, we have*

$$\sup(|a|, |b|, |c|) \leq C_2(\varepsilon)\,\mathrm{rad}(abc)^{1+\varepsilon}.$$

We saw in Exercise 1.13 (Chapter 1) that the analogue of this conjecture in $\mathbb{C}[t]$ holds ("Mason's theorem").

6.8.3 Consequences

These conjectures have such numerous consequences that it is impossible to list them all here. But before mentioning some of them, we recall the famous theorem of S-units.

This result, which is due to C.L. Siegel, is valid in the larger framework of algebraic number fields, but we restrict ourselves to expressing it for the field of rational numbers (for a proof, see [Sil] p. 253).

Given a finite set S consisting of the number -1 and prime numbers $p_1, \ldots, p_n$, let $\langle S \rangle$ denote the multiplicative subgroup of $\mathbb{Q}^*$ generated by S, i.e. the set of rationals equal to $\pm p_1^{\nu_1} \ldots p_n^{\nu_n}$ for arbitrary exponents $\nu_1, \ldots, \nu_n$ in $\mathbb{Z}$.

Theorem 6.8.1 (Siegel) *Let a and $b \in \mathbb{Q}^*$.*
Then the equation

$$ax + by = 1$$

has only a finite number of solutions $(x, y) \in \langle S \rangle \times \langle S \rangle$.

Example 6.8.1

(1) There exists only a finite number of exponents μ, ν in $\mathbb{Z}$ such that $2^\mu \pm 3^\nu = 1$.
(2) The same property is valid for the equation

$$p_1^{\mu_1} \cdots p_m^{\mu_m} \pm q_1^{\nu_1} \cdots q_n^{\nu_n} = 1,$$

where the set $S = \{-1, p_1, \ldots, p_m, q_1, \ldots, q_n\}$ is fixed and where the unknowns

$$\mu_1, \ldots, \mu_m, \qquad \nu_1, \ldots, \nu_n$$

lie in $\mathbb{Z}$.

(a) Bounds for Szpiro's quotient and the *abc* quotient
By the strong Szpiro conjecture, for every $\varepsilon > 0$,

$$\frac{\log |abc|}{\log \operatorname{rad}(abc)} \leq \frac{C_1(\varepsilon)}{\log \operatorname{rad}(abc)} + 3 + \varepsilon.$$

By Siegel's theorem of S-units, the radical of abc cannot remain bounded when the abc relations vary, hence

$$\overline{\lim} \frac{\log |abc|}{\log \operatorname{rad}(abc)} \leq 3 + \varepsilon.$$

Taking $\varepsilon = 1$, we see that there is only a finite number of abc relations for which the Szpiro quotient is greater than 4.

Conjecture (consequence of the generalised Szpiro conjecture) 6.8.5 *There exists an absolute constant K_1' such that for every abc relation, we have*

$$\frac{\log |abc|}{\log \operatorname{rad}(abc)} \leq K_1',$$

and the maximum is reached for a particular (and remarkable!) abc relation.

The present record (1994) for this proportion is held by A. Nitaj of the University of Caen, France. He obtained the value of 4.41901 ... for the relation

$$13 \cdot 19^6 + 2^{30} \cdot 5 = 3^{13} \cdot 11^2 \cdot 31.$$

By the same reasoning, we can also state the following conjecture.

Conjecture (consequence of *abc*) 6.8.6 *There exists an absolute constant K_2' such that for every abc relation, we have*

$$\frac{\log \sup(|a|, |b|, |c|)}{\log \operatorname{rad}(abc)} \leq K_2',$$

and the maximum is reached for a particular (and remarkable!) relation.

The present record for this proportion is held by E. Reyssat of the University of Caen. E. Reyssat obtained the value 1.62991... for the relation

$$2 + 3^{10} \cdot 109 = 23^5.$$

(b) Generalisation of Fermat's equation

Suppose we are given three non-zero pairwise relatively prime integers α, β, γ, such that we cannot have $\alpha + \beta\varepsilon + \gamma\eta = 0$ with $\varepsilon, \eta \in \{1, -1\}$. We consider the equation

$$\alpha x^n + \beta y^n + \gamma z^n = 0, \tag{1}$$

where n denotes an unknown integer ≥ 4.

Conjecture (follows from Szpiro) 6.8.7

(1) *When n is sufficiently large, every primitive solution (a, b, c) of the equation (1) is such that $abc = 0$.*

(2) *There exists only a finite number of quadruples (n, a, b, c) where a, b, c are relatively prime integers and $n \geq 5$ which satisfy equation (1).*

Proof. **(1)** Assume that n is greater than all the exponents of the prime factors of $\alpha\beta\gamma$. Then every solution (a, b, c) of (1) is proportional to a primitive solution (where a, b and c are relatively prime).

(2) If we apply the consequence of Szpiro's conjecture to

$$(A, B, C) = (\alpha a^n, \beta b^n, \gamma c^n),$$

we have

$$\frac{\log|\alpha\beta\gamma| + n \log|abc|}{\log \operatorname{rad}(\alpha\beta\gamma abc)} \leq K_1'.$$

Because $|abc| > 1$, we see that $\log|abc| \geq \log 2$, so n must be bounded.

(3) If n is bounded, then Faltings' theorem applied to each $n \geq 4$ shows that we obtain only a finite number of solutions. □

(c) The Fermat–Catalan Conjecture

It follows from the *abc* conjecture (see Exercises 6.10–6.12) that for fixed positive integers u, v, w, there are only a finite number of triples (abc) of the form $ux^r + vy^s = wz^t$ with $1/r + 1/s + 1/t < 1$.

Example 6.8.2 If $u = v = w = 1$, we know ten triples, namely

$$1 + 2^3 = 3^2,\ 2^5 + 7^2 = 3^4,\ 13^2 + 7^3 = 2^9,\ 17^3 + 2^7 = 71^2,\ 3^5 + 11^4 = 122^2$$
$$17^7 + 76271^3 = 21063928^2,\ 1414^3 + 2213459^2 = 65^7,\ 9262^3 + 15312283^2 = 113^7$$
$$43^8 + 96222^3 = 30042907^2,\ 33^8 + 1549034^2 = 15613^3.$$

(d) Extension to surfaces

Completing equation (1) by a symmetric equation, we obtain the system (Problem 4 of Chapter 1, see also [Hel 1] p. 35–36):

$$\begin{cases} \alpha x^n + \beta y^n + \gamma z^n = 0 \\ \alpha^n x + \beta^n y + \gamma^n z = 0. \end{cases} \tag{2}$$

Conjecture (follows from Szpiro) 6.8.8 *When $n > 5$, the system (2) admits only a finite number of solutions, which are non-trivial integers in $\mathbb{Z}$ such that*

$$(\alpha x, \beta y, \gamma z) = 1.$$

Proof. By Szpiro's conjecture, for every $\varepsilon > 0$, we have

$$\begin{cases} |\alpha\beta\gamma(xyz)^n| \le C(\varepsilon)|\alpha\beta\gamma xyz|^{3+\varepsilon} \\ |(\alpha\beta\gamma)^n xyz| \le C(\varepsilon)|\alpha\beta\gamma xyz|^{3+\varepsilon}. \end{cases}$$

Multiplying these inequalities, we obtain

$$|\alpha\beta\gamma xyz|^{n-5-2\varepsilon} \le C(\varepsilon)^2. \qquad \square$$

COMMENTARY

We explained above how Frey's idea (1985), the publication of Serre's conjectures (1986–87), the Mazur–Ribet theorem (1987), and finally Wiles' exploits (1994) lifted the psychological barrier which had kept the elliptic approach to Fermat's last theorem in a state of hibernation for fifteen years.

As an effect of this liberation, we are currently witnessing an intense flowering of new results and of open questions (enigmas) which makes it hopeless to even attempt to give a coherent synthesis of the current situation.

The first specialised books devoted to this question are beginning to appear, in particular the acts of the Boston Summer Conference of August 1985 [CSS], and the book by Van der Poorten [VDP].

To limit ourselves to currently available literature, we indicate that the notion of a *projective limit* (Sections 6.1 and Section 6.2) can be studied in [Se 1] and [Sil].

An excellent description of the difficult kernel of the subject is given in various articles of the Bourbaki Seminar [Oe] and [Se 4]. The articles of Ribet [Ri 1] and Wiles ([Wi] and [W-T]) should not be neglected but they are by no means easy reading.

For the Serre conjectures, read [Se 4].

Finally, the new results cited in the text are given excellent treatments in the articles by Darmon and Granville [D-G], and in the articles by Ribet [Ri 2] and Darmon-Merel [D-M].

Exercises and Problems for Chapter 6

6.1 Let A be a ring equipped with an ultrametric absolute value $|\ |$, and let $R = \mathrm{M}_{n(A)}$ be the ring of $n \times n$ matrices with coefficients in A.
If $M = (m_{ij}) \in \mathbb{R}$, set

$$||M|| = \sup_{i,j}\{|m_{ij}|\}.$$

Show the following relations:

(i) $||M|| = o \leftrightarrow M = 0,$
(ii) $||M + N|| \leq \sup\{||M||, ||N||\},$
(iii) $||MN|| \leq ||M||\,||N||.$

Give an example showing that the inequality of (iii) can be strict.

6.2 We use the notation of Exercise 6.1, with $A = \mathbb{Z}_p$ for an odd prime p.

(a) Show that if $M \in R$ is such that

$$||M - I|| < 1,$$

then $M \in R^*$ (i.e. M is invertible).
(b) Let G be the set of matrices M such that $||M - I|| < 1$. Show that G is a multiplicative group.
(c) Let q be a prime number different from p, and let $M \in G$ be such that $M \neq I$. Show that $M^q \neq I$.
(d) Show that if $M \in G$ is such that $M \neq I$, then $M^p \neq I$.
(e) Deduce that the only element of finite order in G is the element I.

6.3 We use the notation of Exercise 6.2, and let π denote the canonical homomorphism $\mathbb{Z}_p \to \mathbb{F}_p$ extended to a homomorphism $R \to \mathrm{M}_n(\mathbb{F}_p)$.

(a) Show that $\mathrm{Ker}\,\pi = G$.
(b) Using Exercise 6.2, show that if H is a subfinite group of $GL(\mathbb{Z}_p)$, then

$$\mathrm{Ker}\,\pi \cap H = \{I\}.$$

(c) Deduce that the order of H divides $\#GL(\mathbb{F}_p) = (p^n - 1)(p^n - p) \cdots (p^n - p^{n-1})$.

6.4 Prove lemmas 4.10.1, 4.10.2 and 4.10.3 of Chapter 4 with $\mathbb{Z}_{(p)}$ replaced by $\mathbb{Z}_p$.
Show that the analogue of E_p is a torsion-free group.
Show that the analogue of E_p is of finite index in $E(\mathbb{Q}_p)$ (Lutz' theorem, see [Lut]).
When the curve $\tilde{E}$ reduced modulo p is of genus one, show that the order of the torsion group of $E(\mathbb{Q}_p)$ divides the cardinal of $\widetilde{E}(\mathbb{F}_p)$.

6.5 Let c be a positive integer. We propose to show that the equation $x^n - y^n = c$ has only a finite number of solutions (x, y, n) in $\mathbb{N}^3$.

(i) Assuming that $y > 1$, show that $n\mathrm{Log}x/y < y^{-n/2}$.
(ii) Alan Baker's theory of linear forms of logarithms implies the existence of a constant $\gamma > 0$ such that

$$\mathrm{Log}\,\mathrm{Log}\frac{x}{y} \geq -\gamma\,\mathrm{Log}\,n\,\mathrm{Log}\,y.$$

Deduce from this an upper bound for n.
(iii) Conclude.

6.6 We use the notation of Exercise 6.5 with $S_1 = \langle 2, 3\rangle$. Let E_1 be the set of solutions of $x - y = 1$ in S_1. We propose to show that

$$E_1 = \{(2, 1), (3, 2), (4, 3), (9, 8)\}.$$

Write this equation in the form

$$2^a - 3^b = \pm 1.$$

(i) Assume that a is odd and $b > 0$. Working modulo 3, show that the right-hand side is -1. Then, working in $\mathbb{Z}_3$, show that if $b > 1$, then $a = 6n + 3$. Working modulo 8, show that b is even, and obtain a contradiction modulo 10. Conclude.
(ii) Assume that $a > 0$ is even and $b > 0$, and show by working modulo 3 that the right-hand term is 1. Assume that $a \geq 3$, and by working in $\mathbb{Z}_3$, show that a is divisible by 3.

Show that the equation

$$64^\alpha - 1 = 3^b$$

is impossible if $\alpha \geq 1$. Conclude.

6.7 We use the notation of Exercise 6.5 with $S_2 = \langle 2, 5\rangle$. Let E_2 be the set of solutions of $x - y = 1$ in S_2. Show as in Exercise 6.6 that

$$E_2 = \{(2, 1), (5, 4)\}.$$

6.8 The same question, with $S_3 = \langle 3, 5\rangle$. Show that $E_3 = \phi$.

6.9 We use the notation of the previous exercises, and take $S_4 = \langle 2, 3, 5\rangle$. Let E_4 denote the set of solutions of $x - y = 1$ in S_4.
Show that $E_4 = E_1 \cup E_2 \cup \{(6, 5), (9, 10), (16, 15), (25, 24), (81, 80)\}$ (see [Luc]).

6.10 Consider all the triples $(r, s, t) \in (\mathbb{N}^*)^3$ such that $f(r, s, t) = r^{-1} + s^{-1} + t^{-1} < 1$. Assume that $r \leq s \leq t$, and order the set of these triples according to the lexicographic order, i.e.

$$(r, s, t) < (r', s', t')$$

if and only if $r < r'$ or ($r = r'$ and $s < s'$) or ($r = r'$, $s = s'$ and $t < t'$).

Find the smallest triple (r, s, t) such that $f(r, s, t) < 1$, and compute the minimum of $1 - f(r, s, t)$.

6.11 Deduce from the *abc* conjecture that there exists only a finite number of sextuples $(r, s, t, x, y, z) \in (\mathbb{N}^*)^6$ such that

$$\begin{cases} \frac{1}{r} + \frac{1}{s} + \frac{1}{t} < 1 \\ (x, y, z) = 1 \\ x^r + y^s = z^t. \end{cases}$$

One can use Exercise 6.10.

6.12 Take three numbers u, v, w in $\mathbb{N}^*$, and deduce from the *abc* conjecture that there exists only a finite number of sextuples $(r, s, t, x, h, z) \in (\mathbb{N}^*)^6$ such that

$$\begin{cases} \frac{1}{r} + \frac{1}{s} + \frac{1}{t} < 1 \\ (x, y, z) = 1 \\ ux^r + vy^s = wz^r. \end{cases}$$

6.13 Show that the *abc* conjecture implies the following conjecture:

For every $\varepsilon \in]0, 1/6[$, there exists a constant $c(\varepsilon) > 0$ such that for every pair of relatively prime integers $(x, y) \in (\mathbb{N}^*)^2$, we have

$$|x^3 - y^2| > c(\varepsilon) \max(x^3, y^2)^{1/6-\varepsilon}.$$

6.14 A prime number satisfying the congruence

$$2^{p-1} \equiv 1 \bmod p^2$$

is called a **Wieferich prime**. The Wieferich primes are quite rare since the only $p \leq 3 \cdot 10^{10}$ which are Wieferich primes are 1093 and 3511. However, it is not known whether there exists an infinity of prime numbers which are not Wieferich primes.

Following J. Silverman, we will show that the *abc* conjecture implies a positive answer to the preceding question.

(α) Assume that there exists $n \in \mathbb{N}$ such that $2^n \equiv 1 \bmod p$ and $2^n \not\equiv 1 \bmod p^2$. Show that we have

$$2^{p-1} \not\equiv 1 \bmod p^2.$$

(β) Assume that there exists only a finite number of primes p_i such that $2^{p_i-1} \not\equiv 1 \bmod p_i^2$, and for $n \in \mathbb{N}$, set

$$2^n - 1 = a_n b_n,$$

where a_n lies in the monoid generated by the p_i and b_n does not lie in it.

Using (α), show that the radical of b_n is less than or equal to $b_n^{1/2}$.

(γ) Consider the triple $(1, 2^n - 1, 2^n)$ and let $\varepsilon > 0$. Show that the *abc* conjecture implies that

$$a_n b_n < 2^n \leq C(\varepsilon)\,\mathrm{rad}(2a_n b_n)^{1+\varepsilon}.$$

(δ) Obtain a contradiction, when n tends to infinity, using β and γ.

6.15 We propose to show that the *abc* conjecture implies that if

$$\mathrm{rad}(x+1) = \mathrm{rad}(y+1), \quad \mathrm{rad}(x+2) = \mathrm{rad}(y+2), \quad \mathrm{rad}(x+3) = \mathrm{rad}(y+3),$$

then we have $x = y$ except for a finite number of exceptions (this property is due to M. Langevin).

(α) Assume that $y > x$. Show that the hypothesis implies that

$$\mathrm{rad}((y+1)(y+2)(y+3)) \quad \text{divides} \quad y - x.$$

(β) Applying the *abc* conjecture to the relation

$$1 + (y+1)(y+3) = (y+2)^2,$$

show that y is bounded.

6.16 We say that $n \in \mathbb{N}^*$ is a **powerful number** if $\mathrm{rad}(n)^2$ divides n.

Show that the *abc* conjecture implies that there exists no triple of consecutive powerful numbers (note that if a, b, c are three consecutive numbers, then $b^2 = ac + 1$).

6.17 Let A be a Dedekind domain, and let $n \to \nu_n$ be a sequence of positive integers tending to infinity.

(α) Show that if $x \in A$ is such that the sequence (x^{ν_n}) takes only a finite number of values, then $x \in U(A) \cup \{0\}$, where $U(A)$ denotes the group of units of A.

(β) Generalise the result of (α) to the case where x belongs to the fraction field of A.

(γ) Admit the fact (which is a consequence of Faltings' theorem) that if $r > 3$, then the curve of equation

$$aX^r + bY^r + cZ^r = 0$$

with $(a, b, c) \in (\mathbb{Q}^*)^3$ has only a finite number of points in $\mathbb{P}_2(\mathbb{Q})$. Deduce from (β) that for n large enough, all the solutions of the equation

$$aX^{rn} + bY^{rn} + cZ^{rn} = 0$$

are proportional to elements of $(U(A) \cup \{0\})^3$ (for more details, see [Hel 3]).

6.18 Let F be an algebraically closed field of characteristic p.

(α) Show that if $P \in F[t]$ admits a root $\alpha \in F$ of multiplicity $\nu(\alpha)$ not divisible by p, then the derivative P' admits α as a root of multiplicity $\nu(\alpha) - 1$.

(β) Consider a relation ABC in which A, B and C are relatively prime elements of $F[t]$ such that $A + B + C = 0$, and assume that $ABC \notin F$.

We say that this relation is *separable* if A/B admits a non-zero derivative.

Show that, up to replacing t by t^{p^h}, we can always assume that the relation ABC is separable.

(γ) If $P \in F[t]$, the *tame radical* $R_0(P)$ is the product $\prod_{\substack{P(\alpha)=0 \\ p \nmid \nu(\alpha)}} (t-\alpha) \in F[t]$ in which the α are all the roots of P in F whose multiplicity is not divisible by p.

Following the method of proof of Mason's theorem (Exercise 1.15, Chapter 1), show that if the relation ABC is separable, then

$$\sup\{\deg A, \deg B, \deg C\} < \deg R_0(ABC).$$

(δ) Deduce that if the ABC relation given by

$$a^n + b^n + c^n = 0$$

with $(a, b, c) \in (F[t])^3$ pairwise relatively prime is *separable*, then $n \leq 2$.

6.19 Let $\mathbb{F}_q$ be a finite field of cardinal q, and let $\bar{\mathbb{F}}_q$ be an algebraic closure of $\mathbb{F}_q$.

(α) Show that the map $x \mapsto x^q$ is an automorphism σ of $\bar{\mathbb{F}}_q$, and that

$$\mathbb{F}_q = \{x \in \bar{\mathbb{F}}_q; \sigma(x) = x\}.$$

(β) Let $P \in \mathbb{F}_q[t]$, and set

$$D(P) = \frac{\sigma(P) - P}{\sigma(t) - t}.$$

Show that $D(P) \in \mathbb{F}_q[t]$.

(γ) Show that the map $D : \mathbb{F}_q[t] \to \mathbb{F}_q[t]$ is $\mathbb{F}_q$-linear and that

$$\begin{aligned} D(PQ) &= PD(Q) + D(P)\sigma(Q) \\ &= \sigma(P)D(Q) + D(P)Q. \end{aligned}$$

We say that D is the *Galois derivation* of $\mathbb{F}_q[t]$.

(δ) Show that if $\alpha \in \bar{\mathbb{F}}_q$ is a root of P of multiplicity $\nu(\alpha)$, then

(i) If $\alpha \in \mathbb{F}_q$, α is a root of $D(P)$ of multiplicity $\nu(\alpha) - 1$.
(ii) If $\alpha \notin \mathbb{F}_q$, α is a root of $D(P)$ of multiplicity $\nu(\alpha)$.

(ε) Let $P \in \mathbb{F}_q[t]$. The *rational radical* $R_1(P)$ is the product $\prod_{\substack{P(\alpha)=0 \\ \alpha \in \mathbb{F}_q}} (t-\alpha) \in \mathbb{F}_q[t]$ in which the α run over the roots of P in $\mathbb{F}_q$. Following the method of Exercise 6.18, show that

$$\deg R_1(P) \geq q - (q-2)\sup\{\deg A, \deg B, \deg C\}.$$

(φ) Assume that $q = 2$, and show that ABC has at least two distinct roots in $\mathbb{F}_2$. Could this result be obtained directly?

(η) Redo Problem 5 of Chapter 1 with the radical replaced by the rational radical.

6.20 Limits to optimism
Given a relation (a, b, c), we set

$$\rho(a,b,c) := \frac{\log|abc|}{\log \operatorname{rad}(abc)}, \qquad \tau(a,b,c) := \frac{\log \sup(|a|,|b|,|c|)}{\log \operatorname{rad}(abc)}.$$

(α) Show that $\rho(a,b,c) < 3\tau(a,b,c)$.

(β) We propose to show that

$$\limsup_{(a,b,c)} \rho(a,b,c) \geq 3.$$

For this, we construct a sequence of triples (a_n, b_n, c_n) by the induction relations

$$(a_1, b_1, c_1) = (16, 1, -17)$$
$$a_{n+1} = 4a_n b_n, \quad b_{n+1} = (a_n - b_n)^2, \quad c_{n+1} = -(a_n + b_n)^2.$$

Show that the triple (a_n, b_n, c_n) corresponds to an *abc* relation, and that a_n is always divisible by 16.
Set $\lambda_n := |a_n b_n c_n|/(\operatorname{rad} a_n b_n c_n)^3$, and show that

$$\frac{\lambda_{n+1}}{4\lambda_n} \geq \frac{|c_n|}{|a_n - b_n|} > 1.$$

Deduce that λ_n tends to infinity and conclude.

(γ) Deduce from α and β that one cannot have

$$\limsup_{(a,b,c)} \tau(a,b,c) < 1.$$

(δ) We return to the example of β to show that $\sup(|a|,|b|,|c|)/\operatorname{rad}(abc)$ cannot be universally bounded.
Set $r_n = \operatorname{rad}(a_n b_n c_n)$, $\mu_n = |c_n|/r_n$, $a_n/b_n = tg^2\theta_n$ with $0 < \theta_n < \pi/2$.
Show that $\theta_n \equiv 2\theta_{n-1} \mod \pi$ and that $\mu_{n+1}/\mu_n \geq 1/|\cos 2\theta_n|$.
Deduce that $\mu_{n+1}/\mu_1 \geq 2^n |\sin 2\theta_1| = \log|c_{n+1}|/\log|c_1|\,|\sin 2\theta_1|$.
Conclude.

Problem I

(Nevanlinna Theory)

I. The goal of this first part is to prove the **Poisson–Jensen formula**.

(1) Let f be a holomorphic function in the disk $D = \{z \in \mathbb{C}; |z| < 1\}$, bounded in the closed disk $\bar{D} = \{z \in \mathbb{C}; |z| \leq 1\}$ and continuous on $\bar{D}$ except at a finite number of points. Show that

$$\frac{1}{2\pi}\int_0^{2\pi} f(e^{i\theta})\, d\theta = f(0).$$

(2) Assume now that f is meromorphic in D, continuous and non-zero on the boundary of D and at the origin.

Let $a_1, \dots, a_m$ be the zeros of f in $\bar{D}$, and let $b_1, \dots, b_n$ be its poles (zeros and poles are repeated a number of times equal to their multiplicity). Prove the **Jensen formula**:

$$\log|f(0)| + \sum_{h=1}^{m} \log \frac{1}{|a_h|} - \sum_{\ell=1}^{n} \log \frac{1}{|b_\ell|} = \frac{1}{2\pi} \int_0^{2\pi} \log|f(e^{i\theta})|\, d\theta.$$

For every sufficiently small $\varepsilon > 0$, we can construct a region R_ε in which $f(z)$ has neither poles nor zeros, and we will compute the integral of $\log f(z)/z\, dz$ along the boundary of R_ε.

(3) Suppose now that f is meromorphic in $\mathbb{C}$, and write its Laurent expansion at the origin in the form

$$f(z) = cz^e + \cdots$$

with $c \neq 0$ and $e \in \mathbb{Z}$.

Show that we have the **Poisson–Jensen formula**:

$$\log|c| = \int_{\Gamma_r} \log|f(z)| \frac{dz}{2i\pi z} - \sum_{\substack{0<|a|<r \\ f(a)=0}} \operatorname{ord}(f,a) \frac{r}{|a|} + \sum_{\substack{0<|b|<r \\ f(b)=\infty}} \operatorname{ord}(f,b) \frac{r}{|b|} - e \log r,$$

where Γ_r denotes a circle centred at the origin and of radius r, travelled in the counterclockwise direction. Check that this formula is invariant under the involution $f(z) \leftrightarrow 1/f(z)$.

II. The aim of this part is to define and study the radical and height of a meromorphic function. To the function f, which is meromorphic in $\mathbb{C}$, we associate two **counting functions**:

$$N_f(r,0) = \sum_{\substack{0<|a|<r \\ f(a)=0}} \operatorname{ord}(f,a) \log \frac{r}{|a|} + \begin{cases} 0 & \text{if } e \leq 0 \\ e \log r & \text{if } e > 0 \end{cases}$$

and

$$N_f(r,\infty) = \sum_{\substack{0<|b|<r \\ f(b)=0}} \operatorname{ord}(f,b) \log \frac{r}{|b|} + \begin{cases} 0 & \text{if } e \geq 0 \\ -e \log r & \text{if } e < 0. \end{cases}$$

(1) Show that we have

$$\begin{cases} N_{1/f}(r,0) = N_f(r,\infty) \\ N_{1/f}(r,\infty) = N_f(r,0) \end{cases}$$

and (by the Poisson–Jensen formula):

$$\log|c| + N_f(r,0) = \int_{\Gamma_r} \log|f(z)| \frac{dz}{2i\pi z} + N_f(r,\infty).$$

(2) Following Nevanlinna, we write

$$\int_{\Gamma_r} \log|f(z)| \frac{dz}{2i\pi z} = \int_{\Gamma_r} \log \frac{|f|}{\sqrt{1+|f|}^2} \frac{dz}{2i\pi z} + \int_{\Gamma_r} \log\sqrt{1+|f|^2} \frac{dz}{2i\pi z},$$

and we associate to f two **proximity functions**:

$$m_f(r,0) := \int_{\Gamma_r} \log \frac{|f|}{\sqrt{1+|f|^2}} \frac{dz}{2i\pi z} + \begin{cases} \log \frac{|f(0)|}{\sqrt{1+|f(0)|^2}}, & e \leq 0 \\ \log|c| & e > 0 \end{cases}$$

$$m_f(r,\infty) := \int_{\Gamma_r} \log\sqrt{1+|f|^2} \frac{dz}{2i\pi z} - \begin{cases} \log\sqrt{1+|f(0)|^2}, & e \geq 0 \\ \log|c| & e < 0. \end{cases}$$

Show that we have

$$\begin{cases} m_{1/f}(r,0) = m_f(r,\infty) \\ m_{1/f}(r,\infty) = m_f(r,0), \end{cases}$$

and (by 1),

$$N_f(r,0) + m_f(r,0) = N_f(r,\infty) + m_f(r,\infty).$$

(3) Show that if $|f(z)| \to 0$ when $|z| \to \infty$, then $m_f(r,0)$ tends to $+\infty$ when $r \to +\infty$. Show that if $|f(z)| \to \infty$ when $|z| \to \infty$, then $m_f(r,\infty)$ tends to $+\infty$ when $r \to +\infty$.

(4) Define the **characteristic function** (or **height**) of f by the formula:

$$T_f(r) := N_f(r,\infty) + m_f(r,\infty).$$

Show that if $f = a/b$ where a and b are holomorphic functions on $\mathbb{C}$ having no common zeros, we have

$$T_f(r) = \int_{\Gamma_r} \log\sqrt{|a(z)|^2 + |b(z)|^2} \frac{dz}{2i\pi z} - \log\sqrt{|a(0)|^2 + |b(0)|^2}.$$

III. Finally, let us define two new functions of r:

(i) the **logarithmic radical** S_a of a non-zero holomorphic function a. By definition, we have

$$S_a(r) := \sum_{\substack{0<|x|<r \\ a(x)=0}} \log\frac{r}{|x|} + n_0 \log r,$$

with $n_0 = 0$ if $a(0) \neq 0$ and $n_0 = 1$ otherwise.

(ii) the **ramification counting function** R_f of a non-constant meromorphic function f. By definition, we have

$$R_f(r) := \sum_{\substack{0<|x|<r \\ f(x)\notin\{0,1,\infty\} \\ f'(x)=0}} \operatorname{ord}(f',x)\log\frac{r}{|x|} + \begin{cases} \operatorname{ord}(f',0)\log r & \text{if } f(0) \notin \{0,1,\infty\} \\ 0 & \text{otherwise} \end{cases}$$

The goal of this third part is to give an **abc theorem** for the holomorphic functions in $\mathbb{C}$, and to deduce one of its consequences (see [VF] for more details). Recall Mason's theorem (Exercise 1.15, Chapter 1) on the polynomials of $\mathbb{C}[z]$.

Consider three non-constant relatively prime polynomials a, b, c such that $a + b + a = 0$. Then

$$\sup(\deg a, \deg b, \deg c) < \deg \operatorname{rad}(abc).$$

Now, if a, b, c are three entire non-constant functions on $\mathbb{C}$ having no common zeros and such that $a + b + c = 0$, we have the following result.

***abc* Theorem** *For every r outside a set of finite Lebesgue measure, we have*

$$T_{a:b:c}(r) \leq S_{abc}(r) - R_{a/c}(r) + 2\log T_{a:b:c}(r) + O(\log\log T_{a:b:c}(r)),$$

where $T_{a:b:c}$ is defined by

$$T_{a:b:c}(r) = \int_{\Gamma_r} \log\sqrt{|a(z)|^2 + |b(z)|^2 + |c(z)|^2}\,\frac{dz}{2i\pi z} - \log\sqrt{|a(0)|^2 + |b(0)|^2 + |c(0)|^2}.$$

(1) Consider three non-constant entire functions a, b, c such that $a^n + b^n = c^n$, with $n \in \mathbb{N}$. Show that we can assume that a, b, c have no common zeros; in what follows, we make this assumption.

(2) Set

$$\begin{cases} T(r) = T_{a^n:b^n:c^n}(r) \\ S(r) = S_{a^n b^n c^n}(r) = S_{abc}(r). \end{cases}$$

Show that the *abc* theorem implies that

$$T(r) \leq S(r) + 2\log T(r) + O(\log\log T(r)).$$

(3) Show that

$$\begin{aligned} S(r) &\leq N_{a/b}(r, 0) + N_{b/c}(r, 0) + N_{c/a}(r, 0) \\ &\leq T_{a/b}(r) + T_{b/c}(r) + T_{c/a}(r) + O(1). \end{aligned}$$

(4) Show that

$$T(r) \geq T_{a^n/b^n}(r) + O(1) = nT_{a/b}(r) + O(1).$$

(5) Deduce that if r is sufficiently large, then

$$nT_{a/b}(r) \leq T_{a/b}(r) + T_{b/c}(r) + T_{c/a}(r) + 3\log T(r).$$

(6) Conclude that we must have $n \leq 3$.

Problem 2

(An application of the Langlands–Tunnell theorem due to J.-P. Serre)

A. Preliminaries on $\mathrm{GL}_2(\mathbb{F}_3)$.

(1) $\mathrm{GL}_2(\mathbb{F}_3)$ denotes the group of invertible 2 by 2 matrices with coefficients in $\mathbb{F}_3$. Show that the centre of $\mathrm{GL}_2(\mathbb{F}_3)$ is equal to $\{\lambda I_2;\ \lambda \in \mathbb{F}_3^*\}$.

(2) Let $\mathrm{PGL}_2(\mathbb{F}_3)$ denote the quotient of $\mathrm{GL}_2(\mathbb{F}_3)$ by its centre. Show that the cardinal of this group is equal to 24. Interpreting this group as a group of permutations of the lines of $\mathbb{F}_3^2$, deduce that $\mathrm{PGL}_2(\mathbb{F}_3) \simeq \mathcal{S}_4$.

(3) Show that $\mathrm{GL}_2(\mathbb{F}_3)$ is generated by

$$\alpha = \begin{pmatrix} -\bar{1} & \bar{1} \\ -\bar{1} & \bar{0} \end{pmatrix} \quad \text{and} \quad \beta = \begin{pmatrix} \bar{1} & -\bar{1} \\ \bar{1} & \bar{1} \end{pmatrix}.$$

(4) Note that $\mathbb{F}_3 \simeq \mathbb{Z}[\sqrt{-2}]/\pi$, where π denotes the ideal of $\mathbb{Z}[\sqrt{-2}]$ generated by $1+\sqrt{-2}$. Show that there exists a unique monomorphism φ from $\mathrm{GL}_2(\mathbb{F}_3)$ to $\mathrm{GL}_2(\mathbb{Z}[\sqrt{-2}])$ such that

$$\varphi(\alpha) = \begin{pmatrix} -1 & 1 \\ -1 & 0 \end{pmatrix} \quad \text{and} \quad \varphi(\beta) = \begin{pmatrix} 1 & -1 \\ 1 & 1 \end{pmatrix}$$

and we have

$$\varphi(g) \equiv g \mod \mathbb{P}$$

for all $g \in \mathrm{GL}_2(\mathbb{F}_3)$. Deduce that for every $g \in \mathrm{GL}_2(\mathbb{F}_3)$, we have

$$\overline{\mathrm{trace}(\varphi(g))} = \mathrm{trace}(g), \quad \overline{\det(\varphi(g))} = \det(g),$$

where the overline denotes the class in $\mathbb{F}_3$.

B. In this second part, we will consider a plane elliptic curve E defined over $\mathbb{Q}$, and its 3-torsion group $E[3]$ in $\mathbb{P}^1\mathbb{C}$. We consider the Galois representation

$$G_{\mathbb{Q}} \overset{\rho_3}{\rightarrow} \mathrm{GL}_{\mathbb{F}_3}(E[3])$$

given by the action of the absolute Galois group on $E[3]$.

(1) Show by an example that ρ_3 is not necessarily irreducible.

(2) From now on, we fix a basis for $E[3]$ over $\mathbb{F}_3$, and we identify ρ_3 with a representation of $G_{\mathbb{Q}}$ in $\mathrm{GL}_2(\mathbb{F}_3)$. Show that $\theta = \varphi \circ \rho_3$ is a representation of $G_{\mathbb{Q}}$ in $\mathrm{GL}_2(\mathbb{Z}[\sqrt{-2}]) \subset \mathrm{GL}_2(\mathbb{C})$.

(3) Let σ denote complex conjugation, and show that

$$\det \theta(\sigma) = \pm 1.$$

(4) Using Weil's alternating form, show that

$$\Lambda^2_{\mathbb{F}_3} E[3] \simeq \mu_3,$$

where μ_3 denotes the group of cube roots of unity in $\mathbb{C}$, i.e. a line over $\mathbb{F}_3$.

(5) Can $E[3]$ have an $\mathbb{F}_3$-basis (P, Q) consisting of rational points?
(6) Deduce from (4) that

$$\det \theta(\sigma) \equiv -1 \mod \pi.$$

(7) Show that the eigenvalues of $\rho_3(\sigma)$ are $\overline{1}$ and $-\overline{1}$.
(8) Assume from now on that ρ_3 is irreducible over $\mathbb{F}_3$. Show that every matrix M of $\mathbb{M}_2(\overline{\mathbb{F}}_3)$ which commutes with all the matrices of $\operatorname{Im} \rho_3$ is of the form λI_2 with $\lambda \in \overline{\mathbb{F}}_3$.
(9) Deduce that ρ_3 is absolutely irreducible (i.e. irreducible in $\mathrm{GL}_2(\overline{\mathbb{F}}_3)$).

C. Let us admit that for almost all primes q, we have:

$$\operatorname{trace}_{\rho_3}(\operatorname{Frob}_q) \equiv q + 1 - \#\widetilde{E}(\mathbb{F}_q),$$

where the second term is considered modulo 3. Recall that $\mathcal{S}_4$ is a solvable group. Show that the representation $\theta : G_{\mathbb{Q}} \to \mathrm{GL}_2(\mathbb{C})$ satisfies the following properties:

(1) θ is irreducible (over $\mathbb{C}$);
(2) $\det \theta(\sigma) = -1$, where σ denotes complex conjugation;
(3) $\operatorname{Im} \theta$ is a solvable group.

Remark A famous theorem due to **Langlands** and **Tunnell** states that then there exists

$$f(z) = \sum_{n=1}^{\infty} a_n e^{2\pi i n z} \in S_1\big(\Gamma_0(N), \chi\big),$$

for some N and some χ, such that θ "comes from" f in the sense that

$$\operatorname{trace}\big(\theta(\operatorname{Frob}_q)\big) = b_q$$

for almost all primes q. This was a decisive first step in the proof of Wiles' theorem and the Shimura–Taniyama–Weil conjecture.

APPENDIX

THE ORIGIN OF THE ELLIPTIC APPROACH TO FERMAT'S LAST THEOREM

OUTLINE

The subject of this talk[67] is a historical (and somewhat personal) description of a early step in the succession of results which culminated in Wiles' proof of the so-called Fermat's last theorem.

As is well known, Wiles' proof centres around three main themes: elliptic curves, Galois representations and modular forms. So one may ask where Fermat's last theorem lies; this is the step we want to concentrate upon[68].

A.1 STATEMENT OF THE ELLIPTIC TRICK

Traditionally this trick is presented as follows ([O]). Suppose we are in possession of a **hypothetical** solution of Fermat's equation:

$$\alpha^p + \beta^p + \gamma^p = 0 \tag{1}$$

with $p \geq 5$, $(\alpha, \beta, \gamma) \in \mathbb{Z}^3$ primitive **and** $\alpha\beta\gamma \neq 0$; then we consider the two cubic curves:

$$\begin{cases} Y^2 = X(X - \alpha^p)(X + \beta^p) \\ Y^2 = X(X + \alpha^p)(X - \beta^p). \end{cases} \tag{2}$$

[67] This is the text of a lecture given at Cambridge on 28 November 1995.

[68] Although in a primitive form, almost all the material presented here can be found in [H72]. However, in 1972, the absence of available textbooks on elliptic curves was conducive to ad hoc proofs.

These curves are elliptic **because** $\alpha\beta\gamma \neq 0$ (the solution of Fermat's equation is **non-trivial**) and one of them is semi-stable over $\mathbb{Q}$: let us call it $E_{A,B,C}$ with $A = \alpha^p$, $B = \beta^p$, $C = \gamma^p$.

Denote by $\mathbb{Q}(\zeta_p)$ the field generated over $\mathbb{Q}$ by a primitive p^{th} root of unity.

Then the extension K_p of $\mathbb{Q}(\zeta_p)$ generated by the points of p-torsion of $E_{A,B,C}$ is unramified over $\mathbb{Q}(\zeta_p)$ except at the primes dividing $2p$ in the first case of FLT and at the primes dividing 2 in the second case. Equivalently, we can say that the representation ρ of the absolute Galois group $G_{\mathbb{Q}}$ through the $\mathbb{F}_p$-plane of p-torsion points of $E_{A,B,C}$ is not ramified outside $2p$ and "not much" at p (which implies tamely ramified).

Remark A.1.1 **(1)** It is obvious that both curves (2) are semi-stable everywhere, except perhaps at 2.

To be precise, we must specify which is the better of the two curves in (2).

The recipe is simple: the right curve (i.e. the semi-stable one) is the curve which is congruent to

$$y^2 = x^2(x + 1)$$

modulo 4.

(2) If we dislike hypothetical assertions and want to be "positive", we can work locally over one ℓ-adic field $\mathbb{Q}_\ell$ in which Fermat's equation has a solution, and we get a local version of the story.

(3) Of course I was aware, in 1969, that the curve $E_{A,B,C}$ was truly an "Orlando's mare" (in Orlando Furioso, Orlando's mare is supposedly endowed with every possible virtue except that of existence).

So, I was hoping that Serre could prove the impossibility of the representation ρ, or that someone could prove that the existence of the curve $E_{A,B,C}$ contradicted Beppo Levi's conjecture – the one which was "part of the folklore" ([C] p. 264) at that time and which is now Mazur and Merel's theorem.

Merel's theorem A1.1 *The torsion of an elliptic curve defined over a given number field K is uniformly bounded by a constant C(K).*

A.2 BIRTH OF THE ELLIPTIC TRICK

The elliptic trick, together with the properties of non-ramification of the p-torsion of the elliptic curve, were presented by the author of this talk at the 1969 Journées Arithmétiques de Bordeaux. Moreover, as far as I can see, it was not known before.

(a) The journey from elliptic curves to FLT

It seems that only two people, using different methods, actually made this journey at the end of the 1960s: B.A. Demjanenko and I. We were both working on B. Levi's conjecture over $\mathbb{Q}$. As I was trying to find an "elementary" proof of Manin's theorem limiting uniformly the p-torsion of elliptic curves, I was surprised to discover that the existence of a torsion point of order $2p^h$ (in a certain $E(\mathbb{Q})$) led to a **non-trival** solution of the Fermat equation of degree p^{h-1}, and moreover in the "second case" of FLT (the harder case[69]). I was using tools I had developed some years before in a study of points of order 14 on elliptic curves ([H65a] and [H65b]). These tools were

(i) the model

$$Y^2 = X(X^2 + CX + D),$$

(ii) substitutes for Néron's algebraic groups (which I called Lutz's groups), and
(iii) Tate p-adic elliptic functions.

During this time, Demjanenko was reaching similar results with the model

$$Y^2 = X^4 + AX^2 + B$$

and extensive elementary calculations.

I will state these results in the form given in [H75].

Theorem A.2.1 *If a rational elliptic curve admits a rational point of order* $2p^2$, *then the curve*

$$\begin{cases} x_1^p + x_2^p = x_3^p \\ x_1^p - x_2^p = x_4^p \end{cases}$$

admits $(p-1)/2$ *non-trivial points in* $\mathbb{P}_3(\mathbb{Q})$.

Corollary A.2.1 *Under the hypothesis of theorem A.2.1, the curve*

$$y_1^{2p} - y_2^{2p} = y_3^p$$

admits non-trivial solutions in $\mathbb{P}_2(\mathbb{Q})$.

Theorem A.2.2 *If a rational elliptic curve with completely rational* 2*-torsion admits a point of order* p^2, *then the curve*

$$z_1^{2p} + z_2^{2p} = z_3^{2p}$$

admits $(p-1)/2$ *non-trivial solutions in* $\mathbb{P}_2(\mathbb{Q})$.

Remark A.2.1

(1) These solutions belong to the "second case of FLT".
(2) With the hypothesis of Theorem A.2.2, one gets at the same time non-trivial solutions of $r_1^{4p} - r_2^{4p} = r_3^p$, $s_1^{2p} - s_2^{2p} = 4s_3^{2p}$ and $t_1^{4p} - t_2^{4p} = 4t_3^p$.

[69] Because it is closer p-adically to a trivial solution.

(b) The way back

For several reasons (among them pure logic) I was led to a converse. This converse was presented orally at the "Journées Arithmétiques de Bordeaux, 1969", and in print two years later ([H71], [H72]) when I found a **positive** result.

The idea was simple: starting from a hypothetical non-trivial primitive solution (α, β, γ) of Fermat's equation, and using what I knew of the p-torsion of elliptic curves, I was led to consider $E_{A,B,C}$ curves, with $A = \alpha^p$, $B = \beta^p$, $C = \gamma^p$ and to study their p-torsion points. Looking for a clue to prove the non-existence of this "Orlando's mare", I noticed that these torsion points lay in a field which had so little ramification outside $2p$ that I thought something strange was happening.

Example A.2.1 Consider the curve:

$$X^{2p} - Y^{2p} = 2^4 Z^{2p} \tag{3}$$

and suppose (α, β, γ) is a primitive non-trivial solution such that p divides $\alpha\beta\gamma$. Then the field L_p generated by the $4p$-torsion points of $E_{A,B,C}$ is non-ramified over $\mathbb{Q}(\zeta_{4p})$. Note that we can now prove that p divides γ, see [H90].

It was not until 1971 that I found a way to make a modest but presentable step forward.

Theorem A.2.3 *If equation (3) admits an infinity of primitive solutions in the "second case of FLT", then there is a number field K, depending only on p, where an infinity of non-equivalent elliptic curves admit a complete group of K-rational $2p$-torsion points.*

Using the terminology of Lang ([K-L] p. 195) according to which a curve V defined over $\mathbb{Q}$ is said to be "**Mordellic**" if it has only a finite number of points in any number field, we have:

Corollary A.2.2 *If Fermat's equation of degree p has an infinity of solutions in the second case, then the modular curve $X_1(2p)$ is not Mordellic.*

And this corollary is refuting Mordell's conjecture (which became Falting's theorem in 1983).

Motivations (apart from pure logic)

(1) It is known that the second case of FLT is notoriously more difficult than the first, and that one of the reasons why this is so is that the trivial solutions $(1, -1, 0)$, $(0, 1, -1)$, $(-1, 0, 1)$ are not excluded by the definition of the second case, as they are by the definition of the first. Saying that the cubic $E_{A,B,C}$ was an **elliptic curve** was tantamount to saying that the primitive solution (a, b, c) was **not trivial**. So the consideration of these curves, as **elliptic curves**, was a means of treating both cases on an equal footing (inasmuch as they differ only by this property).

(2) At that time I was making no distinction between the two $E_{A,B,C}$ elliptic curves (which are only isomorphic over $\mathbb{Q}(i)$). In that way a nice symmetry in the correspondence between solutions of Fermat equations and those curves was preserved: $\mathfrak{S}_3$ was operating on both. Later in [F86] Frey considers only the $E_{A,B,C}$ curve which is **semi-stable** at 2 (both are obviously semi-stable everywhere else) using a different terminology. The notion of a semi-stable Abelian variety goes back at least to 1968 (see [S-T]). In view of Diamond's recent improvement of Wiles' result, one may ask if Frey's distinction (which breaks the symmetry) is still necessary.

(3) Kummer's method was the use of an **enlargement** of the ring of integers $\mathbb{Z}$, but this enlargement was of finite dimension over $\mathbb{Q}$. The use of the curve $E_{A,B,C}$ could be compared to an enlargement of $\mathbb{Z}$ of **infinite** dimension: to **a point** of Fermat's curve an **elliptic curve** is associated. But, of course, that type of metaphorical comparison could not be mentioned in front of a French audience.

(4) The use of $E_{A,B,C}$ curves was a means of applying Abel's General Method for solving impossibility problems: "We need to give a form to the problem which is such that it is always possible to solve it, which can be done for every problem. Instead of asking for a relation when we do not know whether it exists or not, we need to ask if such a relation is possible".

One can thus study $E_{A,B,C}$ curves in general (as they do exist), and then wonder if the ones related to a non-trivial solution of the Fermat's equation do exist.

(c) Subsequent evolution

This elliptic approach presented at the end of the sixties, entered a strange period of **hibernation** for more than ten years. On the one hand, it seemed well understood as references in papers of Frey [F77, F82, F89], but noticeably *not* [F86]) testify.

On the other hand, it seems strange that the fuse waited for so long before exploding suddenly in 1985 when Frey conjectured in an unpublished paper[70], that the $E_{A,B,C}$ curves linked with Fermat's last theorem should not be **modular**. At last, a plan to tap the hidden power of these curves had been suggested.

A tentative explanation for this hibernation period should take into account the existence of intense activity in neighbouring areas during the seventies:

1977: G. Terjanian proved the impossibility of $a^{2p} + b^{2p} = c^{2p}$ in the **first case**. [T]
1978: B. Mazur proved his famous result on points of finite order. [M]
1983: G. Faltings proved Mordell's conjecture. [Fa]

A.3 LOCAL STUDY

To understand the introduction of the elliptic curves related to the Fermat's equation, let us go into more technical details.

In my thesis I made a detailed local study of elliptic curves defined over $\mathbb{Q}$ with an hypothetical torsion point of order $2n$ for n odd.

We will now present the main facts of this study in a more modern and general setting.

[70] "Modular elliptic curves and Fermat's conjecture" and later published in his Saraviensis paper [F86].

Take a local field K, complete with respect to an additive valuation v, with ring of integers R. Denote by E an elliptic curve defined over K with minimal Weierstrass equation

$$y^2 + a_1xy + a_3y = x^3 + a_2x^2 + a_4x + a_6, \tag{E}$$

with $a_i \in R$ for all i. We suppose that K is of zero characteristic.

As we are assuming that (E) has a point of order $2n$, with n odd, we denote by (x_0, y_0) a rational point of order 2 on (E) and we use the isomorphism

$$\begin{cases} X = 4(x - x_0) \\ Y = 8y + 4a_1 + 4a_3 \end{cases}$$

to get a new model

$$Y^2 = X(X^2 + CX + D) \tag{A}$$

with C, D in R.

If $v(2) = 0$, then (A) is also a minimal model for the curve. If $v(2) > 0$, equation (A) is no longer a minimal model, but it is rather close to a minimal model.

Remark A.3.1 If $v(2) > 0$ and if $a_1/2$ and $a_3/2$ are in R, then we can modify the isomorphism to get a minimal model of type (A).

Definition A.3.1 *We will say that (A) is a "nearly minimal model".*
Now we suppose that $P = (X, Y)$ is of odd order $n = p^h$ with $p \geq 5$ in $A(K)$, and we make the fundamental assumption:

Hypothesis A.3.1 $v[Y([i]P)]$ *is* **not constant** *when i varies in* $(\mathbb{Z}/n\mathbb{Z})^*$.

To study the situation we use the following well-known facts:

Lemma A.3.1 (Nagell–Lutz) *If $2v(p)/(p-1) < 1$, the abscissa of the points of order p^h in $A(K)$ lie in R.*

Lemma A.3.2 (Kodaira–Néron) *Let $A_0(K)$ be the subgroup of points of $A(K)$ with non-singular reduction. Then either:*

$$[A(K) : A_0(K)] \leq 4$$

or $A(K)/A_0(K)$ is cyclic of order $-v[j(A)]$ and $A(K)$ has split multiplicative reduction.

Now since $v[Y([i])P)]$ is not constant, one of the multiples Q of P reduces to a point of order 2 in $\widetilde{A}(R/\mathfrak{m})$. If $P \in A_0(K)$, then Q is also in $A_0(K)$ and its reduction $\widetilde{Q}$ must be killed by 2 and n, so $\widetilde{Q}$ must be the point at infinity which contradicts lemma A.3.1. But since $p \geq 5$, lemma A.3.2 shows that E has split multiplicative reduction and n divides $-v[j(A)]$ which is > 0. So A is a Tate elliptic curve.

Proposition A.3.1 *Let K be a local field in which 2 is not ramified and p is "not very ramified". Suppose $A(K)$ has a point of order $2n$ with an odd $n \geq 5$, and let P be a point of order n in $A(K)$.*

If $v[Y([i]P)]$ is not constant when i goes through $(\mathbb{Z}/n\mathbb{Z})^*$ then A is K-equivalent to a Tate curve and the v-adic parameter of P has a valuation $\not\equiv 0 \mod v(q)$.

Digression A.3.1 Suppose **all the points of order 2 of** A **are in** $A(K)$ and take Jacobi–Tate theta functions[71]; the cubic is then K-isomorphic to

$$\Theta : \quad y^2 = x_1 x_2 x_3$$

where:

$$\begin{cases} x_1 = \theta_3^2\theta_4^2(0)\dfrac{\theta_2^2}{\theta_1^2}(t), \; x_2 = \theta_4^2\theta_2^2(0)\dfrac{\theta_3^2}{\theta_1^2}(t), \; x_3 = \theta_2^2\theta_3^2(0)\dfrac{\theta_4^2}{\theta_1^2}(t) \quad \text{and} \\ \\ y = \theta_2\theta_3\theta_4(0)\dfrac{\theta_2\theta_3\theta_4}{\theta_1^3}(t). \end{cases}$$

We can always write:

$$\begin{aligned} A : \quad Y^2 &= X_1(X_1 - A)(X_1 + B) \\ &= X_2(X_2 - B)(X_2 + C) \\ &= X_3(X_3 - C)(X_3 + A) \end{aligned}$$

with $A + B + C = 0$ (write $X_3 = X_1 - A, X_2 = X_3 - C$).

Similarly:

$$\begin{aligned} \Theta : \quad y^2 &= x_1(x_1 + \theta_4^4)(x_1 + \theta_3^4) \\ &= x_2(x_2 - \theta_3^4)(x_2 - \theta_2^4) \\ &= x_3(x_3 + \theta_2^4)(x_3 - \theta_4^4) \end{aligned}$$

with Jacobi's relation:

$$\theta_3^4 = \theta_2^4 + \theta_4^4,$$

and we now have :

$$(\theta_2^4, \theta_3^4, \theta_4^4) \sim (16q^{1/2} + \cdots, 1 + \cdots, 1 + \cdots)$$

[71] One must note that q is then a square in $\mathfrak{m}$.

So, looking at the q-expansions of the theta-constants, we see that the relation $A+B+C=0$ is a deformation of Jacobi's relation to a local one

$$16X^n + Y^n = Z^n$$

with $v(X\,Y\,Z) > 0$.

A.4 GLOBAL STUDY

Now we will take an elliptic curve defined over $\mathbb{Q}$ and $K = \mathbb{Q}_\ell$ for all primes ℓ. We assume that $A(\mathbb{Q})$ has a torsion point of order $2n$ with odd $n \geq 5$ and that $P \in A(\mathbb{Q})$ is of order n.

Definition A.4.1 *When ℓ is such that $v_\ell[Y([i]P)]$ is not constant when i varies in $(\mathbb{Z}/n\mathbb{Z})^*$, we will say that ℓ is a* **sensitive prime**.

Lemma A.4.1 *If $n = p$ is a prime number ≥ 5, then p is a* **sensitive prime**.

Proof. This follows from the fact that Hasse's bound for $\widetilde{A}(\mathbb{F}_p)$ forces A to have bad reduction at p.

Again, since p does not divide the order of $\widetilde{A}_0(\mathbb{F}_p)$, $\widetilde{P}$ cannot lie in the identity component of Néron's group. So $[A(\mathbb{Q}_p) : A_0(\mathbb{Q}_p)]$ is divisible by p, we have a Tate curve and the valuation of the p-adic parameter of P is not $\equiv 0 \mod v(q)$. $\square$

Now, to get Fermat-type equations, we will introduce the isogenous curve

$$Y_1^2 = X_1(X_1^2 - 2CX_1 + G) \qquad (A_1)$$

with $G = C^2 - 4D$ and we write

$$\begin{cases} X(t) = V^2(t) \\ Y(t) = V^2(t)U(t) \end{cases} \qquad \begin{cases} X_1(2t) = U^2(t) \\ Y_1(2t) = 2U^2(t)V(2t) \end{cases} \qquad (4)$$

with, for example,

$$V(t) = C_1 \frac{\theta_2}{\theta_1}(t), \qquad U(t) = C_2 \frac{\theta_3\theta_4}{\theta_1\theta_2}(t).$$

The addition formulae for U and V are (in additive notation for arguments[72]):

$$\frac{V(t+s) + V(t-s)}{V(t+s) - V(t-s)} = -\frac{U(t)}{U(s)}$$

and symmetric ones, so we have:

$$\frac{V(4t) + V(2t)}{V(4t) - V(2t)} = -\frac{U(3t)}{U(t)}. \qquad (5)$$

Now we consider $n = p^2$, with $p \geq 5$, and we take into account the local study of $Y([i]P)$ and $Y_1([i]\varphi(P))$ where φ is the isogeny $A \to A_1$ of order 2.

[72] $t+s$ (resp. $t-s$, $2t$, $3t$, ...) means ts (resp. ts^{-1}, t^2, t^3, ...).

Looking at sensitive primes for (A, P) and $(A_1, \varphi(P))$ we get :

Lemma A.4.2 *Suppose there is a point Q of order $p^2 \geq 25$ in $A(\mathbb{Q})$ and consider the functions $U(it)$ and $V(it)$ attached to $[i]Q$.*

Let ℓ be a prime and suppose $v_\ell[U(it)]$ or $v_\ell[V(it)]$ varies when i varies in $(\mathbb{Z}/p^2\mathbb{Z})^*$. Then ℓ is a sensitive prime and the variations of $v_\ell[U(ipt)]$ and $v_\ell[V(ipt)]$ are in $p\mathbb{Z}$.

Proof. Suppose that, contrary to the conclusion, $v_\ell[Y(it)]$ **and** $v_\ell[Y_1(it)]$ are constant. Then the formulae (4) imply that

$$v_\ell(2) + v_\ell[Y^2(it)] - v_\ell[Y_1(it)] = 4v_\ell[V(it)] - v_\ell[V(2i)] = 3C$$

Write $f(i) := v_\ell[V(it)] - C$; we have

$$f(2i) = 4f(i),$$

and since there is an $n \in \mathbb{N}$ such that $2^n \equiv 1 \mod p$, we have $f(i) = 0$, so $v_\ell[V(it)]$ is constant. A similar reasoning works for U. □

Using lemma A.4.2 and Jacobi–Tate theta functions, we can simplify both sides of (5), since the non-sensitive primes are eliminated in the quotients. Now, if we assume that $P = [p]Q$ with $Q \in A(\mathbb{Q})$, the same approach shows that the variations of $v_\ell[V(it)]$ and $v_\ell[U(it)]$ lie in $p\mathbb{Z}$ when ℓ is a sensitive prime. Simplifying the fractions and equating numerators and denominators, we get theorems A.2.1 and A.2.2 (lemma A.4.1 ensures that we are in the "second case" of Fermat's last theorem and that the solution is not p-adically trivial).

Digression in the style of Section 3. We assume that the whole 2-torsion of A is rational, and we look for a Fermat-like equation by a **global deformation of Jacobi's relation**.

(1) Suppose there is a **point P of order** $p \geq 5$ in $A(\mathbb{Q})$. Then we get:

$$2^4aX^p + bY^p + cZ^p = 0$$

where X, Y, Z are made up of the multiplicative contribution of the sensitive primes (so p divides XYZ), 2 does not divide YZ and where a, b, c come from the dead bad primes.

Of course we have

$$(a, X) = (b, Y) = (c, Z) = 1$$

and

$$(X, Y, Z) = (a, b, c) = 1.$$

(2) Suppose there is a **point Q of order** p^2 in $A(\mathbb{Q})$. Similarly we have:

$$2^4aX^{p^2} + bY^{p^2} + cZ^{p^2} = 0.$$

A.5 THE WAY BACK

Proving the impossibility of

$$X^{2p} + Y^{2p} = Z^{2p}, \quad XYZ \neq 0, \qquad (F_{2p})$$

especially in the "second case of Fermat's last theorem", would have been a great achievement in 1971!

This reason explains the particular type of equation which is the object of theorem A.5.1.

Theorem A.5.1 *Let a, b, c be fixed integers of varying signs pairwise relatively prime and not divisible by a $2p^{th}$ power.*

If the curve

$$aX^{2p} + bY^{2p} + cZ^{2p} = 0 \qquad (F)$$

admits an infinity of primitive non-trivial solutions $(\alpha, \beta, \gamma) \in \mathbb{Z}^3$ such that p divides $\alpha\beta\gamma$, then there exists a number field L, depending only on a, b, c and p, in which all the curves $E_{A,B,C}$ would admit a complete group of $2p$-division points.

Instead of proving theorem A.5.1, as in [H71] and [H72], it might be more useful to give some details on Fermat's original equation

$$X^p + Y^p + Z^p = 0, \quad XYZ \neq 0 \qquad (F_p)$$

of prime degree $p \geq 5$.

We will use the ideas of 1969–1971 together with references to [Sil].

Let (α, β, γ) be a primitive solution of (F), and take $A = \alpha^p$, $B = \beta^p$ and $C = \gamma^p$. To fix ideas we write

$$E_{A,B,C}: \quad Y^2 = X(X - A)(X + B)$$

with even $A \equiv 0$ and $B \equiv 1 \bmod 4$.

Lemma A.5.1 *Let ℓ be a prime dividing α and let ζ_p be a primitive p^{th}-root of unity.*

Then the coordinates of the p-division points of $E_{A,B,C}$ lie in $\mathbb{Q}_\ell(\zeta_p, \sqrt[p]{2}, \beta^{1/2})$.

Proof. Since $E_{A,B,C}$ has discriminant $\Delta = 2^4(ABC)^2$, modular invariant

$$j = \frac{c_4^3}{\Delta} = -2^8 \frac{(BC + CA + AB)^3}{(ABC)^2}$$

and since $p \geq 5$, we have $v_\ell(j) < 0$ whenever $v_\ell(\Delta) > 0$.

Since j is a $2p$-th power in $\mathbb{Q}_\ell$, Tate's parameter q is also a $2p$-th power in $\mathbb{Q}_\ell$ and $E_{A,B,C}$ is isomorphic to the Tate curve E_q over an (at most) quadratic extension of $\mathbb{Q}_\ell$ (see [S] p. 359). We know (ibid.) that E_q is characterised by q <u>and</u> by the fact that it has split multiplicative reduction at ℓ.

If $\ell = 2$, E_q is equivalent to $E_{A,B,C}$ over $\mathbb{Q}_2(2^{1/p})$. Since the parameters of the p-torsion points lie in $\langle \zeta_p, q^{1/p} \rangle / \langle q \rangle$, we know (ibid.) that they are in $\mathbb{Q}_2(2^{1/p}, \zeta_p)$. If $\ell \neq 2$, then q is a $2p$-th power in $\mathbb{Q}_\ell$ (because j is also a $2p$-th power). So the parameters of the p-torsion points lie in $\langle \zeta_p, q^{1/p} \rangle \mod \langle q \rangle$, so in $\mathbb{Q}_\ell(\zeta_p)^* \mod \langle q \rangle$. It remains to show that $E_{A,B,C}$ is split multiplicative over $\mathbb{Q}_\ell(b^{1/2})$, but this is obvious since B is a square in this field.

If $\ell = p$ divides α, this result extends to $\mathbb{Q}_p(\zeta_p, \sqrt[p]{2}, b^{1/2})$, but a little more care is necessary to prove that j is a $2p$-th power in $\mathbb{Q}_\ell$. □

Weak form of Theorem A.5.1 *Let K be the field generated over $\mathbb{Q}$ by the p-division points of $E_{A,B,C}$.*

(i) *If p divides $\alpha\beta\gamma$, then K_p is unramified everywhere except over 2 and p, and the ramification index over p is $p - 1$.*
(ii) *If p does not divide $\alpha\beta\gamma$, then K_p is unramified everywhere except over 2 and p, but the curve has good reduction over p.*

Proof.
(i) We saw in lemma A.5.1 that K_p was *not much ramified* over $\mathbb{Q}_\ell$ when ℓ divides $\alpha\beta\gamma$. If ℓ does not divide $p\alpha\beta\gamma$, then it does not divides Δ, and $E_{A,B,C}$ has good reduction modulo ℓ. So the criterion of Néron–Ogg–Shafarevitch ([Si] p. 184) shows that K_p is not ramified over $\mathbb{Q}_\ell$ in this case.
(ii) The only difference concerns the ramification over p, but since we have good reduction at p, we need not consider this point (see [Se,a] p. 274 and [Se,b] p. 191). □

Remark A.5.1 **(1)** The Fermat-like equation:

$$X^p + Y^p = 2Z^p$$

seems simple enough, but it is actually more difficult than Fermat's original equation, since $(X, Y, Z) = (1, 1, 1)$ is an obvious solution for which $E_{A,B,C}$ is elliptic.

(2) Considerations developed in [H70] suggest that the system:

$$\begin{cases} X\xi^p + Y\eta^p + Z\zeta^p = 0 \\ X^p\xi + Y^p\eta + Z^p\zeta = 0 \end{cases}$$

should also be devoid of non-trivial solutions when $p \geq 7$.

(3) This analysis seems a bit more precise than Frey's in some respects.
Consider the Fermat–Jacobi equation of Section 3:

$$16X^{2p} + Y^{2p} = Z^{2p} \qquad (FJ)$$

with $p \geq 5$.

Suppose α, β, γ is a non-trivial primitive solution of (FJ); then p divides α by [H90] and we see, as above, that all the points of p-division of $E_{A,B,C}$ lie in $\mathbb{Q}_\ell(\sqrt{-1}, \zeta_p)$. This differs a bit from Frey's result ([F86] p. 12).

But one cannot jump to the conclusion that all the $4p$-division points of $E_{A,B,C}$ lie in $\mathbb{Q}(\zeta_{4p})$, as Serre made me realize in 1969!

A.6 BIBLIOGRAPHY

[C] Cassels, J.W.S., Diophantine equations with special reference to elliptic curves, *J. London Math. Soc.* **41**, 193–291, 1966.

[D70, a] Demjanenko, B.A., O Totchkax krutchenia Elliptitcheskix Krivix. *Izv. Ak. Nauk, S.S.S.R., Ser. Mat.* **34**, 757–774, 1970

[D70,b] Demjanenko, B.A., O Totchkax kaniechnovo poriadka Elliptitcheskix Krivix. *Mat. Zametki* **7**, (5), 563–567, 1970.

[D71,a] Demjanenko, B.A., O Totchkax kaniechnovo poriadka Elliptitcheskix Krivix. *Acta Arithmetica* **XIX**, 185–194, 1971.

[D71,b] Demjanenko, B.A., O Krutcheni Elliptitcheskix Krivix. *Izv. Ak. Nauk. S.S.S.R., Ser. Mat.* **35**, 280–307, 1971.

[Fa] Faltings, G., Endlichkeitsätze für abelsche Varietäten über Zahlkörpern. *Inv. Math.* **73**, 349–366, 1983.

[F77] Frey, G., Some remarks concerning points of finite order on elliptic curves over global fields. *Ark. f. Mat.* **15**, 1–19, 1977.

[F82] Frey, G., Rationale Punkte auf Fermatkurven und getwisteten Modulkurven. *J. Reine u. Angew. Math.* **33**, 185–191, 1982.

[F86] Frey, G., Links between stable elliptic curves and certain Diophantine equations. *Ann. Univ. Sarav. Math. Ser.* **1**, 1–40, 1986.

[F89] Frey, G., Links between solutions of $A - B = C$ and elliptic curves, Number Theory (Ulm 1987), 31–62; Lecture Notes in Mathematics 1380, Springer-Verlag, New York, 1989.

[H65,a] Hellegouarch, Y., Une propriété arithmétique des points exceptionnels d'ordre pair d'une cubique de genre 1. *C.R. Acad. Sc. Paris* **260**, 5989–5992, 1965.

[H65,b] Hellegouarch, Y., Applications d'une propriete arithmétique des points exceptionnels d'ordre pair d'une cubique de genre 1. *C.R. Acad. Sc. Paris* **260**, 6256–6258, 1965.

[H70] Hellegouarch, Y., Etude des points d'ordre fini des variétés abéliennes de dimension un définies sur un anneau principal. *J. Reine u. Angew. Math.* **244**, 20–36, 1970.

[H71] Hellegouarch, Y., Points d'ordre fini sur les courbes elliptiques, *C.R. Acad. Sc. Paris* **273**, 540–543, 1971.

[H72] Hellegouarch, Y., Courbes elliptiques et équation de Fermat. *Thèse, Besancon,* 1972.

[H75] Hellegouarch, Y., Points d'ordre $2p^h$ sur les courbes elliptiques. *Acta Arith.* **XXVI**, 253–263, 1975.

[H90] Hellegouarch, Y., A generalization of Terjanian's theorem. *C.R. Math. Canada* **XII**, (4), 103–108, 1990.

[K-L] Kubert, D., and Lang, S., Modular Units. Springer, 1981.

[M] Mazur, B., Modular curves and the Eisenstein ideal. *IHES Publ. Math.* **47**, 33–186, 1977.

[O] Oesterlé, J., Nouvelles approches du théorème de Fermat. *Sem. Bourbaki* 1987–88, SMF, 165–186.

[Se,a] Serre, J-P., Propriétés galoisiennes des points d'ordre fini des courbes elliptiques. *Inv. Math.* **15**, 259–331, 1972.

[Se,b] Serre, J-P., Sur les représentations modulaires de degré 2 de $\mathrm{Gal}(\bar{\mathbb{Q}}/\mathbb{Q})$. *Duke Math. J.* **54**, 179–230, 1987.

[S-T] Serre, J-P., and Tate, J., Good reduction of Abelian varieties. *Ann. Math.* **88**, 492–517, 1968.

[Sil] Silverman, J.H., The Arithmetic of Elliptic Curves. Springer–Verlag, Berlin–New York, 1986.

[T] Terjanian, G., Sur l'équation $x^{2p} + y^{2p} = z^{2p}$. *C.R. Acad. Sci. Paris* **285**, 973-975, 1977.

BIBLIOGRAPHY

[Ab] Abel N.H – Oeuvres Complètes, Deuxième éd., éd. Sylow et Lie, Oslo 1881.

[Am] Amice Y. – Nombres $p-$adiques, P.U.F., Paris, 1975.

[Ap] Apostol T.M. – Modular Functions and Dirichlet Series in Number Theory, Springer, Berlin, 1976.

[Ar 1] Artin E. – Galois Theory, Ann Arbor, 1946.

[Ar 2] Artin E. – Quadratische Körper im Gebiet der höheren Kongruenzen I, II Math Zeitschrift 19 (1924), 153–246 (also in Complete Works).

[B-B] Borwein J.M. and Borwein P.B. – Pi and the AGM, John Wiley, New York, 1987.

[B-C] Borevitch Z.I. and Chafarevitch I.R. – Théorie des Nombres, Gauthier-Villars, Paris, 1967.

[B-M] Boyer C.B. and Merzbach U.C. – A History of Mathematics, 2nd edn, John Wiley, New York, 1991.

[Bos] Bost J-B. – Introduction to Compact Riemann Surfaces, Jacobians and Abelian Varieties, in From Number Theory to Physics; Springer, Berlin, 1989.

[Bou 1] Bourbaki N. – Eléments de Mathématique, Algèbre, Chapter 5, Hermann, Paris, 1959.

[Bou 2] Bourbaki N. – Eléments de Mathématique, Algèbre, Chapter 6, Hermann, Paris.

[Bou 3] Bourbaki N. – Eléments de Mathématique, Topologie Générale, Chapter 2, Hermann, Paris, 1965.

[Br] Brassine E. – Précis des Oeuvres Mathématiques de P. de Fermat, J. Gabay, Sceaux, 1989.

[Car] Cartier P. – An Introduction to Zeta Functions, in From Number Theory to Physics, Springer, Berlin, 1989.

[Ca 1] Cassels J.W.S. – Local Fields, L.M.S. Student Texts 3, Cambridge U.P. Cambridge, 1986.

[Ca 2] Cassels J.W.S. – Diophantine equations with special reference to elliptic curves. *London Math. Soc.*, **41**, 193–291, 1966.

[Ca 3] Cassels J.W.S. – Lectures on Elliptic Curves, L.M.S. Student Texts, Cambridge U.P., Cambridge, 1991.

[C-N] Charpentier N. and Nikolski N. Leçons de Mathematiques d'aujourd'hui, Cassini, 2000.

[Co] Cornell G. – Modular Forms and Fermat's Last Theorem, Springer, Berlin, 1997.

[CSS] Modular Forms and Fermat's Last Theorem, Papers from the Instructional Conference on Number Theory and Arithmetic Geometry held at Boston University, Boston, MA, August 9–18, 1995. G. Cornell, J. Silverman, G. Stevens, eds. Springer-Verlag, New York, 1997.

[D] Darmon H. – A proof of the full Shimura-Taniyama-Weil conjoncture is announced. Notices Amer-Math-Soc. **46**(1999), no 11, 1397–1401.

[D-D-T] Darmon H., Diamond F. and Taylor R. – Fermat's last theorem, in Current Developments in Maths, 1995, International Press.

[D-G] Darmon H. and Granville A. – On the equations $z^m = F(x, y)$ and $Ax^p + By^q = Cz^r$. *Bull. London Math. Soc.* **27**, 513–543, 1995.

[D-S] Deligne, Serre J.P. – Formes modulaires de poids 1. *Ann. Sci. E.N.S.* **7**, 507–530, 1974.

[Den] Dénes P. – Über die Diophantische Gleichung $x^\ell + y^\ell = cz^\ell$. *Acta Math.* **88**, 241–251, 1952.

[Des] Descombes R. – Eléments de Théorie des Nombres, P.U.F. Paris, 1986.

[Di 1] Dieudonne J. – Foundations of Modern Analysis, Academic Press, London, 1960.

[Di 2] Dieudonne J. – Abrégé d'Histoire des Mathématiques, Hermann, Paris, 1978.

[D-M] Darmon, H. and Merel. L. Winding quotients and some variants of Fermat's last theorem, *J. Reine Angew. Math.* **490**, 81–100, 1997.

[Ed] Edwards H.M.– Fermat's Last Theorem, Springer, Berlin, 1977.

[Eu] Euler L. – Introduction to Analysis of the Infinite, Vol. I, Springer, Berlin, 1988.

[F-G] Fauvel J. and Gray J. – The History of Mathematics, Macmillan, London, 1987.

[Fu] Fulton W. – Algebraic Curves, Addison-Wesley, New York, 1989.

[Go] Goldstein C. – Un théorème de Fermat et ses lecteurs, Presse Universitaires de Vincennes, 1995.

[Gu] Gunning R.C. – Lectures on Modular Forms, Annals of Math. Studies 48, Princeton University Press, Princeton, 1962.

[Ha] Hartshorne R. – Algebraic Geometry, Springer, Berlin, 1977.

[Hec 1] Hecke E. – Lectures on the Theory of Algebraic Numbers, Springer, Berlin, 1981.

[Hec 2] Hecke E. – Mathematische Werke, Vandenhoeck und Ruprecht, Göttingen, 1983.

[Hel 1] Hellegouarch Y. – Etude des points d'ordre fini des variétés abéliennes de dimension un définies sur un anneau principal. *J. Reine Angew. Math.* **244**, 20–36, 1970.

[Hel 2] Hellegouarch Y. – Points d'ordre fini sur les courbes elliptiques. *C.R. Acad. Sc. Paris* **273**, 540–543, 1971.

[Hel 3] Hellegouarch Y. – Courbes ellliptiques et équation de Fermat, Thèse, Besançon, 1972.

[Hel 4] Hellegouarch Y. – Points d'ordre $2p^k$ sur les courbes elliptiques. *Acta Arith.* **XXVI**, 253–263, 1975.

[Hou 1] Houzel C. – De Diophante à Fermat. Pour la Science, *N*° 219, Paris, janv., 88–96, 1966.

[Hou 2] Houzel C. – Fonctions elliptiques et intégrales abéliennes, in Abrégé d'Histoire des Mathématiques éd. J. Dieudonné, Hermann, Paris 1992, 293–314.

[Hus] Husemöller D. – Elliptic Curves, Springer, Berlin, 1987.

[I] Igusa J.I. – Theta Functions, Springer, Berlin, 1972.

[I-R] Ireland K. and Rosen M. – A Classical Introduction to Modern Number Theory, 2nd edn, Springer, Berlin, 1990.

[J-L] James G. and Liebeck M. – Representations and Characters of Groups, Cambridge Math textbooks, Cambridge, 1993.

[J-S] Jones G.A. and Singerman D. – Complex Functions, Cambridge U.P., Cambridge, 1987.

[Kap] Kaplansky I.– Fields and Rings, 2nd edn, Univ. of Chicago Press, Chicago, 1972.

[Kl] Klein F. – Gesammelte mathematische Abhandlungen, 3 vol., Springer, Berlin, 1921–1923.

[Kn] Knapp A-W. – Elliptic Curves, Princeton U.P., Princeton, 1992.

[Ko] Koblitz N. – Introduction to Elliptic Curves and Modular Forms, Springer, Berlin, 1984.

[K-L] Kubert D.S. and Lang S. – Modular Units, Springer, Berlin, 1981.

[Kuh] Kuhn T.– La structure des révolutions scientifiques, Flammarion, Paris, 1983.

[Kum] Kummer E-E. – Collected Papers, edited by André Weil, Springer, Berlin, 1975.

[L] Lang S. – Algebraic Number Theory, Springer, Berlin, 1994.

[Luc] D. Luçon – Répartition des entiers de la forme $2^a.3^b.5^c$; Revue de Math. Spé., 1987–88, 3, p. 117–119

[Lut] Lutz E. – Sur l',quation $y^2 = x^3 - Ax - B$ dans les corps $p-$adiques. *J. Reine Angew, Math.* **177**, 431–466, 1937.

[M-M] Maillard R. and Millet A. – Géométrie, Classe de Mathématiques, Hachette, Paris, 1951.

[M] Mazur B. – Modular curves and the Eisenstein ideal, *IHES Publ. Math.*, **47**, 33–186, 1977.

[Mi] Miyake T. – Modular Forms, Springer, Berlin, 1989.

[ML] Maclaurin C. – A Treatise of Algebra ... To which is added, an Appendix, concerning the general properties of geometrical lines, 5th edition, London, 1788.

[Mum] Mumford D. – Tata Lectures on Theta, 1 and 2, Birkhaüser, Basle, 1983, 1984.

[N] Nitaj D. – Conséquences et aspects expérimentaux des conjectures *abc* et de Szpiro, Thèse, Caen, 1994.

[Oe] Oesterlé J. – Nouvelles approches du "théorème" de Fermat Sém. Bourbaki, 1987–88, no. 694.

[Og 1] Ogg A-P. – Modular Forms and Dirichlet Series, Benjamin, 1969.

[Og 2] Ogg A-P. – Abelian curves of small conductor, *J. Reine Angew. Math.* **226**, 204–215, 1967.

[Per] Perrin D. – Géométrie Algébrique, Interéditions/CNRS éditions, Paris, 1995.

[Ra] Rademacher H. – Topics in Analytic Number Theory, Springer, Berlin, 1973.

[Rbb] Ribenboim P. – 13 Lectures on Fermat's Last Theorem, Springer, Berlin, 1979.

[Re] Reyssat E. – Quelques aspects des Surfaces de Riemann, Birkhaüser, Basle, 1989.

[Ri 1] Ribet K.A. – On modular representations of Gal($\bar{\mathbb{Q}}$, $\mathbb{Q}$) arising from modular forms. *Invent. Math.* **100**, 431–416, 1990.

[Ri 2] Ribet K.A. – On the equation $a^p + 2b^p + c^p = 0$, (manuscript).

[Ro] Roquette P. – Analytic theory of elliptic functions over local fields, Vandenhoeck and Ruprecht, Göttingen, 1970.

[Sa] Samuel P. – Théorie algébrique des nombres, Hermann, Paris, 1967.

[Sch] Schikhof W.H. – Ultrametric Calculus, Cambridge U.P., Cambridge, 1984.

[Se 1] Serre J-P. – Cours d'Arithmétique, P.U.F., Paris, 1970.

[Se 2] Serre J-P. – Représentations linéaires des groupes finis, $3^{è}$ éd., Hermann, Paris, 1978.

[Se 3] Serre J-P. – Propriétés galoisiennes des points d'ordre fini des courbes elliptiques. *Invent. Math.* **15**, 259–331, 1972.

[Se 4] Serre J-P. – Sur les représentations modulaires de degré 2 de Gal($\bar{\mathbb{Q}}/\mathbb{Q}$). *Duke Math. J.* **54** 179–230, 1987.

[Se 5] Serre J-P. – Corps Locaux, Hermann, Paris, 1962.

[Se 6] Serre J-P. – Travaux de Wiles (et Taylor) Séminaire Bourbaki, 1994–95, no. 803.

[Sh] Shimura G. – Introduction to the Arithmetic Theory of Automorphic Functions, Princeton, 1971.

[Sie] Siegel C-L. – Topics on Complex Function Theory, Wiley, New York, 1971.

[Sil] Silverman J-H. – The Arithmetic of Elliptic Curves, Springer, Berlin, 1986.

[S-T] Silverman J-H. and Tate J. – Rational Points on Elliptic Curves, Springer, Berlin, 1992.

[St-T] Stewart I.N. and Tall D.O. – Algebraic Number Theory, Chapman and Hall, London, 1987.

[Tat] Tate J. – The arithmetic of elliptic curves. *Invent. Math.* **23**, 171–206, 1974.

[Tau] Tauvel P. – Mathématiques Générales pour l'Agrégation, Masson, Paris, 1992.

[Va] Valiron G. – Théorie des Fonctions, Masson, Paris, 1955.

[VDP] Van der Poorten A. – Notes on Fermat's Last Theorem, Wiley, 1995.

[Ve] Verriest G. – Théorie des Equations selon Galois, Gauthier-Villars, Paris, 1939.

[VF] Van Frankenhuysen M. – Hyperbolic Space and the *abc* Conjecture, Thèse, La Hague, 1995.

[Vui] Vuillemin J. – La philosophie de l'algèbre, P.U.F., Paris, 1962.

[VW 1] Van der Waerden B-L. – Modern Algebra I, Ungar, New York, 1953

[VW 2] Van der Waerden B-L. – Geometry and Algebra in Ancient Civilizations, Springer, Berlin, 1983.

[Web] Weber H. – Lehrbuch der Algebra, T. III, Chelsea, New York, 1962.

[Wei] Weil A. – Number Theory, Birkhäuser, Basle, 1984.

[Wi] Wiles A.J. – Modular elliptic curves and Fermat's Last Theorem. *Ann. Maths* **142**, 443–551, 1995.

[W-T] Wiles A.J. and Taylor R. – Ring-theoretic properties of certain Hecke algebras. *Ann. Maths* **142**, 553–572, 1995.

[W-W] Whittaker E.T. and Watson G.N. – Modern Analysis, Cambridge U.P., Cambridge, 1958.

[Z] Zagier D. – Introduction to modular forms, in From Number Theory to Physics, ed. Waldschmidt, Springer, Berlin, 1989.

INDEX